AKITO ARIMA
MEMORIAL VOLUME

AKITO ARIMA
MEMORIAL VOLUME

Editors

Thomas T S Kuo
Stony Brook University, USA

Takaharu Otsuka
University of Tokyo, Japan

KK Phua
World Scientific Publishing Company, Singapore

James Vary
Iowa State University, USA

World Scientific

NEW JERSEY • LONDON • SINGAPORE • BEIJING • SHANGHAI • TAIPEI • CHENNAI

Published by

World Scientific Publishing Co. Pte. Ltd.
5 Toh Tuck Link, Singapore 596224
USA office: 27 Warren Street, Suite 401-402, Hackensack, NJ 07601
UK office: 57 Shelton Street, Covent Garden, London WC2H 9HE

Library of Congress Control Number: 2025939442

British Library Cataloguing-in-Publication Data
A catalogue record for this book is available from the British Library.

AKITO ARIMA MEMORIAL VOLUME

ISBN 978-981-98-1334-6 (hardcover)
ISBN 978-981-98-1335-3 (ebook for institutions)
ISBN 978-981-98-1336-0 (ebook for individuals)

For any available supplementary material, please visit
https://www.worldscientific.com/worldscibooks/10.1142/14318#t=suppl

Desk Editor: Carmen Teo Bin Jie

Typeset by Stallion Press
Email: enquiries@stallionpress.com

https://doi.org/10.1142/9789819813353_fmatter

Preface

T.T.S. Kuo[a], T. Otsuka[b,c], J. Vary[d] and K.K. Phua[e]

[a] *Department of Physics, State University of New York at Stony Brook, New York, NY 11794, USA*
[b] *Department of Physics, University of Tokyo, Hongo, Bunkyo-ku, Tokyo, 113-0033, Japan,*
[c] *RIKEN Nishina Center, 2-1 Hirosawa, Wako, Saitama 351-0198, Japan*
[d] *Department of Physics and Astronomy, Iowa State University, Ames, Iowa 50011, USA*
[e] *World Scientific Publishing Company, Singapore*

It is with great pleasure and years of memories that we write a short introduction to this memorial volume honoring Akito Arima (1930–2020). This book is formed by many enthusiastic contributions by his students, colleagues and friends.

Akito Arima passed away on the 6th of December, 2020. It was very sudden news, and virtually all of us learned the news from broadcast media or later. He looked healthy and was busy until then. Many of us expected that he would remain active at least for ten more years, as he himself kept claiming. Akito Arima was born in Osaka, Japan in 1930 and turned 90 years old in September 2020. Some of his ex-students were planning a special party called SOTSUJYU KAI (celebration for the 90th birthday in Japanese tradition), but the COVID epidemic forced us to postpone it, to our deep regret.

Akito Arima graduated from the University of Tokyo in 1953, and entered the Graduate course of the University of Tokyo. He became a research associate (assistant professor in the present classification) of the Institute of Nuclear Study, the University of Tokyo in 1956, and was granted the Doctor of Science in 1958. He joined the Argonne National Laboratory in 1959 on the Fulbright scholarship program. In 1960, he returned to Tokyo as a lecturer of the Department of Physics, the Faculty of Science (presently the School of Science), the University of Tokyo. He was promoted to associate professor (in 1964) and full professor (in 1975). He has been engaged in research and education all over the world, including guest professors in Princeton University and New York State University at Stony Brook.

From the early days of nuclear physics, he made major contributions. In five years after the birth of the "shell model" by M. G.-Mayer and J. D. Jensen, he and Dr. H. Horie presented the theory of configuration mixing, which turned out to be a milestone of nuclear physics, serving as the backbone of today's many-body theories of nuclei. With his experiences in Argonne, he started a shell-model tradition in Japan, being still developed to date. Moreover, his dedication went beyond nuclear physics: he initiated Japanese efforts towards supercomputing in general, for instance, through his leadership as the director of the Computer Centre, the University of Tokyo.

Besides the development of the shell model in Japan, he worked on a variety of important topics in nuclear physics ranging from the medium/meson effects on magnetic and weak processes, to α clustering and α decay, and to algebraic approaches like the pseudo SU(3) scheme. He certainly enjoyed them all, including the famous imprisonment at the Delta-hole controversy: he was confined in a "jail" in a comic presented in a conference. We are happy that what he advocated was shown, by experiment by Dr. H. Sakai and his collaborators, to be correct later, like many cases of the political prisoners.

Akito is certainly best known for the Interacting Boson Model, proposed with Dr. F. Iachello, on the quadrupole deformation of nuclear shapes. This model covers various shapes in a simple and unified way. Such scientific achievements brought him many honors, including Nishina Memorial Awards, Japanese Academy Awards, Cultural Medal, Bonner Award, the Humboldt Award, and the Region Donour Medal. He was also instrumental in international cooperation, receiving many emeritus professorships and doctoral degrees.

Akito made highly respected efforts for improving the conditions of education and research in Japan. Having this goal in mind, he became a very important political figure: a member of the House of Councilors, the Minister of Education. The "University Poverty Story" campaign, an appeal for raising government spending, is still recalled partly because of its remarkable success. In addition, he was a great poet: he brought a new style to Haiku poems by including memories and impressions received abroad, and was granted the Dakotsu Prize, the highest honor for a Haiku poet. He donated a Haiku to the University of Tokyo. This Haiku is carved on a rock placed in the Hongo campus of the University, and its photo is shown on the back cover of this book. The Haiku reads

銀杏散る
萬巻の書の
頁より
朗人

The direct translation to English may be "Ginko leaves fly out from pages of tens of thousands of books". Ginko leaves change their color to yellow in autumn, and fall from the tree. A large number of such leaves are often blown up to the sky by wind. In the mind of Akito Arima, Ginko leaves symbolize, combined with "books", a great variety of knowledge, and this Haiku actually implies his wish, in our interpretation, that the University of Tokyo delivers such knowledge to broad communities of the whole world. The photo of the carved rock is used for the book cover, paying respects to his wish. This photo is also displayed below.

We are so grateful to Akito Arima for his great contributions to physics research and to society's support for research and education. Honoring such broad and great contributions, the book contains articles on broader subjects.

We indeed wonder whether any scientist after him will be able to speak in person to prime ministers on the promotion of scientific research and higher education in general as he did. We pray once again, with our greatest respects and gratitudes, that his soul may rest in peace, after long hard times.

We sincerely thank the authors of this book for their enthusiastic support. We are very delighted that the family of Akito Arima not only approved the publication of this book but also is pleased with the book. We are grateful to Dr. T. Suzuki for his help in editing the manuscripts and to Dr. N. Yoshida for his help with photo caption. We thank the Musashi Academy of the Nezu Foundation for providing with the photo of Prof. A. Arima used for the cover of this book. We also sincerely thank Carmen Teo Bin Jie, for her excellent help in preparing this book. Last, but not least, the support from World Scientific is gratefully acknowledged.

A stone showing a Haiku poem by Akito Arima (see the text)

https://doi.org/10.1142/9789819813353_fmatter

Contents

https://doi.org/10.1142/9789819813353_fmatter

Memory of Professor Akito Arima

Koichi Yazaki

Professor Akito Arima was my great teacher not only in physics but also in the way of life. I met him first in 1961 when he came back from the United States to take a faculty position at the University of Tokyo. He gave us a course of advanced quantum mechanics. His lecture was clear, attractive and also very practical. Quantum theory of angular momentum and rotational group was one of his favorite subjects and I learned for the first time the beauty and usefulness of symmetries and conservation laws in quantum world. I still keep and use his note he distributed to us during his lectures on this subject including Clebsch-Gordan coefficients and Racah algebra.

I learned a great deal from him since then. He was very good at explaining and solving problems in elegant ways always with their future prospects. His attitude was positive and encouraging. He listened carefully to what his students told him about their ideas in physics and, even if the ideas were crude and undeveloped, he always tried to find their positive aspects and gave encouraging advices.

Professor Arima was very considerate in giving advices not only for research but also for personal problems. Although he was extremely busy after he became the dean of the faculty of science, he always found time to talk to students or friends who wanted to consult with him.

On the other hand, he was also very strict about the way of treating students. I still remember his words when I stupidly tried to avoid the responsibility for a mistake in a paper I wrote with a student. He said to me "If you do it once more, you are not qualified to be a thesis advisor." Since then, I have been careful not to do something similar. His advice was always to the point, useful and sympathetic. I miss him so much.

Professor Arima was also famous for his cultural activities and, as everyone knows, he was one of the leaders in Japanese poem, Haiku. But he was not indifferent to athletic activities and was especially fond of hiking. In spite of his tight schedule, he encouraged us to organize hiking once or twice a year. The picture shown below was a photo taken at one of such occasions, when we went to Mt. Takao located about an hour by train from the center of Tokyo in the autumn of 1983.

Professor Arima was truly a great, kind and pleasant person.

May his soul rest in peace.

Professor Arima and his group in Mt. Takao hiking (taken in 1983). Shown are K. Yazaki, T. Kajino, N. Takahashi, Y. Koike, N. Yoshinaga, T. Shigehara, O. Morimatsu, N. Yoshida, A. Arima, and S. Takeuchi.

(Koichi Yazaki, Professor Emeritus, University of Tokyo,
Research Advisor, RIKEN iTHEMS, Wako, Saitama, Japan)

https://doi.org/10.1142/9789819813353_fmatter

Memorial Episodes of Professor Akito Arima, Related to the RIKEN Radioactive Isotope Beam Factory Project

Yasushige Yano
Executive Director of Nishina Memorial Foundation

In this short note, I would like to reveal my memorial episodes of late Professor Akito Arima (hereafter "Arima-sensei") related to the RIKEN Radioactive Isotope Beam Factory (RIBF) Project.

The RIBF now demonstrates the unparalleled capability in the world to produce unprecedentedly strong radioactive isotope (RI) beams. Since its commissioning in 2007, the RIBF has been creating a large number of excellent papers unravelling mystery of nuclear world with such capability,

Speaking in short, this RIBF could not have been created without Arima-sensei.

The RIBF project was proposed in 1987 by Professor Masayasu Ishihara and me to construct a world-top nuclear physics accelerator facility through greatly upgrading the newly commissioned (at that time) RIKEN Ring Cyclotron Facility. Since then, however, this project had been merely a dream and virtual until Arima-sensei was inaugurated as the new president of RIKEN in 1993.

After Arima-sensei's inauguration, I explained the outline of this RIBF project to him, and he requested me to receive the international evaluation of the project. Then immediately, Arima-sensei formed the international advisory committee consisting of distinguished nuclear physicists over the world, and we received their evaluation. Such an international evaluation of the Japanese project was a rare case in Japan. For us, it was a little nervous, but very exciting experience. As a result, our RIBF project was quite highly evaluated.

This was Arima-sensei's strategy to make the government approve this very expensive project: namely, based on this international evaluation he requested the government to approve the budget for preparatory research of this project. As a result, in 1995, this budget was approved, and we started to construct one of the world's first superconducting sector magnets: this is very key technology of this project.

In 1996, the then government of prime minister Ryutaro Hashimoto declared the austere fiscal policy and accordingly all the "new" projects proposed were postponed to the indefinite future. Arima-sensei, however, strongly claimed that this RIBF project was not "new", but already approved and now running. This claim was so strong and persuasive that eventually the big budget of the whole project was successfully approved in 1997.

In 1998, Arima-sensei left RIKEN and became a politician, but his support of the RIBF as its father or creator has been continued. Arima-sensei's expectation to the RIBF was a definitive contribution to the "energy problem" of mankind: He has continued to ask us to tackle an invention of a method to transmute radioactive waste from nuclear power plants into stable or useful materials. In front of us, sometimes he told us "You should attack this adventure for mankind, otherwise the RIBF will die."

Below is the memorial picture of Arima-sensei. He visited "Nishina Memorial Room", where I explained some of episodes of Dr. Yoshio Nishina.

A. Arima with Y. Yano in the Nishina Memorial Room (see the text)

https://doi.org/10.1142/9789819813353_fmatter

As a Student of Arima

Nobuaki Yoshida
Faculty of Informatics, Kansai University,
2-1-1 Ryozenji-cho, Takatsuki-shi,
569-1095 Japan,
yoshidan@kansai-u.ac.jp

1. Introduction

I feel deep regret at Arima-sensei's death. I studied the interacting boson model (IBM) under the guidance of Prof. Arima. In the following, I will write about some subjects that I studied by him together with memories.

2. Research

One of the most important academic contributions that he made was the IBM. When I became a graduate student of Arima in 1978, it was only a few years after the introduction of the interacting boson model (IBM) with F. Iachello in 1975. The model approximates a pair of nucleons by a boson, thus approximate active nucleons in a nucleus by bosons. In those days, the IBM began to be accepted widely in the field of nuclear physics.

In the following, I give some of the subjects that I myself took part in.

Boson mapping The first subject that he suggested to me was the test of the boson models as an approximation of the nuclear shell model. Using the Ginocchio model, a schematic shell model, I studied how to derive the formulae in the group theory learning the SU(2) algebra from his hand-written notes. The Otsuka-Arima-Iachello method to lead the IBM from the shell model was compared with other methods such as the boson expansion technique. Later he gave me the opportunity to visit Los Alamos to study under Ginocchio.

High-spin states In those days, there were increasing amount of experimental data in high-spin states. He suggested me, as the theme of the thesis, to describe those high-spin states using the IBM. The nuclei were the Ce isotopes, and later, the Ge isotopes were studied, too.

Random interaction In the shell model, the predominance of the 0^+ ground state was observed the two-body interactions taken at random. By Prof. Zhao's proposal, we confirmed that similar 0^+ dominance was observed, too, in the IBM.

Prof. Arima and the author in Peking University in 1987.

β-decay For a long time, Arima was interested in studying the β-decay with the IBM. Although there was not enough time to pursue it in the limited period, I am sure that his suggestions had large effects on the recent β-decay studies using the IBM.

The above subjects are only part of his, even in the field of the IBM. It is hard for me to know all the areas that he covered.

I had opportunities to visit many places in the world with him. Belgium in the Capture Gamma-Ray Symposium in 1987 was one of them. He was interested in the art of that area, and it was the first time for me to know Hieronymus Bosch while visiting museums with him. In the same year, we visited China in the collaboration with Peking University, where many scientists there got interest in the IBM.

3. Memories

Lectures When I was an undergraduate student of the department of physics, Prof. Arima had a course of lectures in nuclear physics. I remember that he gave detailed explanation about the angular-momentum algebra. Usually, he came to the classroom without carrying anything, and derived many mathematical formulas as if he hit them on on the spot.

Excursions The Arima group used to have an excursions in spring and autumn. The walking courses include Hakone, Takao, Nokogiriyama, Miura, the mountains around Chichibu, Oume, the mountains around Fuji, Choshi, and Tanzawa. Those were good memories for me.

New year parties When Arima worked in the university, he invited all the members of the group to his home to have a new year party. In 2013, we again had a new year party around him in restaurants in Tokyo. It lasted as a regular meeting until 2020, where I met him last time.

P. von Brentano (left) and A. Arima (right)

A. Arima (left) and F. Iachello (right)

Akito Arima (courtesy of the Musashi Academy of the Nezu Foundation)

A. Arima in front of Mont-Saint-Michel, France (courtesy of K. Kubodera)

From left to right, H. Feshbach, I. Talmi, F. Iachello and A. Arima

From left to right, T. Otsuka, A. Arima and T. Suzuki, Arima is holding messages from ex-students for his 90th birthday. Taken in September, 2020.

A. Arima (center) and his wife, Hiroko, with some ex-students and their wives. From left to right, M. Oka, K. Takayanagi, N. Yoshida, H. Nakada, A. Arima, Y. Otsuka, T. Otsuka, H. Arima (front, on chair), T. Suzuki, H. Suzuki, and T. Kajino. This photo was taken at the occasion of the celebration for Arima's Order of Culture of Japan.

https://doi.org/10.1142/9789819813353_0001

Effect of Ground-State Shape Deformation on the Collective Excitations in Even-Mass Nd Isotopes

M. Abdullah*, D. Das*, S. Bagchi*,‡, and M. N. Harakeh†,§

*Department of Physics,
Indian Institute of Technology (Indian School of Mines), India
†Nuclear Energy Group, ESRIG, University of Groningen,
9747 AA Groningen, the Netherlands
‡sbagchi@iitism.ac.in
§m.n.harakeh@kvi.nl

Using inelastic α-scattering at extremely forward angles, including $0°$, the strength distributions of the Isoscalar Giant Monopole Resonance (ISGMR) have been measured in $^{142,146-150}$Nd (even A) isotopes. The nuclear ground-state shape changes from almost spherical to prolate in going from ^{142}Nd to ^{150}Nd. Thus, we can study the effect of ground-state shape deformation on the strength distribution of giant resonances. The optical potential parameters are extracted from the α-elastic scattering off ^{142}Nd.

1. Introduction

Giant resonances are the high-frequency collective vibrational modes in nuclei.[1] In a hydrodynamical model, an isoscalar excitation can be referred to as in-phase oscillation of proton and neutron fluids whereas an isovector excitation is the out-of-phase oscillation of the above nuclear fluids. Among the collective modes, the Isoscalar Giant Monopole Resonance (ISGMR) is referred to as "breathing mode" and the Isoscalar Giant Dipole Resonance (ISGDR) is referred to as "squeezing mode". Both of these modes are of crucial importance because their excitation energies are directly related to the incompressibility of nuclear matter. Nuclear incompressibility is a key input to the equation-of-state of nuclear matter, which in turn is crucial to understand some cosmological phenomena such as radii and masses of neutron stars, and supernova explosions.

Isoscalar giant resonances have been studied well for spherical nuclei but for the deformed nuclei, very few experimental data are available. However, the deformation effect on Isovector Giant Dipole Resonance (IVGDR) has been studied well.[2] In spherical nuclei, giant-resonance strength does not split because of directional symmetry. However, in the case of deformed nuclei, where the projection onto the symmetry axis, K, is a good quantum number, the giant-resonance strength splits into the different K components. Nevertheless, the strength of ISGMR, which has only a $K = 0$ component, splits in deformed nuclei into two parts because of the coupling of the ISGMR with the $K = 0$ component of the Isoscalar Giant Quadrupole Resonance (ISGQR).[3] The splitting in the ISGMR strength distribution has been studied only in the deformed Sm isotopes,[4] in the fission decay of the ^{238}U nucleus[5] and in light nuclei, e.g., the ^{24}Mg nucleus[6] and the ^{28}Si nucleus.[7] The deformation in a nucleus also affects the ISGQR strength distribution resulting in its splitting into three K components. In spherical ^{144}Sm, there is a small high-energy component of the ISGQR strength distribution. The fraction of the Energy-Weighted Sum Rule (EWSR) of this high-energy component increases with the increase in nuclear ground-state deformation.[4] For ISGDR, a "bi-modal" strength distribution has been observed for medium-mass to heavy-mass nuclei.[8] There is no change in the high-energy component of the "bi-modal" ISGDR strength distribution but the width and the EWSR fraction of the low-energy part increase with the deformation. Similarly, the strength distribution of the High-Energy Octupole Resonance (HEOR), seems to be changing with nuclear deformation as reported by Morsch *et al.*[9] In this chapter, we will discuss the preliminary results of inelastic α-particle scattering off Nd isotopes for studying the effect of ground-state shape deformation on the strength distribution of giant resonances.

2. Experimental Setup and Details

The work, undertaken in this chapter, was performed at the Research Center for Nuclear Physics (RCNP) in Osaka, Japan. Inelastic scattering of 386-MeV α particles was measured at the Ring Cyclotron Facility of RCNP, Osaka University. Self-supporting enriched metallic thin $^{142,146-148}$Nd foils (having thicknesses around 5 mg/cm^2) were employed as targets. Inelastically scattered α particles were momentum analyzed with the high-resolution magnetic spectrometer Grand Raiden (GR)[10] and focused onto the focal-plane detector system comprised of two Multi-Wire Drift Chambers (MWDCs)[4] and two plastic scintillator detectors. The MWDCs allow measurement of the horizontal and vertical positions of the α particle on the focal plane. The two plastic scintillators downstream of MWDC were used as the trigger of the data-acquisition system as well as to identify the scattered α-particles by energy-loss measurements. In the double-focusing mode of the spectrometer, vertical focusing was utilized to eliminate the instrumental background resulting from the multiple scattering inside the chamber.[4,6] The schematic setup of the GR spectrometer and the related detector arrangement is shown in Fig. 1.

To study the isoscalar giant resonances, a beam of α-particles was used because both its spin and isospin are zero. The α particles were accelerated to an energy of 386 MeV. To keep the beam halo to a minimum, the α particles were extracted in a single turn from the cyclotron. Elastic scattering data were taken at GR angles from 0° to 20.2°, to determine the optical-model parameters (OMPs), which are required to calculate the angular distributions of giant resonances of different multipolarities in the Distorted-Wave Born Approximation (DWBA) framework. Inelastic scattering data were taken at seven different GR angular settings, from 0° to 9.5°.

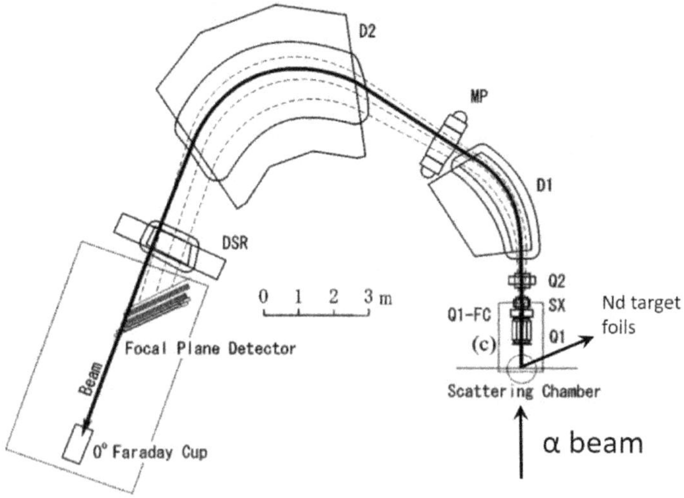

Fig. 1. Experimental and GR spectrometer setups for the study of giant resonances in Nd isotopes.

The measurement at 0°, corresponds to an average angle of 0.75° due to the collimator size (used in the experiment) and the horizontal collimation, used in the offline analysis, having a width of 1.6°. Measurements at forward angles, where the angular distributions are characteristic of multipolarities, are critical. Data were also taken with a ^{24}Mg target (having a thickness of around 2.5 mg/cm^2) at each of the GR angle settings for precise energy calibrations through momentum analysis. The energy resolution of the beam was around 175 keV, which was reasonable to study the structure of giant resonances.

3. Results and Discussions

3.1. *Optical Potential*

To explain the elastic and inelastic scattering data, an optical potential was used, which was obtained by fitting the elastic scattering data. An optical potential of the Woods-Saxon type was utilized for both the real and imaginary parts. This is represented as:

$$U(r) = V(r) + iW(r) \tag{1}$$

$$V(r) = -\frac{V_\circ}{1 + \exp(\frac{r-R}{a})} \tag{2}$$

where $V(r)$ is the real volume potential and is responsible for elastic scattering. Similarly for the imaginary part $W(r)$, which is responsible for the absorption and it arises because of the couplings to all non-elastic channels.

Therefore, we have six optical potential parameters: three real parameters $(\mathrm{V}_{0r}, \mathrm{R}_r, \mathrm{a}_r)$ and three imaginary parameters $(\mathrm{W}_{0i}, \mathrm{R}_i, \mathrm{a}_i)$. The optical-potential

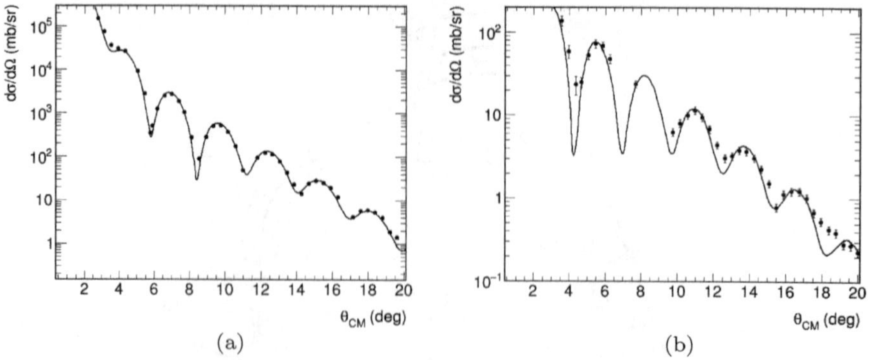

Fig. 2. Differential cross sections for α elastic scattering off ^{142}Nd at 386 MeV (a) and for excitation of the 2^+ excited state (b). The solid lines are results of optical-model and DWBA calculations obtained from the CHUCK program.

Table 1. Optical-model parameters from χ^2minimization using CHUCK.

E_α (MeV)	V_{0r} (MeV)	W_{0i} (MeV)	R_r (fm)	R_i (fm)	a_r (fm)	a_i (fm)
386	79.24	26.01	6.61	7.56	0.792	0.625

parameters were obtained from fitting of the $\alpha + {}^{142}$Nd elastic scattering data using the χ^2-minimization technique.[11] The coupled-channel code CHUCK was used for DWBA calculation.[12] The best fit to the elastic scattering cross section is shown in Fig. 2(a) and the parameters obtained are listed in Table 1. The Coulomb radius R_c was determined by $r_c \times A^{1/3}$ where A is the mass number of the target nucleus and r_c (Coulomb radius parameter) was fixed at 1.25 fm. The error bars are within the data points.

Using the known B(E2) value[13] and the OMPs thus obtained, the differential cross section for the 1.576 MeV 2^+ state was calculated in the DWBA framework and compared with the experimental data, which are shown in Fig. 2(b). An excellent agreement has been observed except at larger angles, suggesting the goodness of the obtained OMPs. The data at large angles were taken towards the end of the beam time when frequent tripping of the high voltages, applied to the MWDCs, occurred. Since the mass numbers of the other Nd isotopes are almost similar to ^{142}Nd, the same OMPs will be used for the other Nd isotopes as well.

3.2. *Energy Calibration*

For each Nd measurement angular setting, a measurement on ^{24}Mg was performed for calibration because of the well-known low-lying discrete states of ^{24}Mg and the spectrum was compared with the high-resolution reference spectra obtained

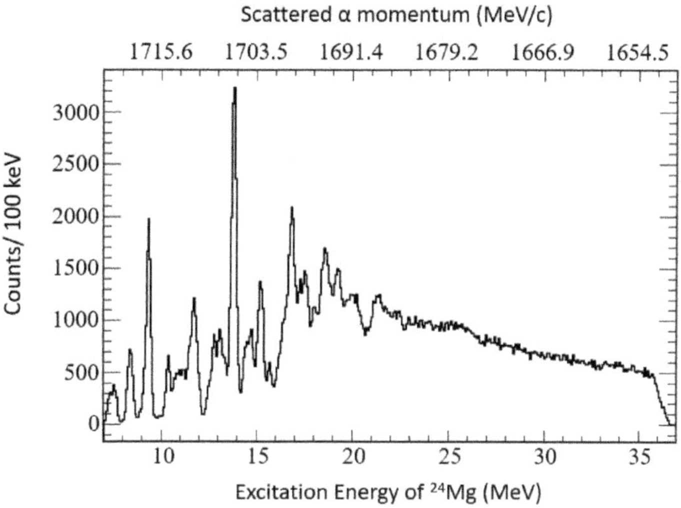

Fig. 3. 0° calibration spectrum for ^{24}Mg(α, α'), showing both the calibrated excitation energy of ^{24}Mg (bottom horizontal axis) as well as the momentum of the scattered α-particle (top horizontal axis).

from Ref.[14] Calibration parameters were obtained using the kinematical relations between the incident α particle, the target nucleus, and the scattered α' particle. Figure 3 shows a calibration spectrum of ^{24}Mg at the zero-degree setting of the GR spectrometer.

3.3. *Excitation-Energy Spectra of* $^{142,146\text{–}150}$*Nd*

If the kinetic energies and scattering angles of recoil α-particles are known in the laboratory frame, the excitation energy of $^{142,146-150}$Nd (even A) can be determined using the missing-mass technique. The excitation-energy spectra of Nd isotopes were extracted after particle identification, subtraction of instrumental background, and ion-optical corrections in offline analysis. Figure 4 shows the excitation-energy spectra of Nd isotopes at 0° GR angle.

3.4. *Subtraction of Contamination*

In the case of each Nd isotope, hydrogen and oxygen contaminants were present with substantial effective thicknesses. In the case of hydrogen contamination, α elastic scattering off hydrogen has a larger cross section than that off the Nd targets and henceforth, it is straightforward to identify and remove the affected data points. However, in the case of ^{16}O, the excited states were extended into the giant-resonance region of Nd isotopes. Discrete peaks of oxygen at 11.520 MeV (2^+) and 12.049 MeV (0^+) were observed in the spectra. Before further analysis, the contribution of scattering from oxygen should be subtracted from the experimental

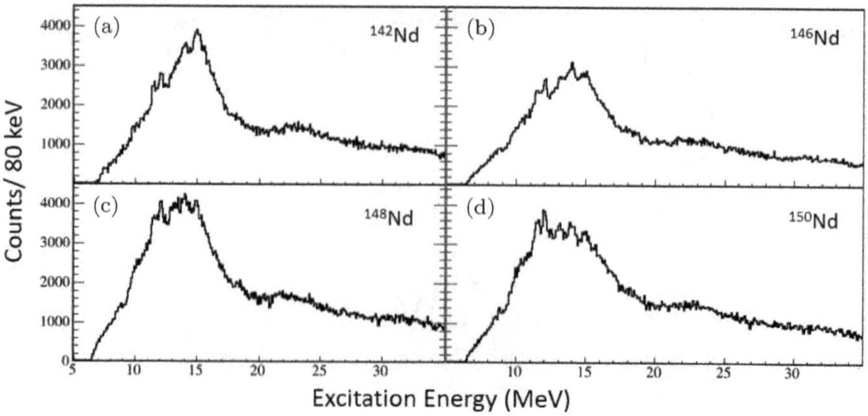

Fig. 4. Excitation-energy spectra of (a) ^{142}Nd, (b) ^{146}Nd, (c) ^{148}Nd and (d) ^{150}Nd, at GR angle ($\theta_{GR} = 0°$) obtained in inelastic α scattering. See text for further details about the data reduction. The narrow discrete peaks are due to oxygen contamination.

Nd spectra. In order to subtract this contamination, high-resolution $^{16}O(\alpha, \alpha')$ spectra from Ref. 15 were used. These spectra were measured using the same α beam energy, at the same scattering angles and using a SiO_2 target. As the resolution in the reference ^{16}O spectra is much higher than the resolution in the present experiment, it was essential to convolute the reference spectra with a Gaussian smearing function to match the resolutions before subtraction. The smearing function is given as follows:

$$\sigma(E_x, \Theta_{avg}) = \frac{1}{\sqrt{4\pi\omega^2}} \int_{-\infty}^{\infty} dE' \exp\left(\frac{(E_x - E')}{2\omega^2}\right) \sigma(E', \Theta_{avg}) \tag{3}$$

where ω is the smearing width, and its value was typically of the order of 140 keV. To get a scaling factor, the ratio of the areas of the relevant peaks was taken in our experimental spectra and the reference spectra. Figure 5 shows the excitation-energy spectrum of ^{142}Nd with oxygen contamination (a) and, after subtracting the oxygen contamination (b). The same procedure was followed for other Nd isotopes and GR angles to subtract the oxygen contamination. Figure 6 shows the $^{142,146-150}$Nd excitation-energy spectra after subtracting the oxygen contamination.

3.5. *Multipole-Decomposition Analysis and Contribution of Isovector Giant Dipole Resonance (IVGDR)*

In the Multipole-Decomposition Analysis (MDA) procedure, the experimental differential cross section of a particular energy interval in the excitation-energy spectrum is fitted with a linear combination of theoretically calculated DWBA differential cross sections for several multipolarities at that particular energy interval.

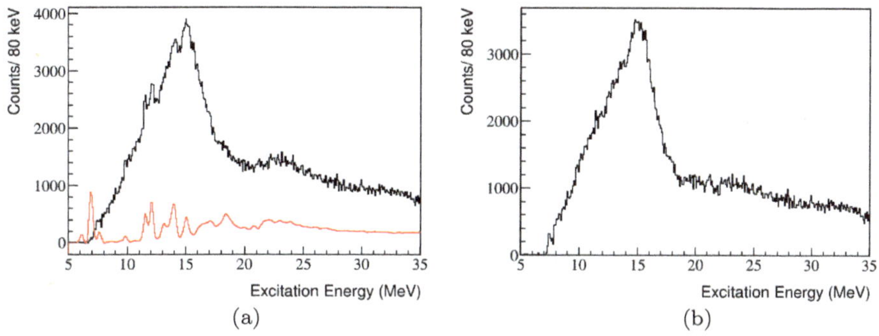

Fig. 5. (Color online) (a) Excitation-energy spectrum of ^{142}Nd including the contribution from the ^{16}O contamination. The reference oxygen spectrum[15] (resolution matched) is shown in red color. (b) Excitation-energy spectrum of ^{142}Nd after subtraction of the oxygen contamination.

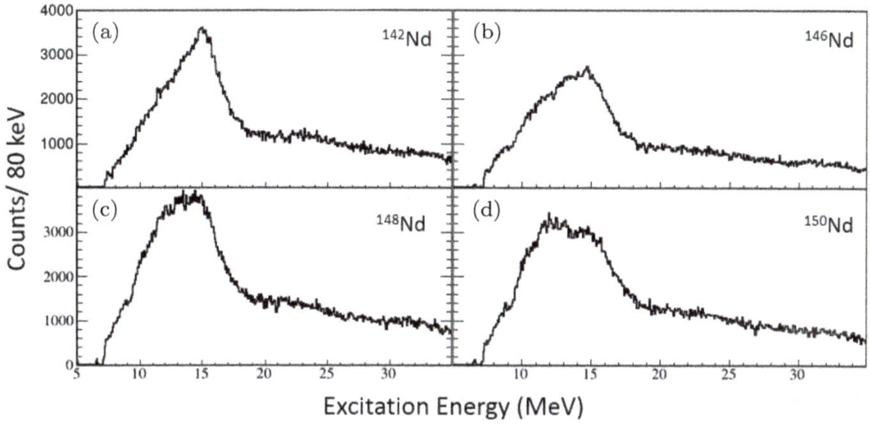

Fig. 6. Excitation-energy spectra of (a) ^{142}Nd, (b) ^{146}Nd, (c) ^{148}Nd and (d) ^{150}Nd, at GR angle $\Theta_{GR} = 0°$ after subtraction of the oxygen contamination. See text for details of the subtraction procedure.

In the case of DWBA calculations, 100% of the Energy-Weighted Sum Rule (EWSR) was considered. The fitting function is given as:

$$\frac{d^2\sigma^{exp}(\theta_{C.M.}, E_x)}{d\Omega dE_x} = \sum_{\lambda} A_\lambda(E_x) \frac{d^2\sigma_\lambda^{DWBA}(\theta_{C.M.}, E_x)}{d\Omega dE_x} \tag{4}$$

where $\frac{d^2\sigma^{exp}(\theta_{C.M.}, E_x)}{d\Omega dE_x}$ is the experimental differential cross section at a particular excitation-energy interval and $\frac{d^2\sigma_\lambda^{DWBA}(\theta_{C.M.}, E_x)}{d\Omega dE_x}$ is the differential cross section obtained from the DWBA calculation at that particular energy interval. The $A_\lambda(E_x)$ coefficients correspond to the fractions of the EWSR for the multipolarity λ exhausted within a particular excitation-energy bin. DWBA cross sections for isoscalar modes are included in the MDA up to $\lambda_{max} = 4$. The whole region of excitation energy $E_x = 10 - 32$ MeV was binned into 400 keV to reduce the statistical fluctuations and the MDA will be performed in 400 keV excitation-energy intervals.

Although the isoscalar giant resonances are excited by the scalar, isoscalar α probe, it should be noted that Coulomb excitation can lead to contributions from the isovector modes and thus the α-particles can couple to the IVGDR mode with a non-negligible effect. The contribution of IVGDR has to be subtracted before performing MDA to extract the strength distribution of the isoscalar giant resonances. In the excitation-energy region, 15–16 MeV, the strength of the IVGDR was estimated to be around 13.5%, 10.4%, 10.5%, and 9.5% of the cross section in that interval for $^{142,146-150}$Nd (even A), respectively, according to the experimental data on photo-neutron cross sections and the fitted Lorentzians to the data points.[16] From the MDA analysis, the strength distributions of different isoscalar giant resonances as functions of excitation energy will be extracted. This will allow the study of the effect of the ground-state shape deformation on the strength distributions.

3.6. *Summary and Outlook*

Inelastic α-particle scattering off $^{142,146-150}$Nd (even A) was performed at RCNP, Osaka University to study the effect of the ground-state shape deformation on the strength distributions of isoscalar giant resonances. A significant amount of oxygen contamination was found in the targets of the Nd isotopes, the contributions of which were subtracted before performing the MDA. IVGDR contributions are estimated for the excitation energies of interest and they will be subtracted before performing the MDA analysis, which is in progress.

Inelastic α-particle scattering was performed on strongly deformed ^{154}Sm both at RCNP[4] and at TAMU.[17] A splitting of the ISGMR strength in ^{154}Sm was observed in both the TAMU and RCNP data with different strengths assigned to the low-energy component in the two experiments. Although this "disagreement" could arise from the bombarding energies at the two different accelerator facilities, it is worth addressing this issue with another strongly deformed nucleus, such as ^{150}Nd in our case. In addition, the finite nuclear incompressibility (K_A) will be studied as a function of the neutron-to-proton asymmetry from which K_τ (the asymmetry term in the nuclear incompressibility[1]) will be inferred as done in Refs. 18, 19 where the atomic numbers of the nuclei are below 60. The detailed results can be found in Ref. 20.

Acknowledgements

The authors gratefully thank the staff of the RCNP Ring Cyclotron Facility for providing a high-quality and halo-free α beam. SB is thankful for the local support during the experiment. The full author list is given in Ref. 21. This work has been supported by the IIT (ISM) seed grant (Grant No.: FRS(154)/2021-2022/Physics) under the faculty research scheme. MA is indebted to MHRD, Government of India, for the financial support.

References

1. M. N. Harakeh and A. van der Woude, *Giant Resonances: Fundamental High Frequency Modes of Nuclear Excitation*. Oxford Science Publications (2001).
2. B. L. Berman and S. C. Fultz, *Rev. Mod. Phys.* **47**, 713 (1975).
3. U. Garg *et al.*, *Phys. Rev. Lett.* **45**, 1670 (1980).
4. M. Itoh *et al.*, *Phys. Rev. C* **68**, 064602 (2003).
5. S. Brandenburg *et al.*, *Phys. Rev. Lett.* **49**, 1687 (1982).
6. Y. K. Gupta *et al.*, *Phys. Lett. B* **748**, 343 (2015).
7. T. Peach *et al.*, *Phys. Rev. C* **93**, 064325 (2016).
8. B. K. Nayak *et al.*, *Phys. Lett. B* **637**, 43 (2006).
9. H. P. Morsch *et al.*, *Phys. Lett.* **119 B**, 311 (1982).
10. M. Fujiwara *et al.*, *Nucl. Instrum. Meth. Phys. Res. A* **422**, 484 (1999).
11. D. Das, M.Sc. thesis, IIT (ISM) Dhanbad, unpublished.
12. P. D. Kunz, Program CHUCK unpublished.
13. B. Pritychenko *et al.*, *At. Data Nucl. Data Tables* **114**, 371 (2017).
14. T. Kawabata, *Private Communication.*
15. M. Itoh, *Private Communication.*
16. V. A. Plujko *et al.*, *At. Data Nucl. Data Tables* **85**, 1 (2018).
17. D. H. Youngblood *et al.*, *Phys. Rev. C* **69**, 034315 (2004).
18. K. B. Howard *et al.*, *Phys. Lett. B* **801**, 135185 (2020).
19. D. Patel *et al.*, *Phys. Lett. B* **718**, 447 (2012).
20. M. Abdullah, PhD thesis, IIT (ISM) Dhanbad, India; in preparation.
21. M. Abdullah, D. Das, S. Bagchi, M. N. Harakeh, H. Akimune, T. Doi, Y. Fujikawa, M. Fujiwara, T. Furuno, U. Garg, K. B. Howard, Y. Hijikata, M. Itoh, K. Inaba, S. Ishida, N. Kalantar-Nayestanaki, T. Kawabata, S. Kawashima, K. Kitamura, N. Kobayashi, Y. Matsuda, A. Nakagawa, S. Nakamura, K. Nosaka, S. Okamoto, S. Ota, R. Pramanik, S. Roy, S. Weyhmiller, Z. Yang, in preparation.

https://doi.org/10.1142/9789819813353_0002

Relations Between Three-Particle Interactions in Nuclear Matter to Observable Quantities

Wolfgang Bentz[*,‡] and Ian C. Cloët[†,§]

[*] *Department of Physics. School of Science, Tokai University*
4-1-1 Kitakaname, Hiratsuka-shi, Kanagawa 259-1292, Japan
[†] *Physics Division, Argonne National Laboratory*
Argonne, Illinois 60439, USA
[‡] *bentz@keyaki.cc.u-tokai.ac.jp*
[§] *icloet@anl.gov*

In the first part of this paper, we use the framework of the Fermi liquid theory to derive model-independent relations between the slope parameters of the symmetry energy and of the incompressibility in nuclear matter to three-particle interaction parameters. Based on these relations, we present simple estimates and compare with the empirical information. In the second part, we discuss the general structure of the three-particle scattering amplitude in nuclear matter, and use methods similar to the Bethe-Brueckner-Goldstone theory to show how three-particle cluster diagrams emerge naturally in the Fermi liquid theory.

1. Introduction

Among the basic physical quantities which play a critical role in shaping the structure of nuclei and neutron stars are the symmetry energy (a_s)[1,2] and the incompressibility (K)[3,4] of nuclear systems. In nuclei, the density dependence of the symmetry energy and its associated pressure determine the neutron skin and the excitation modes of isovector characters, in particular the electric dipole oscillations.[5] In neutron stars, the symmetry pressure determines the radius of the star, which is indeed known to be strongly correlated to the neutron skin thickness of heavy nuclei.[6] In heavy ion collisions, the same physical quantity determines the isospin distribution among the reaction products.[7] On the other hand, the density dependence of the incompressibility and its associated pressure determine the nuclear excitation modes of isoscalar character, in particular the monopole oscillations,[8] and play a critical role to determine the maximum mass of a neutron star.[9] In heavy ion collisions, the same physical quantity has critical impact on the collective flow of the produced particles.[10]

The dependencies of the symmetry energy and the incompressibility on the baryon density (ρ) are usually expressed in terms of the slope parameters $L \equiv 3\rho \frac{\mathrm{d}a_s}{\mathrm{d}\rho}$ and $J \equiv 3\rho \frac{\mathrm{d}K}{\mathrm{d}\rho} - 12\,K$, where the latter one is usually called the skewness.[4] Because the quantities a_s and K in turn are defined in terms of the second derivatives of the energy density w.r.t. $\rho^{(3)} = \rho^{(p)} - \rho^{(n)}$ and $\rho = \rho^{(p)} + \rho^{(n)}$, respectively, it follows that L and J are related to the third derivatives of the energy density,[a] and therefore, by extension of Landau's hypothesis,[11,12] to the $\ell = 0$ moments of the three-nucleon in-medium forward scattering amplitude.[13,14] The first purpose of this article is to present these relations in an exact (model independent) form. Because of the long history of studies on three-particle interactions in nuclear matter,[15,16] it is desirable to know such model independent relations based on first principles. We will present semi-quantitative discussions on these relations in comparison to empirical information. The second purpose of our work is to discuss the physical content of the in-medium three-particle amplitude, the $\ell = 0$ moments of which enter into the model independent relations mentioned above. We will discuss how far the structure of the three-particle amplitude can be specified by using only its definition, and illustrate how further assumptions, similar to the ones used in the Bethe-Brueckner-Goldstone (BBG) theory,[17] can be used to derive more detailed expressions.

For these two purposes we will use the general field theoretic formulation of the Fermi liquid theory due to Landau and Migdal.[18,19] Two important merits of this approach are that it exploits general symmetries like gauge invariance and Galilei invariance to derive relations between interaction parameters (Landau-Migdal parameters) and observable quantities which are in principle exact, and that it naturally leads to the shell structure in finite systems. Nowadays several more merits are well known: The theory can be derived from the renormalization group,[20] it has a close relationship to the more popular Skyrme approach,[21] it can be extended to relativistic field theory,[22,23] and it has deep impacts on nuclear electromagnetic properties.[24,25] Last but not least, the theory played an important role to promote the exchange of people, ideas, and understanding between very different societies of the world.

The layout of the paper is as follows: In Sec. 2 we will discuss our relations between the quantities L and K mentioned above and the three-nucleon s-wave forward scattering amplitudes. Semi-quantitative estimates will be presented in comparison to the empirical information. In Sec. 3 we will discuss the physical content of the three-particle scattering amplitude, and make connection to the BBG theory. In Sec. 4 we summarize our results.

2. Basic Relations and Semi-Quantitative Discussions

In this Section we summarize the main points in the derivation of the model independent relations between the slope parameters of the symmetry energy and of the incompressibility to the three-nucleon s-wave interaction parameters, and present semi-quantitative estimates in comparison to empirical values. More detailed discussions can be found in Refs. 13, 14.

---

[a]The precise relations are given by $L = \frac{3}{2}\rho^2 \frac{\partial^3 E}{\partial\rho\, \partial^2 \rho^{(3)2}} + 3a_s$ and $J = 27\rho^2 \frac{\partial^3 E}{\partial \rho^3} - 9K$, where E is the energy density.

2.1. *Model Independent Relations*

We start from the standard relationships[26] for the second derivatives of the energy density (E) w.r.t. the isoscalar and isovector densities defined in the previous Section:[b]

$$\frac{\partial^2 E}{\partial \rho^2} = \frac{\pi^2}{2p_F M^*} + f_0 \equiv \frac{K}{9\rho}, \tag{1}$$

$$\frac{\partial^2 E}{\partial \rho^{(3)2}} = \frac{\pi^2}{2p_F M^*} + f_0' \equiv \frac{2a_s}{\rho}. \tag{2}$$

Here $M^* \equiv M^*(k = p_F)$ is the Landau effective mass in the isospin symmetric limit (p_F denotes the Fermi momentum in this limit), which is defined via the quasiparticle velocity as usual[26] by $\partial\varepsilon_k/(\partial k) \equiv k/M^*(k)$, where ε_k is the quasiparticle energy. The isoscalar and isovector combinations f_0 and f_0' are defined by[19]

$$f_0 = \frac{1}{2}\left(f_0^{(pp)} + f_0^{(pn)}\right), \quad f_0' = \frac{1}{2}\left(f_0^{(pp)} - f_0^{(pn)}\right), \tag{3}$$

where $f_0^{(\tau_1\tau_2)}$ are the $\ell = 0$ moments of the two-particle forward scattering amplitudes

$$\frac{1}{2\ell+1} f_\ell^{(\tau_1\tau_2)}(k_1, k_2) = \int \frac{d\Omega_2}{4\pi}\left(\hat{\boldsymbol{k}_1}\cdot\hat{\boldsymbol{k}_2}\right)^\ell f^{(\tau_1\tau_2)}(\boldsymbol{k}_1, \boldsymbol{k}_2), \tag{4}$$

at $k_1 = k_2 = p_F$. Here the label τ distinguishes between protons ($\tau = p$) and neutrons ($\tau = n$).

As we explained in the previous Section, in order to obtain the slope parameters of the incompressibility and of the symmetry energy, we have to take the derivatives of (1) and (2) w.r.t. ρ. In the following we summarize the basic relations which are used in the calculation:

- The variations of the quasiparticle energies $\varepsilon^{(\tau)}(k)$ w.r.t. the background densities are generally given by

$$\frac{\delta\varepsilon^{(\tau_1)}(k)}{\delta\rho^{(\tau_2)}} = f_0^{(\tau_1\tau_2)}(k, p_F^{(\tau_2)}), \tag{5}$$

  where $p_F^{(\tau)}$ are the Fermi momenta for protons and neutrons. By taking the derivative of Eq. (5) w.r.t. the momentum k and setting $k = p_F$, we obtain in the isospin symmetric limit

$$\frac{\partial M^*}{\partial \rho} = -\frac{M^{*2}}{2p_F}\frac{\partial f_0}{\partial p_F} + \frac{\pi^2}{2p_F^2}\frac{\partial M^*}{\partial p_F}, \tag{6}$$

$$\frac{\partial \Delta M^*}{\partial \rho^{(3)}} = -\frac{M^{*2}}{p_F}\frac{\partial f_0'}{\partial p_F} + \frac{\pi^2}{p_F^2}\frac{\partial M^*}{\partial p_F}, \tag{7}$$

[b]Here and in the following, all derivatives w.r.t. $\rho^{(3)}$ are defined at $\rho^{(3)} = 0$, although this is not indicated explicitly in order to simplify the notation.

where $\Delta M^* \equiv M^{*(p)} - M^{*(n)}$. We note that the above relations include both the variations of the background densities, as expressed by the first terms on the r.h.s. originating from Eq. (5), and of the external momenta of the particles. We defined[13]

$$\frac{\partial f_0}{\partial p_F} \equiv \left[\left(\frac{\partial}{\partial k_1} + \frac{\partial}{\partial k_2}\right) f_0(k_1, k_2)\right]_{k_1=k_2=p_F},$$

and similarly for f_0', and $\partial M^*/\partial p_F \equiv (\partial M^*(k)/\partial k)_{k=p_F}$.

- In analogy to Eq. (5), the variations of the $\ell = 0$ two-particle forward scattering amplitudes for external momenta k_1, k_2 w.r.t. the background densities are given by the $\ell = 0$ moments of the three-particle forward scattering amplitudes, denoted by h_0 (h_0') for the isoscalar (isovector) case, where the third particle is on the Fermi surface ($k_3 = p_F$). If all three particles are taken on the Fermi surface, we obtain[14]

$$\frac{\partial f_0}{\partial \rho} = h_0 + \frac{\pi^2}{2p_F^2}\frac{\partial f_0}{\partial p_F}, \qquad \frac{\partial f_0'}{\partial \rho} = h_0' + \frac{\pi^2}{2p_F^2}\frac{\partial f_0'}{\partial p_F}. \tag{8}$$

The first terms on the r.h.s. of the above relations come from the variations of the background densities, and the second terms from the variations of the external particle momenta. The isoscalar and isovector combinations h_0 and h_0' are defined by

$$h_0 = \frac{1}{4}\left(h_0^{(ppp)} + 3h_0^{(ppn)}\right), \qquad h_0' = \frac{1}{4}\left(h_0^{(ppp)} - h_0^{(ppn)}\right). \tag{9}$$

They are the $\ell = 0$ moments of the three-particle forward scattering amplitudes

$$\frac{1}{2\ell+1} h_\ell^{(\tau_1\tau_2\tau_3)}(k_1, k_2, k_3) = \int \frac{d\Omega_2}{4\pi} \int \frac{d\Omega_3}{4\pi} \left(\hat{\boldsymbol{k}}_1 \cdot \hat{\boldsymbol{k}}_2\right)^\ell h^{(\tau_1\tau_2\tau_3)}(\boldsymbol{k}_1, \boldsymbol{k}_2, \boldsymbol{k}_3), \tag{10}$$

at $k_1 = k_2 = k_3 = p_F$. The three-particle amplitudes on the r.h.s. of Eq. (10) are generally defined as the variations of the two-particle amplitudes w.r.t. the quasiparticle distribution functions:[c]

$$h^{(\tau_1\tau_2\tau_3)}(\boldsymbol{k}_1, \boldsymbol{k}_2, \boldsymbol{k}_3) = \frac{\delta f^{(\tau_1\tau_2)}(\boldsymbol{k}_1, \boldsymbol{k}_2)}{\delta n_{\boldsymbol{k}_3}^{(\tau_3)}}. \tag{11}$$

- Relations which follow from Galilei invariance play a very important role in the Fermi liquid theory. The best known relation follows from the requirement that under a Galilei transformation the single particle energies transform in the same way as a Hamiltonian in classical mechanics.[27] In the isospin symmetric limit,

[c] For a change of the Fermi momentum by $\delta p_F^{(\tau)}$ we have $\delta n_k^{(\tau)} = \delta p_F^{(\tau)}\,\delta(k - p_F^{(\tau)})$, and for a Galilei transformation to a system with velocity $\boldsymbol{u} \equiv \boldsymbol{q}/M$ we have $\delta n_{\boldsymbol{k}}^{(\tau)} = -\left(\hat{\boldsymbol{k}} \cdot \boldsymbol{q}\right)\delta(k - p_F^{(\tau)})$.

this gives the relation

$$\frac{k}{M^*(k)} + \frac{2p_F^2}{3\pi^2} f_1(k, p_F) = \frac{k}{M}, \tag{12}$$

where M is the free nucleon mass. For $k = p_F$ this becomes the familiar Landau effective mass relation.[26] A less well known relation follows from the requirement that under a Galilei transformation the two-particle scattering amplitudes are invariant. At the Fermi surface and in the isospin symmetric limit, this gives[13,14]

$$p_F \frac{\partial}{\partial p_F}\left(f_0 + \frac{1}{3} f_1\right) + \frac{4}{3} f_1 + \frac{4p_F^3}{3\pi^2} h_1 = 0, \tag{13}$$

and a similar relation with primes attached to all amplitudes. We can convert Eq. (13) to a more useful form by using Eq. (12) and its derivative w.r.t. k at the Fermi surface, to obtain

$$p_F \frac{\partial f_0}{\partial p_F} = \frac{3\pi^2}{p_F} \frac{M - M^*}{M M^*} - \frac{\pi^2}{M^{*2}} \frac{\partial M^*}{\partial p_F} - \frac{4p_F^3}{3\pi^2} h_1. \tag{14}$$

This relation is used to eliminate the derivative $\partial f_0 / \partial p_F$ from the final expressions.

By using the above prerequisites, the derivative of Eq. (1) is calculated by using Eqs. (6), (8), (12), and (14). The derivative of Eq. (2) is obtained by using the same formulas, and in addition Eq. (7) to eliminate the derivative $\partial f_0' / (\partial p_F)$ in favor of the dimensionless parameter

$$\mu \equiv \rho \frac{\partial}{\partial \rho^{(3)}} \left(\frac{\Delta M^*}{M}\right). \tag{15}$$

Defining the dimensionless three-particle interaction parameters by[d]

$$H_\ell = \frac{4p_F^4 M^*}{3\pi^4} h_\ell, \quad H_\ell' = \frac{4p_F^4 M^*}{3\pi^4} h_\ell', \tag{16}$$

the final results can be expressed as follows:

$$J = -9K + \frac{9p_F^2}{M} \times \left[\left(-3 + \frac{8}{3}\frac{M}{M^*}\right) - \frac{4}{3}\frac{M p_F}{M^{*2}}\frac{\partial M^*}{\partial p_F} + \frac{M}{M^*}(H_0 - H_1)\right], \tag{17}$$

$$L = 3a_s - \frac{p_F^2}{2M} \times \left[\left(1 - \frac{2}{3}\frac{M}{M^*}\right) + \mu\left(\frac{M}{M^*}\right)^2 - \frac{M}{M^*}\left(H_0' - \frac{1}{3}H_1\right)\right]. \tag{18}$$

One should note that Eqs. (17) and (18) are exact (model independent) relations.

[d]The prefactor in these relations is the prefactor for the dimensionless two-particle parameters $(2p_F M^*/\pi^2)$ multiplied by the density $\rho = 2p_F^3/(3\pi^2)$.

2.2. *Qualitative Discussions*

As we discussed in Sec. 1, there have been many recent analyses on the empirical values of J and L by using data on heavy nuclei, astrophysical observations, and heavy ion collisions. For definiteness we refer here to the values given in Fig. 4 of Ref. 28 for J, and in Fig. 20 of Ref. 4 for L, at normal nuclear matter density ($\rho = 0.155\ \text{fm}^{-3}$):

$$J = -255 \pm 245\,\text{MeV}, \quad L = 59 \pm 16\,\text{MeV}. \tag{19}$$

We mention that these values are consistent with most of the other analyses mentioned in Sec. 1. Concerning the nucleon effective mass, most of the investigations during the last decades[4, 29] are consistent with the range $M^*/M = 0.85 \pm 0.15$ at the Fermi surface, and indicate that $M^*(k)$ has a pronounced peak very close to the Fermi surface,[30, 31] i.e., $\partial M^*/\partial p_F \simeq 0$ or slightly positive.

Consider first Eq. (17). Using the canonical value of the incompressibility at normal nuclear matter density,[4] $K = 250$ MeV, the lower limit of J ($J > -500$ MeV) indicates that the three-particle term $\frac{M}{M^*}(H_0 - H_1)$ is positive and larger than unity. Considering next Eq. (18), the canonical value of the symmetry energy at normal nuclear matter density is $a_s = 32$ MeV, and the empirical range of the parameter μ defined by Eq. (15) reported in Ref. 4 is 0.27 ± 0.25. If μ is taken to be close to the central value of this range ($\mu \simeq 0.27$), the sum of the first two terms in $[\ldots]$ in the expression (18) for L is $\simeq 0.6$, almost independent of M^*, within the range of M^* given above.[e] The empirical slope parameter L in Eq. (19) then implies that the three-particle term $\frac{M}{M^*}\left(H_0' - \frac{1}{3}H_1\right)$ is negative, with a magnitude smaller than unity.

If we further assume that the $\ell = 1$ moment of the three-particle parameter plays only a subordinate role, that is, $H_1 \ll H_0$ and $H_1 \ll 3H_0'$, — and the simple calculations explained below support this assumption — we can expect that the isoscalar three-particle interaction parameter H_0 is positive and larger than unity, while the isovector one H_0' is negative and smaller than unity in magnitude. The uncertainties in the values of K and a_s do not change this qualitative conclusion.[13, 14]

In order to see how these expectations compare with simple semi-quantitative calculations, we consider the leading term in the Faddeev series for the three-particle amplitude h, shown in Fig. 1a. In this diagram, t denotes the two-particle t-matrix, which is the off-forward generalization of the amplitude f of the Fermi liquid theory. We call the contribution of the leading term the 2-particle correlation (2pc) piece. Its expression reads

$$h^{(2\text{pc})}(\boldsymbol{k}_1, \boldsymbol{k}_2, \boldsymbol{k}_3) = \sum_4 \delta_{\boldsymbol{k}_1+\boldsymbol{k}_2, \boldsymbol{k}_3+\boldsymbol{k}_4} \times P\,\frac{|\langle 12|\,\hat{t}\,|34\rangle_a|^2}{\varepsilon_3 + \varepsilon_4 - \varepsilon_1 - \varepsilon_2} + (1 \leftrightarrow 3) + (2 \leftrightarrow 3). \tag{20}$$

[e] In the literature,[4, 32] it is often assumed that $\mu \simeq \frac{2}{3}\frac{M^*}{M}F_1'$, where $F_1' = (2p_F M^*/\pi^2)\, f_1'$ is the dimensionless isovector $\ell = 1$ Landau-Migdal parameter. The exact relation, however, is more complicated.[14]

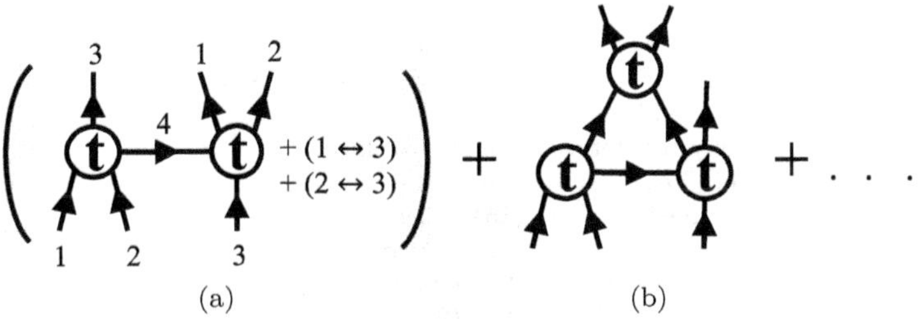

Fig. 1. First two terms in the Faddeev series for the three-particle scattering amplitude h. Lines with arrows denote nucleons, and t denotes the two-particle t-matrix.

In this schematic notation, $1 \sim 4$ represent the momenta $\boldsymbol{k}_1 \sim \boldsymbol{k}_4$ as well as the associated spin and isospin components, though and average over the spin components of $1, 2, 3$ is assumed implicitly. The sum represents momentum integration and summation over spin and isospin components of 4, the δ symbol represents a momentum conserving δ-function, P denotes the principal value, $\langle 12|\,\hat{t}\,|34\rangle_a$ is the antisymmetrized two-particle t-matrix element, and ε_i are the quasiparticle energies. An example for the three-particle correlation (3pc) term of h, which is the next-to-leading term in the Faddeev series, is shown in Fig. 1b. We will derive its form in the next Section.

In order to get a rough estimate of $h^{(2\text{pc})}$, we assume that the matrix elements $\langle 12|\,\hat{t}\,|34\rangle_a$ can be replaced by the $\ell = 0$ part of an effective interaction of the Landau-Migdal type,[19] which corresponds to an approximation of the scattering matrix by the scattering length decomposed into the various spin-isopin channels:

$$\begin{aligned} {}_a\langle 34|\,\hat{t}\,|12\rangle &= f_0\ (\delta_{31}\cdot\delta_{42}) + f_0'\ (\boldsymbol{\tau}_{31}\cdot\boldsymbol{\tau}_{42}) \\ &\quad + g_0\ (\boldsymbol{\sigma}_{31}\cdot\boldsymbol{\sigma}_{42}) + g_0'\ (\boldsymbol{\sigma}_{31}\cdot\boldsymbol{\sigma}_{42})\ (\boldsymbol{\tau}_{31}\cdot\boldsymbol{\tau}_{42}), \end{aligned} \tag{21}$$

where the notation indicates that the spin and isospin operators are defined to act in the particle-hole channel. With the further assumption that the quasiparticle energies can be approximated as $\varepsilon_i = \frac{\boldsymbol{k}_i^2}{2M^*}$, where M^* is in the range given above, the angular averages of Eq. (10) can be carried out analytically,[13] and show that the contribution of H_1 in Eqs. (17) and (18) is indeed small. In this approximation, the dimensionless isoscalar and isovector three-particle $\ell = 0$ amplitudes take the very simple forms

$$H_0^{(2\text{pc})} = \ln 2 \cdot \frac{1}{4}\,(F_0^2 + 3F_0'^2 + 3G_0^2 + 9G_0'^2), \tag{22}$$

$$H_0'^{(2\text{pc})} = \ln 2\ \cdot \frac{1}{4}\left(\frac{1}{3}F_0^2 + \frac{4}{3}F_0F_0' - \frac{1}{3}F_0'^2 + G_0^2 + 4G_0G_0' - G_0'^2\right), \tag{23}$$

where the dimensionless Landau-Migdal parameters are defined as $F_0 = (2p_F M^*/\pi^2)\,f_0$ and similar for the other cases. For illustrative purposes, we show in Table 1 the results for three sets of the extended Skyrme interaction,[33] which are all consistent with the assumptions made at the beginning of this Subsection.

Table 1. Values of various physical quantities entering Eqs. (17) and (18) at nuclear matter saturation density. Sets 1 ∼ 3 correspond to the results for the extended Skyrme interactions[33] eMSL07, eMSL08, eMS09, respectively. The values for μ, defined by Eq. (15), given in this Table refer to the approximate expression given in footnote e. $H_0{}^{(2pc)}$ refers to Eq. (22), and $H_0'{}^{(2pc)}$ refers to Eq. (23).

Set	1	2	3
M^*/M	0.7	0.8	0.9
μ	0.233	0.229	0.360
$H_0^{(2pc)}$	0.528	0.762	1.095
$H_0'^{(2pc)}$	−0.063	−0.101	−0.135

We see that the 2pc contribution to H_0 is positive definite, which is consistent with our expectations, but tends to be smaller than unity, which gives us a hint that 3pc terms like shown in Fig. 1b may be necessary to explain the empirical values of the skewness. The 2pc contribution to H_0', on the other hand, is negative — mainly because of the terms $-\frac{1}{3}F_0'^2$ and $-G_0'^2$ in Eq. (23) — and smaller in magnitude than unity, which is consistent with our expectations. This result thus provides no hint that the 3pc term is necessary to explain the empirical values of the slope parameter of the symmetry energy.

3. Physical Content of the Three-Particle Amplitude

As we have seen in the previous Section, the $\ell = 0, 1$ moments of the three-particle amplitude, Eq. (11), which we express here schematically as[f] $h(1,2,3) = \delta f(1,2)/(\delta n_3)$, are related to observable quantities. It is the aim of this Section to gain some insight into the physics contained in this quantity, and in methods to calculate it by using certain approximations.

The two-particle amplitude has the schematic form[26] $f(1,2) = Z_1 Z_2 t(1,2)$, where the Z-factors are the residues of the single particle propagators at the quasi-particle poles, and $t(1,2)$ is the on-shell value of the two-particle t-matrix in the forward limit, which satisfies the Bethe-Salpeter (BS) equation in the particle-hole channel with the two-body kernel given by $K^{(2)}(1,2) = i\delta\Sigma(1)/(\delta S(2))$,[34] where Σ is the skeleton part of the self energy and S is the single particle propagator. The variation $\delta/(\delta n_3)$ of the Z-factors and of the energy variables in $t(1,2)$ produces retardation terms which will be denoted by $h^{(\mathrm{prod})}$, and have the form of symmetrized products of functions depending only on pairs of variables. For the derivation we refer to Ref. 14, and just quote the result for the case when all particles

[f] To simplify the formulas of this Section, we denote the momenta, spin and isospin projections of the external particles by 1, 2, 3, and refer to the case of isospin symmetric nuclear matter.

are on the Fermi surface:

$$h^{(\mathrm{prod})}(1,2,3) = \frac{1}{2} f(1,2) \left((Z\,\Sigma'') + \frac{\partial}{\partial \varepsilon_F} \right) (f(1,3) + f(2,3))$$
$$+ (1 \leftrightarrow 3) + (2 \leftrightarrow 3),$$

where Z is the values of the Z-factor at the Fermi surface, Σ'' is the second derivative of Σ w.r.t. the energy variable at the Fermi surface, and

$$\frac{\partial f(1,2)}{\partial \varepsilon_F} \equiv Z^2 \left[\left(\frac{\partial}{\partial \varepsilon_1} + \frac{\partial}{\partial \varepsilon_2} \right) t(1,2) \right]_{\varepsilon_i = \varepsilon_F} .$$

The variation $\delta/(\delta n_3)$ of the off-shell two-particle t-matrix gives a contribution $\tilde{h}(1,2,3) \equiv Z_1 Z_2\, \delta t(1,2)/(\delta n_3)$, taken on the energy shells of particles 1 and 2. This can be calculated by straight forward variation of the BS equation, which has the schematic form $t = K^{(2)} - iK^{(2)}\, S\, S\, t$, and one obtains the structure shown in Fig. 2. The particle-hole irreducible three-body kernel which appears there is given by $K^{(3)}(1,2,3) = i\delta^2\Sigma(1)\,(\delta S(2)\delta S(3))$.[35] Typically, one finds that the loop diagrams shown in Fig. 2 represent "direct terms", with the corresponding exchange terms contained in $K^{(3)}$. The total three-particle amplitude can then be expressed as $h(1,2,3) = h^{(\mathrm{prod})}(1,2,3) + \tilde{h}(1,2,3)$.

In order to calculate the two-body and three-body kernels $K^{(2)}$ and $K^{(3)}$ from their definitions, one needs to model the functional dependence of the self energy Σ on the single-particle propagator S. One model which has been widely used in the

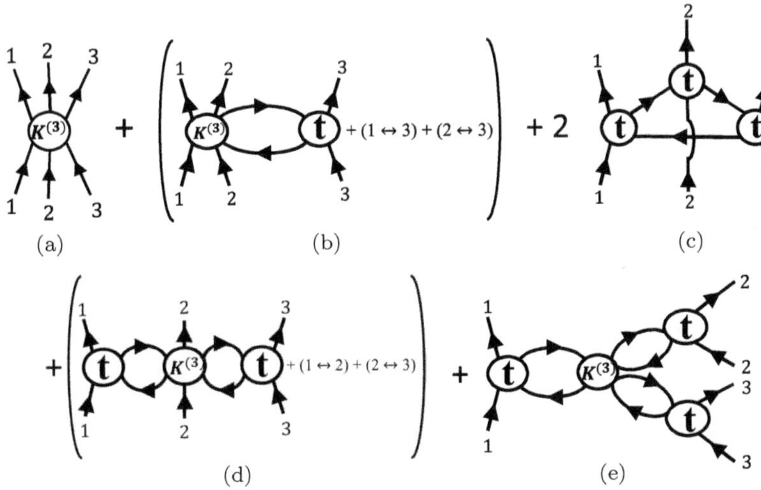

Fig. 2. Graphical representation of the three-particle amplitude $\tilde{h}(1,2,3)$. The three-particle kernel $K^{(3)}$, defined in the main text, is irreducible in the particle-hole channel, and contains terms of second, third and higher orders in the two- particle t-matrix (denoted as t). The products of pole parts of propagators with the same momenta ($S^2_{\mathrm{pole}}(k)$ or $S^3_{\mathrm{pole}}(k)$) in these diagrams are defined as $S^2_p(k) + S^2_h(k)$ or $S^3_p(k) + S^3_h(k)$, without products of particle and hole parts.

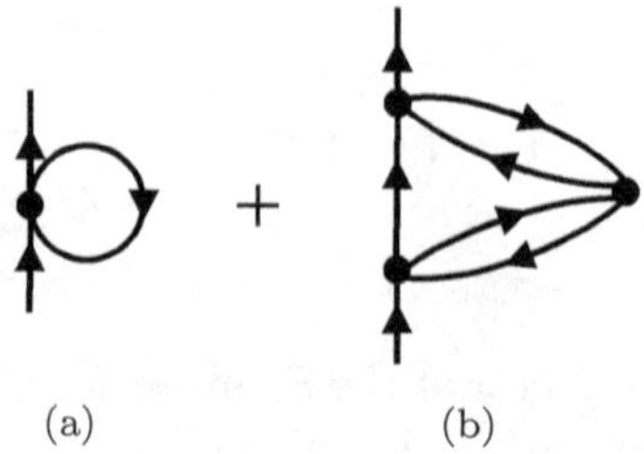

Fig. 3. Skeleton diagrams for the self energy to first order (a) and to third order (b) in the ladder t-matrix. Note that there is no second order contribution. The two-particle t-matrix in the ladder approximation is denoted by a black dot.

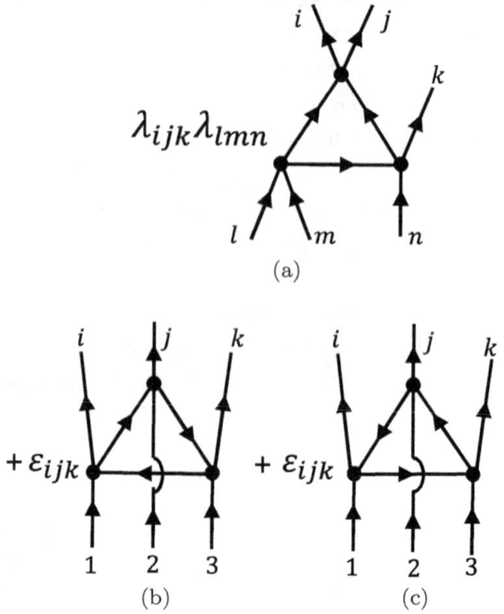

Fig. 4. Contribution of third order in the ladder t-matrix to the three-particle amplitude $\tilde{h}$. The diagram (a) is the second (next-to-leading) term in the Faddeev series. λ_{ijk} is unity for even permutations of (123) and zero otherwise, and an independent sum over (ijk) and (lmn) is implied. The diagrams (b) and (c) arise from the interaction between the three given particles and the particles in the Fermi sea, and do not appear in the usual Faddeev series. ε_{ijk} is the usual antisymmetric tensor, and a sum over $i, j, k = 1, 2, 3$ is implied. The two-particle t-matrix in the ladder approximation is denoted by a black dot.

literature since the works of Brueckner, Day, Bethe and others,[17,36,37] is to express Σ in terms of the two-particle t-matrix calculated in the ladder approximation to the BS equation, which is essentially the same as Brueckner's G-matrix. We have recently developed a scheme[14] to expand Σ in powers of the ladder t matrix, and to carry out the required functional derivatives to get the kernels $K^{(2)}$ and $K^{(3)}$. The self energy up to the third power of the ladder t-matrix is shown graphically in Fig. 3.

Figure 3a is the standard result in Brueckner approximation, and Fig. 3b describes the rescattering of the hole line in Fig. 3a, and the RPA-type correlations of third order in the ladder t-matrix. By calculating $K^{(3)}$ from these diagrams and inserting the result into Fig. 2, one obtains the term $h^{(2\mathrm{pc})}$ of Fig. 1a or Eq. (20) as the second order term, and the contributions shown in Fig. 4 as the third order terms.

The diagram Fig. 4a is the next-to-leading term of the in-medium Faddeev series, with n and k being the spectators of the first and last interactions. The term with $(ijk) = (lmn)$ comes from Fig. 2b, with the three-particle kernel $K^{(3)}$ of second order in the ladder t-matrix shown by Fig. 1a. The other terms with $(ijk) \neq (lmn)$ come from the third order contributions to the three-particle kernel $K^{(3)}$. The diagrams in Figs. 4b and 4c arise from the interaction of the three given particles on the Fermi surface with the particles in the Fermi sea, i.e., they are of four-body nature. These diagrams do not appear in the usual Faddeev series, because there it is assumed from the outset that the first and last interactions occur among the three given particles, and not between one of them and a background particle. The contribution where two intermediate lines in Figs. 4b and 4c have the same momenta come again from Fig. 2b, and the contribution where all three intermediate lines have the same momenta comes from Fig. 2c. Those terms where all three intermediate lines in Figs. 4b and 4c have different momenta come from the third order contributions to the three-particle kernel $K^{(3)}$.

4. Summary

The motivation for the first part of our work was the rapidly expanding interest in the slope parameters of the symmetry energy and the incompressibility of nuclear matter, two physical quantities which have decisive impact on the structure of nuclei and neutron stars. By using the Fermi liquid theory of Landau and Migdal, we derived two model independent relations for these observables, Eqs. (17) and (18). The crucial new point about these relations is that they connect three-body interaction parameters, where the isoscalar s-wave parameter H_0 and its isovector counterpart H_0' play the dominant roles, to observables of nuclear matter. By using empirical information about several other quantities in those relations, we concluded that H_0 must be positive and larger than unity, and H_0' must be negative and smaller in magnitude than unity. We made a semi-quantitative estimate of these two parameters by using the leading term of the Faddeev series and simple two-body contact interactions. We found that H_0 is indeed positive but tends to be smaller than unity, which gives us a hint that higher order terms in the Faddeev series, in particular the three-body cluster of Fig. 1b, give an important contribution to the skweness of nuclear matter. The results for H_0' were in the range of expectations (negative and smaller in magnitude than unity), which gives no clear hint about the role of three-body cluster for the slope parameter of the symmetry energy.

The motivation for the second part of our work was to get more understanding about the physics contained in the three-particle amplitudes in nuclear matter.

Fig. 5. Photo taken in summer 1963 in front of the Physics Building at Argonne National Laboratory. From left to right: A. W. Martin, A. Saperstein, K. C. Wali, H. J. Lipkin, M. Peshkin, W. D. McGlinn, F. Coester, J. Monahan, K. Hiida, A. S. Green, G. E. Brown, A. Arima, K. Tanaka, D. Kurath, M. A. Melvin, N. Rosenzweig, K. Ekstein, S. Tani, M. N. Hack, B. Kursunoglu, T. Sebe, R. A. Ferrell, D. R. Inglis, B. M. Udaonkar, S. Fallieros, S. P. Pandya. Reprinted with kind permission of Argonne National Laboratory.

By using only its definition in the Fermi liquid theory, we expressed it as the sum of retardation terms ($h^{(\mathrm{prod})}$), which come from the energy dependence of the self energy and/or the two-particle t-matrix, and the part $\tilde{h}$ of Fig. 2, which is characterized by the particle-hole irreducible three-particle kernel $K^{(3)}$. In order to get more detailed expressions, we followed the method of the Bethe-Brueckner-Goldstone (BBG) theory and used the ladder approximation to the two-body Bethe-Salpeter equation for the t-matrix as a building block. By expanding all quantities up to the third power of the ladder t-matrix, we found that the Fermi liquid theory naturally leads not only to the first few terms in the Faddeev series (see Fig. 1), but also to medium modifications of four-body nature, induced by the interaction of the three particles on the Fermi surface with the particles in the Fermi sea. It would be very interesting to see how large these terms are in comparison to the usual Faddeev terms, and which kind of medium induced three-body processes appear in higher orders.

Acknowledgments

The authors would like to dedicate their work to Professor Akito Arima, the great physicist, teacher, and friend. Shortly after one of us (W.B.) joined the Nuclear Theory Group at the University of Tokyo 40 years ago, Prof. Arima suggested me to study the connection between the work of the Tokyo group on nuclear electromagnetic properties, which started with the famous Arima-Horie Effect for magnetic moments in 1954, and the Fermi liquid theory. With his help, I could see many similar physical ideas, expressed in a very different and independent language. When Prof. A. B. Migdal visited the Arima group in 1989, he suggested me to work out the connection between Arima's collective bosons and Landau's quasiparticles. Although I did not succeed to accomplish this job, I could get a flavor of Prof. Arima's deep thoughts about different methods to describe nuclear phenomena. I wish to thank him for his generous guidance and support over so many years.

I.C., who is now working at the Nuclear Theory Group in Argonne, would like to contribute a photo, which was taken in the front of the Physics Building in summer 1963, a few years after Prof. Arima's stay as a researcher at Argonne. We can imagine how comfortable and happy he felt among so many active people and friends. With his experiences in Argonne, he started the shell-model tradition in Japan, being still developed to date.

Finally, the authors would like to express their thanks to the Editors and Publishers of this Memorial Book for their kind cooperation.

References

1. M. Baldo and G. F. Burgio, The nuclear symmetry energy, *Prog. Part. Nucl. Phys.* **91**, 203–258 (2016). doi: 10.1016/j.ppnp.2016.06.006.
2. C. J. Horowitz and J. Piekarewicz, Neutron star structure and the neutron radius of Pb-208, *Phys. Rev. Lett.* **86**, 5647 (2001). doi: 10.1103/PhysRevLett.86.5647.
3. J. P. Blaizot, Nuclear compressibilities, *Phys. Rept.* **64**, 171–248 (1980). doi: 10.1016/0370-1573(80)90001-0.
4. B.-A. Li, B.-J. Cai, L.-W. Chen, and J. Xu, Nucleon effective masses in neutron-rich matter, *Prog. Part. Nucl. Phys.* **99**, 29–119 (2018). doi: 10.1016/j.ppnp.2018.01.001.
5. G. Colo, U. Garg, and H. Sagawa, Symmetry energy from the nuclear collective motion: constraints from dipole, quadrupole, monopole and spin-dipole resonances, *Eur. Phys. J. A* **50**, 26 (2014). doi: 10.1140/epja/i2014-14026-9.
6. J. Piekarewicz and F. J. Fattoyev, Neutron rich matter in heaven and on Earth (2019). doi: 10.1063/PT.3.4247.
7. V. Baran, M. Colonna, V. Greco, and M. Di Toro, Reaction dynamics with exotic beams, *Phys. Rept.* **410**, 335–466 (2005). doi: 10.1016/j.physrep.2004.12.004.
8. J. M. Pearson, The incompressibility of nuclear matter and the breathing mode, *Phys. Lett. B* **271**, 12–16 (1991). doi: 10.1016/0370-2693(91)91269-2.
9. A. W. Steiner, J. M. Lattimer, and E. F. Brown, The neutron star mass-radius relation and the equation of state of dense matter, *Astrophys. J.* **765**, L5 (2013). doi: 10.1088/2041-8205/765/1/L5.
10. P. Danielewicz, R. Lacey, and W. G. Lynch, Determination of the equation of state of dense matter, *Science* **298**, 1592–1596 (2002). doi: 10.1126/science.1078070.

11. L. D. Landau, The theory of a Fermi liquid, *Sov. Phys. JETP.* **3**, 920 (1956). URL http://www.jetp.ac.ru/cgi-bin/e/index/e/3/6/p920?a=list.
12. L. D. Landau, Oscillations in a Fermi liquid, *Sov. Phys. JETP.* **5**, 101 (1957). URL http://www.jetp.ac.ru/cgi-bin/e/index/e/5/1/p101?a=list.
13. W. Bentz and I. C. Cloet, Skewness of nuclear matter and three-particle correlations, *Phys. Rev. C* **100**(1), 014303 (2019). doi: 10.1103/PhysRevC.100.014303.
14. W. Bentz and I. C. Cloet, Slope parameter of the symmetry energy and the structure of three-particle interactions in nuclear matter, *Phys. Rev.* **C105**(1), 014320 (2022).
15. H. A. Bethe, Three-body correlations in nuclear matter, *Phys. Rev.* **138**, B804–B822 (1965). doi: 10.1103/PhysRev.138.B804.
16. B. D. Day, Three-body correlations in nuclear matter, *Phys. Rev. C* **24**, 1203–1271 (1981). doi: 10.1103/PhysRevC.24.1203.
17. H. A. Bethe, Theory of nuclear matter, *Ann. Rev. Nucl. Part. Sci.* **21**, 93–244 (1971). doi: 10.1146/annurev.ns.21.120171.000521.
18. L. D. Landau, On the Theory of the Fermi liquid, *Sov. Phys. JETP.* **8**, 70 (1959). URL http://www.jetp.ac.ru/cgi-bin/e/index/e/8/1/p70?a=list.
19. A. B. Migdal, *Theory of Finite Fermi Systems and Applications to Atomic Nuclei.* New York, Wiley (1967).
20. R. Shankar, Renormalization group approach to interacting fermions, *Rev. Mod. Phys.* **66**, 129–192 (1994). doi: 10.1103/RevModPhys.66.129.
21. J. Speth, S. Krewald, F. Grümmer, P. G. Reinhard, N. Lyutorovich, and V. Tselyaev, Landau–Migdal vs. Skyrme, *Nucl. Phys. A* **928**, 17–29 (2014). doi: 10.1016/j.nuclphysa.2014.03.023.
22. G. Baym and S. A. Chin, Landau theory of relativistic Fermi liquids, *Nucl. Phys. A* **262**, 527–538 (1976). doi: 10.1016/0375-9474(76)90513-3.
23. W. Bentz, A. Arima, H. Hyuga, K. Shimizu, and K. Yazaki, Ward identity in the many body system and magnetic moments, *Nucl. Phys. A* **436**, 593–620 (1985). doi: 10.1016/0375-9474(85)90550-0.
24. A. Arima, K. Shimizu, W. Bentz, and H. Hyuga, Nuclear magnetic properties and Gamow-Teller transitions, *Adv. Nucl. Phys.* **18**, 1–106 (1987).
25. W. Bentz and A. Arima, The Relation between the photonuclear E1 sum rule and the effective orbital g factor, *Nucl. Phys. A* **736**, 93–125 (2004). doi: 10.1016/j.nuclphysa.2004.02.004.
26. J. Negele and H. Orland, *Quantum Many-particle Systems.* Westview Press (1998). ISBN 0738200522. URL http://www.worldcat.org/isbn/0738200522.
27. G. Baym and C. Pethick, *Landau Fermi-Liquid Theory: Concepts and Applications.* WILEY-VCH Verlag (2004). ISBN 9780471824183.
28. B.-J. Cai and L.-W. Chen, Constraints on the skewness coefficient of symmetric nuclear matter within the nonlinear relativistic mean field model, *Nucl. Sci. Tech.* **28**(12), 185 (2017). doi: 10.1007/s41365-017-0329-1.
29. C. Mahaux, P. F. Bortignon, R. A. Broglia, and C. H. Dasso, Dynamics of the shell model, *Phys. Rept.* **120**, 1–274 (1985). doi: 10.1016/0370-1573(85)90100-0.
30. J. P. Blaizot and B. L. Friman, On the nucleon effective mass in nuclear matter, *Nucl. Phys. A* **372**, 69–89 (1981). doi: 10.1016/0375-9474(81)90087-7.
31. E. N. E. van Dalen, C. Fuchs, and A. Faessler, Momentum, density, and isospin dependence of the symmetric and asymmetric nuclear matter properties, *Phys. Rev. C* **72**, 065803 (2005). doi: 10.1103/PhysRevC.72.065803.
32. S. Goriely, N. Chamel, and J. M. Pearson, Further explorations of Skyrme-Hartree-Fock-Bogoliubov mass formulas. XII: Stiffness and stability of neutron-star matter, *Phys. Rev. C* **82**, 035804 (2010). doi: 10.1103/PhysRevC.82.035804.

33. Z. Zhang and L.-W. Chen, Extended skyrme interactions for nuclear matter, finite nuclei, and neutron stars, *Phys. Rev. C* **94**, 064326 (Dec, 2016). doi: 10.1103/PhysRevC.94.064326. URL https://link.aps.org/doi/10.1103/PhysRevC.94.064326.
34. G. Baym and L. P. Kadanoff, Conservation laws and correlation functions, *Phys. Rev.* **124**, 287–299 (1961). doi: 10.1103/PhysRev.124.287.
35. J. Speth, Transition probabilities and static moments of excited states in even-even nuclei, *Zeitschrift für Physik A Hadrons and Nuclei.* **239**(3), 249–265 (1970). ISSN 0939-7922. doi: 10.1007/BF01398115.
36. K. A. Brueckner and C. A. Levinson, Approximate reduction of the many-body problem for strongly interacting particles to a problem of self-consistent fields, *Phys. Rev.* **97**, 1344–1352 (1955). doi: 10.1103/PhysRev.97.1344.
37. B. D. Day, Elements of the Brueckner-Goldstone theory of nuclear matter, *Rev. Mod. Phys.* **39**, 719–744 (1967). doi: 10.1103/RevModPhys.39.719.

https://doi.org/10.1142/9789819813353_0003

The Quenching of Spin- and Spin-Isospin-Dependent Operators in Nuclear Structure: Present Microscopic Approach

L. Coraggio and N. Itaco

Dipartimento di Matematica e Fisica, Università degli Studi della Campania "Luigi Vanvitelli", viale Abramo Lincoln 5 - I-81100 Caserta, Italy and Istituto Nazionale di Fisica Nucleare, Complesso Universitario di Monte S. Angelo, Via Cintia - I-80126 Napoli, Italy

The need of reliable nuclear matrix elements for the $0\nu\beta\beta$ decay has restarted a new interest about the quenching of coupling constants in spin- and spin-isospin-dependent operators in order to reproduce experimental results connected with $M1$ transitions and GT decays. The content of this work is focussed to review briefly the present status of microscopic studies tackling such a problem. Akito Arima has been a pioneer in this field, that is tied to the search of new physics beyond the Standard Model.

1. Introduction

Among the few items which may reveal "new physics" beyond the Standard Model, there is the detection of neutrinoless double-β decay ($0\nu\beta\beta$). In fact, the observation of such a rare process would be the evidence of a lepton number violation, and, moreover, it would provide fundamental informations about the nature and properties of the neutrino.[1,2]

The link between $0\nu\beta\beta$ decay and the structure of atomic nuclei is the expression of the half-life of the process, which may be written as

$$\left[T_{1/2}^{0\nu}\right]^{-1} = G^{0\nu} \left|M^{0\nu}\right|^2 \left|f(m_i, U_{ei})\right|^2, \tag{1}$$

where $G^{0\nu}$ is the so-called phase-space factor (or kinematic factor),[3,4] $M^{0\nu}$ is the nuclear matrix element directly related to the wave functions of the parent and grand-daughter nuclei, and $f(m_i, U_{ei})$ accounts for the physics beyond the standard model through the neutrino masses m_i and their mixing matrix elements U_{ei}. The explicit form of $f(m_i, U_{ei})$ depends on the adopted model of $0\nu\beta\beta$ decay, and, if we consider the mechanism of the light-neutrino exchange, $f(m_i, U_{ei})$ has the following expression:

$$f(m_i, U_{ei}) = g_A^2 \frac{\langle m_\nu \rangle}{m_e},$$

where g_A is the axial coupling constant, m_e is the electron mass, and $\langle m_\nu \rangle = \sum_i (U_{ei})^2 m_i$ is the effective neutrino mass.

The above expressions shows that a reliable calculation of $M^{0\nu}$ plays a crucial role to obtain important informations about the neutrino effective mass, and also to evaluate the sensitivity of experimental apparatuses to detect $0\nu\beta\beta$ decay, since by combining $M^{0\nu}$ with neutrino mixing parameters[5] and limits on $\langle m_\nu \rangle$ from current experiments, it is possible to estimate the half-life an experiment should measure in order to be sensitive to a particular value of the neutrino effective mass.[1]

Since the $0\nu\beta\beta$ decay is ruled by the Gamow-Teller (GT) spin-isospin-dependent operator, the predictiveness of nuclear structure calculations can be tested by reproducing observables that are related to spin- and spin-isospin-dependent operators, namely magnetic dipole transitions and β-decay amplitudes with neutrino emission.

Actually, it is well known that most of nuclear structure calculations overestimate the experimental values of quantities connected to $M1$ and GT transitions, and this defect is usually treated by quenching both the spin factor g_s and the axial coupling constant g_A by a factor $q < 1$.[6–8] The need of introducing effective values of g_s, g_A can be traced back to two main sources:

(1) nucleons are not point-like particles, and their quark structure needs to be accounted for by considering the effects of meson-exchange currents.
(2) All nuclear models operate a truncation of the nuclear degrees of freedom, in order to allow the diagonalization of the nuclear Hamiltonian. Consequently, effective Hamiltonians and decay operators must be considered to account for the neglected configurations in the nuclear wave function.[9]

Akito Arima has been a forerunner in the study of the derivation of effective spin- and spin-isospin-dependent operators within the shell model (SM) framework, a scientific interest he started when visiting Stony Brook as a professor in the first half of 1970s,[10,11] and which he expanded through 1980s,[12–14] up to a fundamental review paper on this topic.[15] As a matter of fact, he exploited both sources of the renormalization of shell-model $M1$ and GT transition operators, namely the role of meson-exchange currents as well as the derivation of effective operators accounting for the truncation of the Hilbert space to the SM space.[15]

We want to celebrate his fundamental contribution to this research topic, which is nowadays the subject of a renewed interest matching nuclear structure with beyond-standard-model physics, presenting the results we have obtained calculating SM effective operators for $M1$- and β-decay processes in the region of fp-shell nuclei, considering ^{40}Ca as a doubly-closed core, as well as the SM effective Hamiltonian $H_{\rm eff}$.

To this end, we start from a nuclear potential based on chiral perturbation theory (ChPT),[16,17] that consists of a high-precision two-nucleon (2N) potential derived within the ChPT at next-to-next-to-next-to-leading order (N^{3}LO),[18] and a three-nucleon (3N) component at N^{2}LO in ChPT.[19,20] Then, by way of the many-body perturbation theory,[21–24] we derive consistently the SM effective Hamiltonian for the fp-shell and the one-body decay operators, which account for the truncation of the dimension of the configuration space to the one spanned by the four $0f1p$ proton and neutron orbitals outside doubly-closed ^{40}Ca.

At this stage of our research project, we are not including, for the $M1$ and GT operators, the contributions arising from meson-exchange currents.[9,25] As a matter of fact, ChPT provides a consistent way to construct both nuclear forces and electroweak currents which account for the composite structure of the nucleons,[26–28] an approach that is grounded in the effective field theory (EFT) and that we are currently working out for shell-model calculations.

Our goal is to compare with experiment the results obtained with a SM calculation, employing a H_{eff} derived from a realistic potential. This comparison will be done using both single-particle (SP) matrix elements of the bare $M1$ and GT operators and the one of the SM effective operators, that have been caluclated consistently with the derivation of H_{eff}. This operation should evidence the ability of a theoretical approach to the renormalization of decay operators to provide agreement with data without resorting to quenching factors q.

The focus will be spotted on SP decay properties of one-valence-particle nuclei ^{41}Ca and ^{41}Sc, as well as on the two-neutrino double-β ($2\nu\beta\beta$) decay and GT-strength distribution in ^{48}Ca, that is one of the nuclei which are the candidates for the detection of $0\nu\beta\beta$-decay.

This work is organized as follows: in the following section we sketch out the fundamentals of the theoretical framework where our calculations are grounded, whose details can be found in Refs. 24, 29, 30. In section 3, they are reported the results of the calculated dipole moments of ^{41}Ca and ^{41}Sc, as well as those of the β-decay of $^{41}\text{Sc}\rightarrow^{41}\text{Ca}$, and compare them with available data. They are a direct test of the quality of the effective decay operators, that we have employed to calculate dipole moments of ^{48}Ti, reduced transition probabilities $B(M1)$ in ^{48}Ca, GT transition-strength distributions of ^{48}Ca, and the nuclear matrix element $M^{2\nu}$ for the $2\nu\beta\beta$-decay of ^{48}Ca, which have all experimental counterparts.

Finally, in the last section we report a summary of our work and draw the conclusions which could be inferred from the comparison with data.

2. Outline of the Theory

As mentioned in the Introduction, we derive the effective SM Hamiltonian and the effective operators starting from a realistic nuclear potential consisting in a high-precision 2N potential,[18] and a 3N term,[19,20] derived within the ChPT at next-to-next-to-next-to-leading order (N^{3}LO) and at next-to-next-to-leading order (N^{2}LO), respectively. The two components of the potential (2N and 3N) share the same non-local regulator function and some low-energy constants (LECs) entering their formal expression. As a matter of fact, the c_1, c_3, and c_4 LECs appearing in the two-pion exchange term of the N^{2}LO 3N potential are the same that enter the N^{3}LO 2N force, and their values are therefore determined by means of the renormalization procedure employed to fit the nucleon-nucleon (NN) data.[17]

However, to fully constrain the N^2LO 3N force it is necessary to fix the values of two extra LECs, c_D and c_E, characterizing the one-pion exchange and the contact term, respectively. These values cannot be fitted on two-body observables, but need to be determined by reproducing observables in systems with mass $A > 2$.

We adopt the values $c_D = -1$ and $c_E = -0.34$, that have been determined through no-core shell model (NCSM) calculations,[19,29] where a set of observables in light p-shell nuclei strongly sensitive to the c_D value has been employed to fix its value, while c_E has been constrained to reproduce the binding energies of the $A = 3$ systems.

A detailed description of the calculation of our 3N potential matrix elements in the harmonic-oscillator (HO) basis can be found in Appendix of Ref. 32.

Once the nuclear potential has been chosen, the next step is, as mentioned before, to derive H_{eff} in the model space spanned by the $0f_{7/2}, 0f_{5/2}, 1p_{3/2}, 1p_{1/2}$ proton and neutron orbitals outside doubly-closed ^{40}Ca.

To this end, we derive H_{eff} by way of the time-dependent perturbation theory.[22,30,31] Namely, H_{eff} is expressed through the Kuo-Lee-Ratcliff (KLR) folded-diagram expansion in terms of the vertex function $\hat{Q}$ box, which is composed of irreducible valence-linked diagrams.[21,32] We include in the $\hat{Q}$ box one- and two-body Goldstone diagrams through third order in the 2N potential, and up to first order in the 3N one.

In Fig. 1 we report the one- and two-body diagrams that enter the $\hat{Q}$ box at first order in 3N potential.

For the sake of clarity, we report their explicit expressions in terms of angular-momentum coupled, antisymmetrized but not normalized three-body matrix elements (3BME) $\langle [(j_a j_b)_{J_{ab}}, j_c]_J | V_{3N} | [(j_d j_e)_{J_{de}}, j_f]_J \rangle$:

$$\langle j_a | 1\text{b}_{3N} | j_a \rangle = \sum_{\substack{h_1,h_2 \\ J_{12} J}} \frac{\hat{J}^2}{2\hat{j}_a^2} \langle [(j_{h_1} j_{h_2})_{J_{12}}, j_a]_J | V_{3N} | [(j_{h_1} j_{h_2})_{J_{12}}, j_a]_J \rangle, \tag{2}$$

$$\langle (j_a j_b)_J | 2\text{b}_{3N} | (j_c j_d)_J \rangle = \sum_{h,J'} \frac{\hat{J'}^2}{\hat{J}^2} \langle [(j_a j_b)_J, j_h]_{J'} | V_{3N} | [(j_c j_d)_J, j_h]_{J'} \rangle. \tag{3}$$

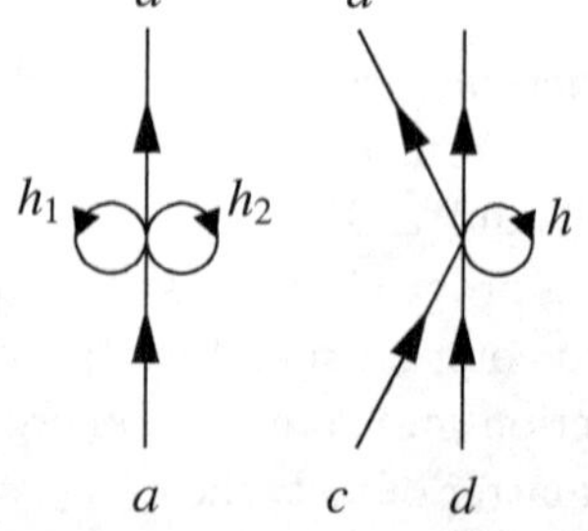

Fig. 1. First-order one- and two-body diagrams with a three-body-force vertex. See text for details.

The above expressions in Eqs. (2) and (3) correspond to the coefficients of the one-body and two-body terms, respectively, arising from the normal-ordering decomposition of the three-body component of a many-body Hamiltonian.[33]

It is worth to note that the effective Hamiltonian for one valence-nucleon systems ^{41}Sc, ^{41}Ca provides the SP energies for the SM calculations, then the two-body matrix elements (TBME) of the SM Hamiltonian are obtained by way of a subtraction procedure from H_{eff} for nuclei with two valence nucleons outside the doubly-closed ^{40}Ca.

Actually, nuclei such as ^{48}Ca and ^{48}Ti, which are the subjects of our SM calculations whose results are reported in Section 3, own more than two valence nucleons, so effective many-body forces — arising from the interaction via the two-body force of the valence nucleons with excitations outside the model space[34] — have to be included in the perturbative expansion of H_{eff}, besides the chiral N^{2}LO force.

To this end, we also account for the contributions of three-body diagrams calculated at second order in perturbation theory.[35] Actually, present SM codes cannot manage three-body SM Hamiltonians, then we have derived a density-dependent two-body contribution at one-loop order from the three-body diagrams, summing over the partially-filled model-space orbitals.[24]

The SP energies and TBME that have been employed in our present study are the same that have been derived for a study of the contribution of chiral three-body forces to the monopole component of H_{eff}.[36]

The main goal of this paper is, however, the derivation of the SP matrix elements of SM effective $M1$ and GT operators, which are connected to measurable quantities such as dipole moments, $B(M1)$ strengths, GT-strength distributions, and the nuclear matrix element of the $2\nu\beta\beta$ decay $M^{2\nu}$.

The derivation of SM effective operators within a perturbative approach dates back to the earliest attempts to employ realistic potentials for SM calculations,[9–11, 34, 37–39] and here we have followed a procedure, that has been introduced by Suzuki and Okamoto,[23] which provides decay operators consistently with the approach pursued to obtain H_{eff}. As a matter of fact, it is based on perturbative expansion of a vertex function $\hat{\Theta}$ box, analogously with the derivation of H_{eff} in terms of the $\hat{Q}$ box.

Our procedure has been reported in details in Ref. 24, and applied in previous investigations about effective spin- and spin-isospin dependent operators.[40–42]

Also the expansion of the vertex fuction $\hat{\Theta}$ box has been carried out at third-order in perturbation theory, and in Fig. 2 they are reported all the diagrams up to second order for a one-body operator Θ.

Now, it is in order a brief description of the calculation of $M^{2\nu}$ and the running sums of the GT strengths ($\Sigma B(p,n)$), whose results we report in Section 3.

Two-neutrino double-β decays are the occurrence of two single-β decay transitions inside a nucleus, and the expression of the GT component of the nuclear matrix

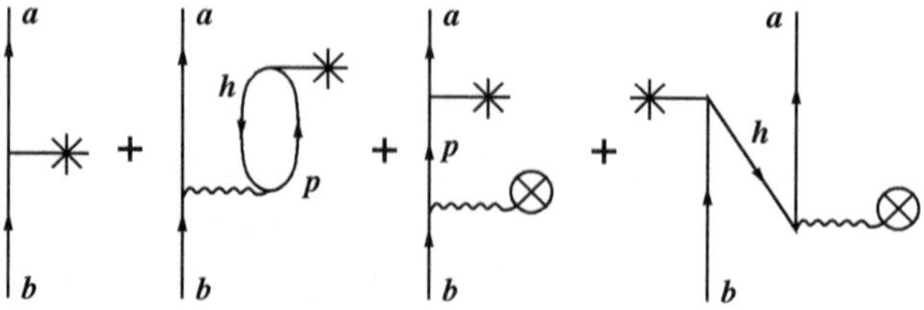

Fig. 2. One-body second-order diagrams included in the perturbative expansion of $\hat{\Theta}$ box. The asterisk indicates the bare operator Θ, the wavy lines the two-body NN interaction, the circle with a cross inside accounts for the self-energy contribution (see Ref. 24).

element $M^{2\nu}$ is the following

$$M^{2\nu}_{\rm GT} = \sum_n \frac{\langle 0^+_f || (\vec{\sigma}\tau^-)_{\mathcal{I}} || 1^+_n \rangle \langle 1^+_n || (\vec{\sigma}\tau^-)_{\mathcal{I}} || 0^+_i \rangle}{E_n + E_0}, \quad (4)$$

where the subscript $\mathcal{I}$ indicates we are employing the matrix elements of either the bare or the effective one-body GT operators.

In Eq. 4, E_n is the excitation energy of the $J^\pi = 1^+_n$ intermediate state, and $E_0 = \frac{1}{2} Q_{\beta\beta}(0^+) + \Delta M$, where $Q_{\beta\beta}(0^+)$ and ΔM are the Q value of the transition and the mass difference of the parent and daughter nuclear states, respectively. The index n runs over all possible intermediate states induced by the given transition operator.

We have adopted for our calculations the Lanczos strength-function method,[43] that is the most efficient way to obtain $M^{2\nu}$, by including a number of intermediate states that is sufficient to provide the needed accuracy for its calculation.

As regards the calculation of $\Sigma B(p,n)$, obtained with both bare and effective GT operators, we remind that the GT strength is defined as follows:

$$B(p,n) = \frac{\left| \langle \Phi_f || \sum_j \vec{\sigma}_j \tau^-_j || \Phi_i \rangle \right|^2}{2J_i + 1}, \quad (5)$$

where indices i, f refer to the parent and grand-daughter nuclei, respectively, and the sum is over all interacting nucleons.

The single-β decay GT strengths, defined by the above equation, can be obtained experimentally by way of intermediate-energy charge-exchange reactions, since the β-decay process is forbidden for ^{48}Ca $\rightarrow$ ^{48}Sc. As a matter of fact, the GT strength can be extracted from the GT component of the cross section at zero degree, following the standard approach in the distorted-wave Born approximation (DWBA):[44,45]

$$\frac{d\sigma^{GT}(0^\circ)}{d\Omega} = \left(\frac{\mu}{\pi \hbar^2} \right)^2 \frac{k_f}{k_i} N^{\sigma\tau}_D |J_{\sigma\tau}|^2 B(p,n), \quad (6)$$

where $N^{\sigma\tau}_D$ is the distortion factor, $|J_{\sigma\tau}|$ is the volume integral of the effective NN interaction, k_i and k_f are the initial and final momenta, respectively, and μ is the reduced mass.

3. Results and Comparison with Experiment

We start the comparison of our results with data considering the magnetic dipole moments of ^{41}Sc and ^{41}Ca, which are reported in Table 1. The calculated values reported in columns I and II are obtained with the bare magnetic-dipole operator and the effective one, respectively.

The comparison between experimental and calculated magnetic dipole moments μ is a direct test for the theoretical $M1$ operator because of the relationship:

$$\mu_j = \sqrt{\frac{4\pi}{3}} \sqrt{\frac{j}{(j+1)(2j+1)}} \langle j||M1||j\rangle \,, \tag{7}$$

where j denotes the total angular momentum of the SP state, and $\langle j||M1||j\rangle$ is the diagonal one-body reduced matrix element of the $M1$ transition operator, that in our calculations may be the bare or the effective one.

As can be seen, the effective $M1$ operator provides a larger quenching factor for neutrons than protons, being $q = 0.77$ for ^{41}Ca and $q = 0.86$ for ^{41}Sc. As regards the comparison with the experimental counterparts,[46,47] the action of the effective operator does not lead to a real improvement for ^{41}Sc, while it leads a better result with respect to the one calculated with bare operator for ^{41}Ca. Actually, it should be pointed out that the discrepancy between the theoretical value of the dipole moment, as obtained with the bare $M1$ operator, and experiment is as a percentage larger for ^{41}Ca than ^{41}Sc.

The experimental GT $\log ft_{\text{GT}}$ for the β^+-decay of ^{41}Sc, from its ground state to the one of ^{41}Ca, has been measured in Ref. 49 and its value is 3.736 ± 0.014. The results obtained with bare and effective GT-decay operators are 3.472 and 3.572, respectively. Both calculated values underestimate the experimental one, a flaw that evidence the need of a quenching of the matrix elements of the GT operator, the result obtained with the effective GT operator being closer to the datum.

So far, we have considered only the case of nuclei with only one-valence particle outside doubly-closed ^{40}Ca, but now we move to focus our attention to the ^{48}Ca and ^{48}Ti nuclear systems, since they are those which are involved in the $0\nu\beta\beta$ decay of ^{48}Ca, a process that is nowadays the subject of a renewed experimental interest[48] as well as of theoretical investigations with nuclear shell model and *ab initio* methods.[49–56]

Table 1. Calculated (I and II) and experimental magnetic dipole moments (in μ_N) of ^{41}Sc and ^{41}Ca.

Nucleus	J^π	I	II	Expt
^{41}Sc				
	$\frac{7}{2}^-_1$	+5.793	+5.001	+5.431[46]
^{41}Ca				
	$\frac{7}{2}^-_1$	−1.913	−1.468	−1.595[47]

As mentioned in the Introduction we consider spectroscopic properties that deal with $M1$ and GT transitions which have experimental counterparts, so that we can provide an assessment of our theoretical approach to the calculation of SM effective operators, without resorting to empirical quenching factors q. We made already a similar analysis in a previous work,[41] but the starting point of the derivation of SM effective operators was the high-precision CD-Bonn 2N potential,[57] whose non-perturbative high-momentum components was integrated out by way of the so-called $V_{\text{low}-\text{k}}$ approach.[58]

First, in Fig. 3, we show the experimental[59,60] and calculated low-energy spectra of ^{48}Ca and ^{48}Ti. Next to the arrows, that are proportional to the $B(E2)$ strengths, we report the explicit experimental[59,60] and calculated $B(E2)$s in $e^2\text{fm}^4$, that have been obtained with SM effective operators and free proton and neutron charges $e_p = e$, $e_n = 0$.

As can be seen, we reproduce the observed shell-closure of the neutron $0f_{7/2}$ orbital in ^{48}Ca and the agreement between the experimental and calculated spectra, as well as the $B(E2)$s, is quite satisfying. We point out that we have shown that our H_{eff}, derived from chiral 2N and 3N potentials — including also effective 3NF forces (see previous section) — are able to provide a successful description of the shell-evolution of isotopic chains belonging to the fp region.[36]

In Table 2 the low-energy experimental and calculated observables related to the $M1$ operator are reported, the legenda being the same as in Table 1.

From the inspection of the above table, it can be seen that the calculated $B(M1)$ transition rates compare well with the observed value for ^{48}Ca. In particular, the values obtained employing the effective shell-model $M1$ operator (II) are quenched with respect to that calculated with the bare operator (I), and in a better agreement with experiment.

As regards the magnetic moments, data are available for ^{48}Ti, and they are underestimated by the theory. However, the contribution due to the effective

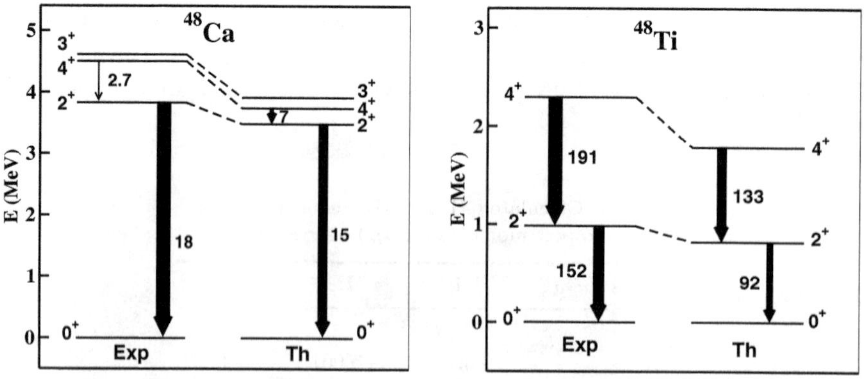

Fig. 3. Experimental and calculated spectra of ^{48}Ca and ^{48}Ti. $B(E2)$ strengths (in $e^2\text{fm}^4$) are also reported (see text for details).

Table 2. Experimental and calculated $B(M1)$ strengths (in μ_N^2) and magnetic dipole moments (in μ_N) of ^{48}Ca and ^{48}Ti. We report those for the observed states in Fig. 3 (see text for details).

Nucleus	$J_i \rightarrow J_f$	I	II	$B(M1)_{\mathrm{Expt}}$
^{48}Ca				
	$3_1^+ \rightarrow 2_1^+$	0.090	0.044	0.023 ± 0.004[61]
Nucleus	J^π	**I**	**II**	μ_{Expt}
^{48}Ti				
	2_1^+	0.26	0.34	$+0.78 \pm 0.04$[59]
	4_1^+	1.0	1.1	$+2.2 \pm 0.5$[59]

Table 3. Experimental[62] and calculated $M^{2\nu}$ of the $2\nu\beta\beta$ decay (in MeV^{-1}) for ^{48}Ca. The same notation of Table 1 is used (see text for details).

Decay	I	II	$M^{2\nu}_{\mathrm{Expt}}$
^{48}Ca→^{48}Ti	0.057	0.047	0.042 ± 0.004

operator points in the right direction, leading to a slightly better agreement with experiment.

Now, we start to examine the results for GT transitions.

In Table 3 we report the observed and calculated values of $M^{2\nu}$ for the $2\nu\beta\beta$ decay of ^{48}Ca into ^{48}Ti. Calculated $M^{2\nu}$s both overerestimate the experimental one, but we have a better agreement by employing the effective operator, this value being 10% larger than the experimental one, since that obtained with the bare one (I) exceeds the experimental $M^{2\nu}$ by 35%.

In this context, it should be mentioned that in our previous work,[41] the $M^{2\nu}$ calculated values slightly underestimated the experimental one, but the different outcome with respect to those reported in Table 3 should be related to the fact that the wave functions with chiral 2N and 3N forces are able to reproduce the closure property of ^{48}Ca, at variance with the result obtained employing a $V_{\text{low-}k}$ derived from CD-Bonn 2N potential.[41]

Finally, in Fig. 4, the calculated $\Sigma B(p,n)$ for ^{48}Ca is shown as a function of the excitation energy, and compared with the data reported with a red line.[63] The results obtained with the bare operator (I) are drawn with a blue line, while those obtained employing the effective GT operator is plotted using a black line.

It can be seen that the distribution obtained using the bare operator (I) strongly overestimates the observed one, while the one provided by the effective GT operator (II) is closer to experiment.

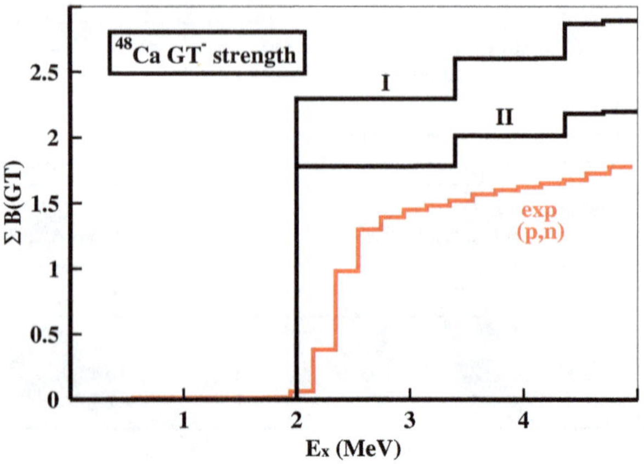

Fig. 4. Running sums of the ^{48}Ca $B(p,n)$ strengths as a function of the excitation energy E_x up to 5 MeV (see text for details).

The results obtained with the SM effective operator, as reported in Table 3 and in Fig. 4, evidence that the contribution to the renormalization of the GT transition operator, which accounts for the truncation of the Hilbert to the SM model space, goes to the right direction to improve the comparison with data. It is worth pointing out that an investigation about the role of meson-exchange currents to GT transitions is in order, this further renormalization availing space to further improvements.

4. Summary and Conclusions

In this work, the results of SM calculations of spin and spin-isospin dependent quantities in the region of fp-shell nuclei have been presented. Our goal has been to show how both the approach to the derivation of shell-model effective operators — the Hamiltonian and decay operators — and the construction of nuclear forces rooted in the fundamental symmetries of QCD have evolved to reproduce successfully observables without resorting to empirical parameters, adjusted to reproduce the experiment.

Such an endeavour aims to bridge the calculation of reliable nuclear wave functions with the research that investigates the existence of new physics, beyond the present limits of the Standard Model, and then to crossover different open questions about the understanding of quantum systems.

As a part of the community of the researchers who are involved in these kind of investigations, we are deeply benefitting from standing on the shoulders of giants such as Akito Arima, who paved the way which opens before us. It is our committment to take advantage of his efforts and heritage, in order to contribute to the progress of our knowledge of the physics of atomic nuclei.

References

1. F. T. Avignone, S. R. Elliott, and J. Engel, *Rev. Mod. Phys.* **80**, 481 (2008).
2. J. D. Vergados, H. Ejiri, and F. Šimkovic, *Rep. Prog. Phys.* **75**, 106301 (2012).
3. J. Kotila and F. Iachello, *Phys. Rev. C* **85**, 034316 (2012).
4. J. Kotila and F. Iachello, *Phys. Rev. C* **87**, 024313 (2013).
5. M. Tanabashi *et al.*, *Phys. Rev. D* **98**, 030001 (2018).
6. H. Ejiri, *Nucl. Phys. A* **396**, 181 (1983).
7. G. Martínez-Pinedo, A. Poves, E. Caurier, and A. P. Zuker, *Phys. Rev. C* **53**, R2602 (1996).
8. J. Barea, J. Kotila, and F. Iachello, *Phys. Rev. C* **91**, 034304 (2015).
9. I. S. Towner, *Phys. Rep.* **155**, 263 (1987).
10. H. Hyuga and A. Arima, *J. Phys. Soc. Jpn. Suppl.* **34**, 538 (1973).
11. K. Shimizu, M. Ichimura, and A. Arima, *Nucl. Phys. A* **226**, 282 (1974).
12. A. Arima, *Prog. Part. Nucl. Phys.* **11**, 53 (1984).
13. T. Cheon, K. Shimizu, and A. Arima, *Phys. Lett. B* **138**, 345 (1984).
14. W. Bentz, A. Arima, H. Hyuga, K. Shimizu, and K. Yazaki, *Nucl. Phys. A* **436**, 593 (1985).
15. A. Arima, K. Shimizu, and W. Bentz, *Adv. Nucl. Phys.* **18**, 1 (2003).
16. E. Epelbaum, W. Glöckle, and U.-G. Meissner, *Nucl. Phys. A* **747**, 362 (2005).
17. D. R. Entem and R. Machleidt, *Phys. Rep.* **503**, 1 (2011).
18. D. R. Entem and R. Machleidt, *Phys. Rev. C* **66**, 014002 (2002).
19. P. Navrátil, V. G. Gueorguiev, J. P. Vary, W. E. Ormand, and A. Nogga, *Phys. Rev. Lett.* **99**, 042501 (2007).
20. P. Navrátil, *Few-Body Syst.* **41**, 117 (2007).
21. T. T. S. Kuo, J. Shurpin, K. C. Tam, E. Osnes, and P. J. Ellis, *Ann. Phys. (NY).* **132**, 237 (1981).
22. M. Hjorth-Jensen, T. T. S. Kuo, and E. Osnes, *Phys. Rep.* **261**, 125 (1995).
23. K. Suzuki and R. Okamoto, *Prog. Theor. Phys.* **93**, 905 (1995).
24. L. Coraggio and N. Itaco, *Frontiers in Physics.* **8**, 345 (2020).
25. A. Baroni, L. Girlanda, A. Kievsky, L. E. Marcucci, R. Schiavilla, and M. Viviani, *Phys. Rev. C* **94**, 024003 (2016).
26. T. S. Park, D. P. Min, and M. Rho, *Phys. Rep.* **233**, 341 (1993).
27. S. Pastore, L. Girlanda, R. Schiavilla, M. Viviani, and R. B. Wiringa, *Phys. Rev. C* **80**, 034004 (2009).
28. A. Baroni, L. Girlanda, S. Pastore, R. Schiavilla, and M. Viviani, *Phys. Rev. C* **93**, 015501 (2016).
29. P. Maris, J. P. Vary, and P. Navrátil, *Phys. Rev. C* **87**, 014327 (2013).
30. T. T. S. Kuo and E. Osnes, *Lecture Notes in Physics.* vol. 364, Springer-Verlag, Berlin (1990).
31. L. Coraggio, A. Covello, A. Gargano, N. Itaco, and T. T. S. Kuo, *Ann. Phys. (NY).* **327**, 2125 (2012).
32. T. T. S. Kuo, S. Y. Lee, and K. F. Ratcliff, *Nucl. Phys. A* **176**, 65 (1971).
33. M. Hjorth-Jensen, M. P. Lombardo, and U. van Kolck, eds., *Lecture Notes in Physics.* vol. 936, Springer, Berlin (2017).
34. P. J. Ellis and E. Osnes, *Rev. Mod. Phys.* **49**, 777 (1977).
35. A. Polls, H. Müther, A. Faessler, T. T. S. Kuo, and E. Osnes, *Nucl. Phys. A* **401**, 124 (1983).
36. Y. Z. Ma, L. Coraggio, L. De Angelis, T. Fukui, A. Gargano, N. Itaco, and F. R. Xu, *Phys. Rev. C* **100**, 034324 (2019).
37. H. A. Mavromatis, L. Zamick, and G. E. Brown, *Nucl. Phys. A* **80**, 545 (1966).

38. H. A. Mavromatis and L. Zamick, *Nucl. Phys. A* **104**, 17 (1967).
39. P. Federman and L. Zamick, *Phys. Rev.* **177**, 1534 (1969).
40. L. Coraggio, L. De Angelis, T. Fukui, A. Gargano, and N. Itaco, *Phys. Rev. C* **95**, 064324 (2017).
41. L. Coraggio, L. De Angelis, T. Fukui, A. Gargano, N. Itaco, and F. Nowacki, *Phys. Rev. C* **100**, 014316 (2019).
42. L. Coraggio, N. Itaco, G. De Gregorio, A. Gargano, R. Mancino, and F. Nowacki, *Phys. Rev. C* **105**, 034312 (2022).
43. E. Caurier, G. Martínez-Pinedo, F. Nowacki, A. Poves, and A. P. Zuker, *Rev. Mod. Phys.* **77**, 427 (2005).
44. C. D. Goodman, C. A. Goulding, M. B. Greenfield, J. Rapaport, D. E. Bainum, C. C. Foster, W. G. Love, and F. Petrovich, *Phys. Rev. Lett.* **44**, 1755 (1980).
45. T. N. Taddeucci, C. A. Goulding, T. A. Carey, R. C. Byrd, C. D. Goodman, C. Gaarde, J. Larsen, D. Horen, J. Rapaport, and E. Sugarbaker, *Nucl. Phys. A* **469**, 125 (1987).
46. M. Tadanori, N. Yoichi, M. Kensaku, T. Kunihiko, K. Atsushi, O. Takashi, O. Akira, and I. Mitsuyoshi, *Nucl. Phys. A* **516**, 365 (1990).
47. E. Brun, J. J. Kraushaar, W. L. Pierce, and W. J. Veigele, *Phys. Rev. Lett.* **9**, 166 (1962).
48. K. Tetsuno *et al.*, *J. Phys. Conf. Ser.* **1468**, 012132 (2020).
49. E. Caurier, J. Menéndez, F. Nowacki, and A. Poves, *Phys. Rev. Lett.* **100**, 052503 (2008).
50. R. A. Sen'kov and M. Horoi, *Phys. Rev. C* **88**, 064312 (2013).
51. Y. Iwata, N. Shimizu, T. Otsuka, Y. Utsuno, J. Menéndez, M. Honma, and T. Abe, *Phys. Rev. Lett.* **116**, 112502 (2016).
52. J. Menendez, *J. Phys. G.* **45**, 014003 (2018).
53. L. Coraggio, A. Gargano, N. Itaco, R. Mancino, and F. Nowacki, *Phys. Rev. C* **101**, 044315 (2020).
54. J. M. Yao, B. Bally, J. Engel, R. Wirth, T. R. Rodríguez, and H. Hergert, *Phys. Rev. Lett.* **124**, 232501 (2020).
55. S. Novario, P. Gysbers, J. Engel, G. Hagen, G. R. Jansen, T. D. Morris, P. Navrátil, T. Papenbrock, and S. Quaglioni, *Phys. Rev. Lett.* **126**, 182502 (2021).
56. A. Belley, C. G. Payne, S. R. Stroberg, T. Miyagi, and J. D. Holt, *Phys. Rev. Lett.* **126**, 042502 (2021).
57. R. Machleidt, *Phys. Rev. C* **63**, 024001 (2001).
58. S. Bogner, T. T. S. Kuo, L. Coraggio, A. Covello, and N. Itaco, *Phys. Rev. C* **65**, 051301(R) (2002).
59. Data extracted using the NNDC On-line Data Service from the ENSDF database, file revised as of March 9, 2022.
60. Data extracted using the NNDC On-line Data Service from the XUNDL database, file revised as of March 8, 2022.
61. J. R. Vanhoy, M. T. McEllistrem, S. F. Hicks, R. A. Gatenby, E. M. Baum, E. L. Johnson, G. Molnár, and S. W. Yates, *Phys. Rev. C* **45**, 1628 (1992).
62. A. Barabash, *Universe.* **6**, 159 (2020).
63. K. Yako *et al.*, *Phys. Rev. Lett.* **103**, 012503 (2009).

https://doi.org/10.1142/9789819813353_0004

In Memoriam Akito Arima
The Electron Neutrino Mass by Electron Capture

Amand Faessler

Institute of Theoretical Physics, University of Tuebingen, Germany
faessler@uni-tuebingen.de

Akito Arima was a good friend of mine and even as Japanese Minister of Science he visited me in Tuebingen. Akito has exceptional high achievements in science (He developed the Interacting Boson Approximation (IBA)) and in science administration as Rector of the University of Tokyo and as Japanese Minister of Science. Here I present in his honor some results of my work on neutrino masses. The electron neutrino mass can be determined by electron capture. One expects the largest influence of the neutrino mass on this decay for a small Q value of Q = 2.8 keV for $^{163}_{67}$Ho + e $\rightarrow$ $^{163}_{66}$Dy + ν. The energy of the Q value is distributed to the emitted neutrino and the excitation of the Dy atom. The Dy recoil can be neglected. Thus the energy difference between the Q value and the upper end of the electron deexcitation spectrum is the electron neutrino mass. The electron wave functions are calculated selfconsistently by the Dirac-Hartree-Fock approach for the bound and the continuum states. To extract the neutrino mass from the spectrum is only possible, if the background is reduced relative to the present situation. The analysis presented here shows, that the determination of the electron neutrino mass by electron capture is difficult, but seems not to be impossible.

1. In Memoriam Akito Arima

I met Akito Arima first in 1963, when I was visiting Ray Sheline at the Florida State University in Tallahassee. He came down from Oak Ridge to collaborate with Takeshi Udagawa. His expertise in Nuclear Structure and Nuclear reactions and his friendly manor impressed me immediately. Later when I was at the University of Tuebingen / GERMANY he visited me several times even during his time as Rector of the Tokyo University Today and even during his time as Japanese Minister of Science. He also invited me several times to the University of Tokyo, once at the same time, when President Theis of the University of Tuebingen was also on a short visit at Today.

2. The Neutrino Mass

The neutrino is one of the least known particles, although it plays a central role in weak interactions. We do not know the mass of the three neutrinos: electron, muon and tauon neutrinos (and antineutrinos).[1–9,11] We do not know their hierarchy. The nature of the neutrinos, Dirac (neutrino different from the antineutrino) or Majorana (neutrino identical with the corresponding antineutrino) is still unknown.[12–16]

The topic of this contribution is the theoretical side of the determination of the electron neutrino mass by electron capture in $^{163}_{67}$Ho as developed by the Tuebingen group and coworkers.[5,17–20] For the capture

$$^{163}Ho + e \rightarrow ^{163}Dy + \nu \tag{1}$$

the Q value obtained by a Penning trap measurement by Eliseev et al.[21] (see also[22–24]) is:

$$Q = 2.833(30_{stat})(15_{sys})\ keV \tag{2}$$

This means, that electron capture in ^{163}Ho is only possible from electron orbits bound less than Q = 2.8 keV. So capture is only possible from $3s_{1/2}, M1$ with a binding energy 2.128 keV and less bound electron states. In addition the electron orbits in Ho, from which an electron can be captured must have an overlap with the nucleus to react with a proton. The energy of the Q value after capture is distributed over the excitation energy of the Dy atom (one-, two-, three-hole and higher excitations) and the emitted neutrino. The sum of the deexcitation energy can be measured by a bolometer. The maximum bolometer energy yields then the minimum neutrino energy, which is the electron neutrino rest mass. Thus the neutrino mass can be determined by the difference of the Q-value and the upper end of the bolometer spectrum. The excitation energy of the Dy* is calculated in the sudden approximation. The excitation of the Dy* is then given by the overlap with Ho* with an electron hole in the capture orbit with the different excited configurations $|D>$ in the daughter Dy*.

According to a publication by Intemann and Polock[10] DeRujula and Lusignoli[11] use non-relativistic pure Coulomb wave functions for the electrons in Ho, which do not allow capture from $np_{1/2}$. The effect of the partial screening of the nuclear Coulomb potential by the electron cloude is considered by effective charges. Since in reference[11] the perturbation approach is used to calculate the Dy* wave functions, this requires orthonormal basis functions. Thus Ref. 11 uses for each capture process, e. g. from $3s_{1/2}$, the same screening factor for all single electron orbits needed in the perturbation approach for this special capture process. Starting from these "bad" wave functions for Ho Ref. 11 calculates as proposed by[10] the bound and continuum electron wave functions for Dy* with the perturbation:

$$H'(r) = +1/r - \int d^3r_1\ |\varphi_{3s/4s}(\vec{r}_1)|^2/|\vec{r} - \vec{r}_1|. \tag{3}$$

The first term takes into account, that in Dy one has one proton less compared to Ho. The second term originates from the electron hole in the $3s_{1/2}$ or the $4s_{1/2}$ state, which are the only hole states considered in Ref. 11. We use here selfconsistent Dirac-Hartree-Fock wave functions calculated specifically for Ho* and Dy*.[25–27] The continuum electron wave functions in Yb* are determined[28] in the selfconsistent Z-1 = 65 electron potential orthogonal to all 65 bound electrons in Yb.

3. Electron Capture

The spectrum of the the decay of the excited $^{163}Dy^*$ after electron capture in ^{163}Ho Refs. 4, 11 and 5 assuming Lorentzian shapes for the decay resonances are:

$$\frac{d\Gamma}{dE_c} \propto \sum_{i=1,\ldots,N_\nu} (Q-E_c)\cdot U^2_{e,i}\cdot\sqrt{(Q-E_c)^2-m^2_{\nu,i}}$$
$$*\left(\sum_{f=f'}\lambda_0 B_f \frac{\Gamma_{f'}}{2\pi}\frac{1}{(E_c-E_{f'})^2+\Gamma^2_{f'}/4}\right.$$
$$+\sum_{f=f';p'<F;q'_b>F}\lambda_0 B_{f,p'<F;q'_b>F}\frac{\Gamma_{f',p'}}{2\pi}\frac{1}{(E_c-E_{f',p'})^2+\Gamma^2_{f',p'}/4}$$
$$\left.+\int dk_{q'}\lambda_0 B_{f,p'<F;q'_c>0}\frac{\Gamma_{f',p',q'}}{2\pi}\frac{1}{(E_c-E_{f',p',q'})^2+\Gamma^2_{f',p',q'}/4}\right) \quad (4)$$

The factor in front of the bracket is the same as for the single beta decay. It reflects the phase space of the emitted neutrino (1). The three expressions in the bracket of Eq. (4) describe the decay of the 1-hole f' excitations in Dy*, of the 2-hole excitations f', p' with shake-up of p' to q' into a bound orbit and of 2-holes f', p' states with one electron p' moved to q' into the continuum (shake-off). The last term has as the others a dimensionless strength $dk_{q'}\cdot B_{f',p'<F,q'>o}$. The transformation to an integral over the energy yields the factor in front of the third term, which can be deduced from Eq. (11). Different neutrino mass eigenstates $|i>$ are mixed into the electron neutrino flavor states by $U^2_{e,i}$. The Q-value is given by Eq. (2) from the ECHo collaboration.[22,23,29–31] The excitation energy of the final Dy* is E_c. The energy $Q-E_c$ is carried away by the neutrino. B_f, $B_{f,p'<F;q'_b>F}$ and $B_{f,p'<F;q'_c>0}$ are the overlap and exchange corrections for the 1-hole, the bound 2-hole and the shake-off 2-hole states. λ_0 contains the nuclear matrix element squared.[32] Eq. (4) yields the spectrum in arbitrary units and must be normalized at one energy to the data. We use the N1, $4s_{1/2}$ peak for the normalisation. $E_{f'}$, $E_{f',p'}$ and $E_{f',p';q'>0}$ are the 1-hole, the 2-hole shake-up and the 2-hole shake-off excitation energies in Dysprosium (see Refs. 9 and 12). $\Gamma_{f'}$, $\Gamma_{f',p'}$ and $\Gamma_{f',p';q'>0}$ are the widths of the one- and two-hole states and the two-hole states with shake-off in Dysprosium.[17–20]

The overlap and exchange corrections B_f, $B_{f,p'<F;q'_b>F}$ and $B_{f,p'<F;q'_c>0}$ are calculated in second quantization with Dirac-Hartree-Fock (DHF) electron wave functions. This yields automatically the antisymmetrization and the exchange terms. The antisymmetrized two-hole state in Dy with shake-off is for example:

$$|A'_{p',f':q'>0}>=a'^{\dagger}_1 a'^{\dagger}_2\cdots a'^{\dagger}_{f'-1}a'^{\dagger}_{f'+1}\cdots a'^{\dagger}_{p'-1}a'^{\dagger}_{p'+1}\cdots a'^{\dagger}_Z a'^{\dagger}_{q'>0}|0> \quad (5)$$

The probability to form a two-hole shake-off state is proportional to ($|G>$ is the DHF ground-state of Holmium):

$$P_{f',p';q'>0}=|<A'_{f',p';q'}|a_i|G>|^2 \quad (6)$$

4. Electrons in the Dy Continuum

The selfconsistent Coulomb field for the shake-off electron is in atomic units:

$$V_{shake\text{-}off}(r) = \frac{-66}{r} + \sum_{k\ occupied\ e} g'_k \int d^3\, r \cdot \frac{|\varphi_k(\vec{r}\,')|^2}{|\vec{r} - \vec{r}\,'|} \tag{7}$$

$g'_k s$ are the number of bound electrons in the selfconsistent occupied orbits $|k> = |n, \ell, j>$ of Dy. To determine the potential for the shake-off electrons in Dy one needs the occupied selfconsistent Dirac-Hartree-Fock upper $P_k(r)$ and lower $Q_k(r)$ components of the selfconsistent electron wave functions.

$$\varphi_k(\vec{r}) * r = (P_k(r); Q_k(r)) \tag{8}$$

$$V_{shake\text{-}off} \approx -\frac{66}{r} + \frac{65}{r} \cdot [1 - exp(-a \cdot r)]\ [1/(length = a.u.)]; \quad a = 3.4 \tag{9}$$

We adjust an analytic expression (9) to the DHF potential (see Fig. 1). The selfconsistent Dy potential for the 66th. electron in the continuum is shown in Fig. 1 for Dy. For the normalization of the continuum Dirac wave function we follow here the recommendation of Perger and Karrighattam on their page 394.[34] In the asymptotic expression (10) η is the Sommerfeld parameter, Δ the Coulomb phase shift and δ takes into account deviations from a pure Coulomb potential. The electron energy in the continuum can be due to energy conservation not larger than the Q-value of 2.8 keV minus the excitation energy of the two hole state.

$$\int_{r=0\ to\ \infty} dr \cdot 2 \cdot \sin\left(kr - \ell \cdot \frac{\pi}{2} - \eta \ln 2kr + \Delta + \delta\right)$$
$$\cdot 2 \cdot \sin\left(k'r - \ell \cdot \frac{\pi}{2} - \eta \ln 2kr + \Delta + \delta\right) \approx 2\pi \cdot \delta(k - k'). \tag{10}$$

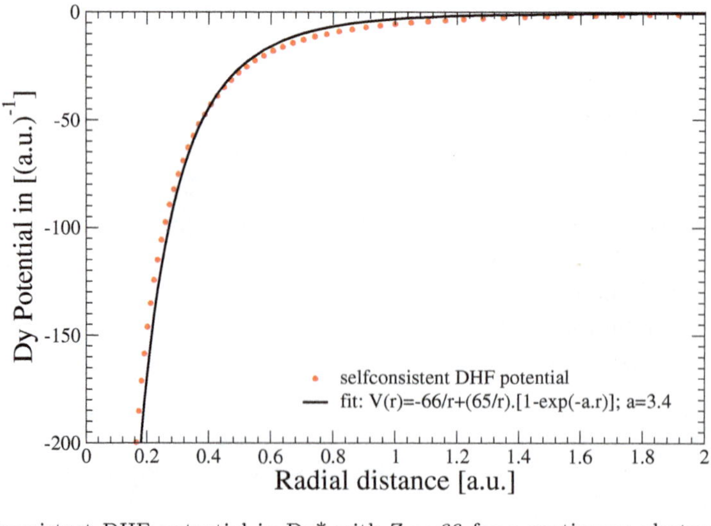

Fig. 1. Selfconsistent DHF potential in Dy* with Z = 66 for a continuum electron (No 66) of 65 electrons (dimension: 1/length = atomic units) in the core and analytic approximation with a = 3.4 (See Eq. (9)).

The wave number is connected with the relativistic and non-relativistic energies by the equations:

$$E_{rel}^2 = c^2\hbar^2k^2 + m^2c^4 \to c^2k^2 + c^4 \quad (in\ atomic\ units); \quad E_{n-rel} = \frac{1}{2}k^2;$$
$$k = \alpha\sqrt{E_{n-rel}(E_{n-rel} + 2\cdot c^2)}; \quad with\ c = 1/\alpha = 137.035999\ in\ [atomic\ units]. \tag{11}$$

5. Electrons Emitted into the Continuum

The overlap between the bound Ho states $ns_{1/2}$ and $np_{1/2}$ with $n \geq 3$ and the continuum wave functions in Dy squared gives the probability for shake-off. We restrict this work to s-wave shake-off, which is expected to be the largest contribution. The s-wave shake-off is given by the overlap $< n \geq 3, s_{1/2}, Ho|E,\ s, Dy >$ squared. In the spectrum with 1-hole, 2-hole and shake-off in Fig. 2 the shake-off contribution is hardly visible. Here and also in Ref. 11 only the decay width of the 2-hole states are included. Three hole states can be neglected.[19] The electron in the continuum has an escape width, which is not included.

Figure 2 shows the logarithmic spectrum of the 1-hole, the 2-hole and the s-wave shake-off contributions. The shake-off spectrum can hardly be seen on this scale in the total spectrum.

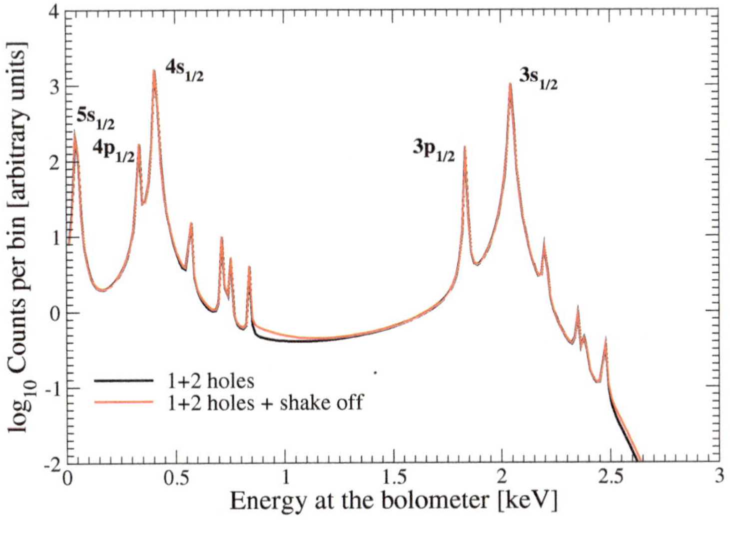

Fig. 2. Theoretical results in arbitrary units of the sum of the one- and two-hole deexcitations compared to the sum of the one-, two-hole and the shake-off deexcitation as measured by the bolometer spectrum (4). The arbitrary units are adjusted to the experimental N1, $4s_{1/2}$ 1-hole peak. The nature of the one hole states are indicated. The two-hole peaks are by about two orders of magnitudes smaller than the one hole peaks. Shake-off can almost not been seen.

6. Conclusions

Akito Arima has exceptionally high achievements in Science (He developed among many other topics the Interacting Boson Approach) and in Science administration as Rector of the University of Tokyo and as Japanese Minister of science. In this contribution I summarize my theoretical work on the determination of the electron neutrino mass by electron capture[5,9,17–20] in $^{163}_{67}$Ho to $^{163}_{66}$Dy. The energy of the Q-value $= 2.833 \pm 0.05\ keV$ is distributed to the excitation of Dy and the emission of a neutrino. The recoil of ^{163}Ho can be neglected. The total deexcitation energy as X-rays and Auger electrons of Dy can be measured by a bolometer. The upper end of the spectrum determines the minimum energy of the neutrino. Thus the difference of Q minus this upper limit of the deexcitation spectrum is the rest mass of the electron neutrino. We describe the bound states in Ho and Dy by the Dirac-Hartree-Fock approach[25–27] even including occupations in Dy due to different hole states. The s-wave continuum wave functions in Dy are determined with the Dirac equation in the selfconsistent potential.[28] (See Fig. 1).

Finally we want to stress, that the accuracy needed to extract the neutrino mass can not be obtained by theoretical calculation alone. One must fit simultaneously the neutrino mass, the Q-value, the excitation energy of the highest resonance, their width and their strength at the upper end of the spectrum to extremely accurate data. For this the present background of the experimental spectrum has to be drastically reduced.[35] If the background can be reduced by one to two orders, it should be enough to adjust only the two parameters, the Q-value and the neutrinos mass, to the upper end of the spectrum for the determination of the electron neutrino mass.

References

1. G. Drexlin, V. Hannen, S. Mertens, and C. Weinheimer, *Adv. High Energy Phys.* **2013**, 293986 (2013) and arXiv 13077.0101v1 (2013).
2. A. Faessler, E. Huster, O. Kraft, and F. Krahn, *Z. Phys.* **238**, 352 (1970).
3. E. W. Otten and C. Weinheimer, *Rep. Prog. Phys.* **71**, 086201 (2008).
4. A. De Rujula, arXiv 1305.4857v1 [hep-ph] 21 May 2013, unpublished. A. De Rujula and M. Lusignoli, *Phys. Lett.* **118 B**, 429 (1982).
5. A. Faessler, L. Gastaldo, and F. Šimkovic, *J. Phys. G* **42**, 015108 (2015) and arXiv: 1407.6504 [nuclo-th] (2014).
6. R. G. H. Robertson, arXiv: 1411.2906v1 [nucl-15 th] (11 Nov 2014) and *Phys. Rev. C* **91**, 035504 (2015).
7. C. Velte *et al.*, *Euro Phys. J.* **C79**, 15 and 1026 (2019).
8. F. Terranova, *A Modern Primer in Particle and Nuclear Physics.* Oxford Univ. Press. ISBN978-01928455252 (2021).
9. A. Faessler, arXiv1501.04338 and *Phys. Rev. C* **91**, 045505 (2015).
10. R. L. Intemann and F. Polock, *Phys. Rev.* **157**, 41 (1967).
11. A. De Rujula and M. Lusignoli, *JHEP* **1605**, 015 (2016); arXiv: 1601.04990 [hep-ph].
12. A. Faessler, R. Hodak, S. Kovalenko and F. Šimkovic, arXiv 1304.5032v3 (2013) and "European Physical Journal Web of Conferences".
13. A. Faessler, *Nucl. Phys. Proc. Suppl.* **188**, 20–24 (2009).

14. F. Šimkovic, A. Faessler, V. Rodin, P. Vogel, and J. Engel, *Phys. Rev.* **C77**, 045503 (2008).
15. P. Vogel and J. Engel, *Phys. Rev.* **C77**, 045503 (2008).
16. F. Avignone, S. R. Elliott, and J. Engel, *Rev. Mod. Phys.* **80**, 481–516 (2008).
17. A. Faessler, L. Gastaldo, and F. Šimkovic, arXiv: 1611.00325 [nucl-th] and *Phys. Rev. C* **95**, 045502 (2017).
18. A. Faessler and F. Simkovic, *Phys. Scripta*, **91**, 043007 (2016). (2015).
19. A. Faessler, Ch. Enns, L. Gastaldo, and F. Šimkovic, *Phys. Rev. C* **91**, 064302 (2015).
20. A. Faessler and F. Šimkovic, *Phys. Rev.* **C91**, 045505 (2015).
21. S. Eliseev, K. Blaum, M. Block, S. Chenmarek, H. Dorrer, Ch. E. Duellmann, C. Enss, P. E. Filiamin, L. Gastaldo, M. Goncharov, U. Koester, F. Lautenschlaeger, Yu. N. Novikov, A. Rischka, R. X. Schuessler, L. Schweighard, and A. Tuerler, *Phys. Rev. Lett.* **115**, 062501 (2015).
22. P. C.-O. Ranitzsch, J. P. Porst, S. Kempf *et al.*, *Jour. of Low Temp. Phys.* **167**, 1004 (2012) and private communication.
23. K. Blaum, A. Doerr, C. E. Duellmann, K. Eberhardt, S. Eliseev, C. Enss, A. Faessler, A. Fleischmann, L. Gastaldo, S. Kempf, M. Krivoruchenko, S. Lahiri, M. Matai, Yu. N. Novikov, P. C. Ranitzsch, F. Šimkovic, Z. Scusc, and M. Wegner, arXiv 1306.2655v1 [physics.ins-det] (2013), unpublished.
24. P. C. Ranitzsch, C. Hassel, M. Wegner, Kempf, A. Fleischmann, C. Enss, L. Gastaldo, A. Herlert, and K. Johnston, arXiv: 1409.007v1 (2014) unpublished.
25. I. P. Grant, *Adv. Phys.* **19**, 747 (1970).
26. J. P. Desclaux, *Comp. Phys. Com.* **9**, 31–45 (1975).
27. A. L. Ankudinov, S. I. Zabinsky, and J. J. Rehr, *Comp. Phys. Com.* **98**, 359–364 (1996).
28. F. Salvat, J. M. Fernandez-Varea, and W. Williamson, *Comp. Phys. Com.* **90**, 151–168 (1995).
29. J. U. Anderson *et al.*, *Phys. Lett.* **B398**, 72 (1982).
30. F. Gatti *et al.*, *Phys. Lett.* **B398**, 415 (1997).
31. G. Audi, A. H. Wapstra, and C. Thibault, *Nucl. Phys.* **A279**, 337 (2003).
32. W. Bambynek, H. Behrens, M. H. Chen, B. Crasemann, M. L. Fitzpatrick, K. W. D. Ledingham, H. Genz, M. Mutterer, and R. L. Intemann, *Rev. Mod. Phys.* **49**, 109 (1977).
33. G. P. Williams, "X-Ray Data Booklet", Lawrence Berkeley National Laboratory, Table 1-1.
34. W. F. Perger and Vasan Karighattam, *Comp. Physics Com.* **66**, 392 (1991).
35. L. Gastaldo, private information.

https://doi.org/10.1142/9789819813353_0005

Akito Arima and Pseudospin Symmetry

Joseph N. Ginocchio
Fellow, Los Alamos National Laboratory
Los Alamos, NM, 87544, USA
j.ginocchio@comcast.net

We review the relativistic origins of the pseudspin doublets that Akito Arima observed along with others 53 years ago. We examine the consequences of this symmetry on magnetic dipole and Gamow Teller transitions in nuclei and on anti-nucleon scattering from nuclei. We then review the evidence for pseudospin symmetry in quark dynamics. We then study the spin wavefunctions produced from the transformation from pseudospin symmetric wavefunctions. We then give some personal remembrances of Akito.

1. Introduction

Over fifty years ago Akito and his collaborators noticed a quasi degeneracy of certain single particle states. The states in question have quantum numbers $(n\, \ell_j, n'\ell'_{j'})$ where $n' = n - 1, \ell' = \ell + 2, j' = j + 1$ and n, ℓ, j are the radial, orbital angular momentum, and total angular momentum quantum numbers, respectively;[1,2] for example, $1s_{1/2}$ and $0d_{3/2}$. These quasi-degeneracies persist in measurements in nuclei far from stability.[3] Subsequently pseudospin doublets in deformed nuclei were discovered.[4] Since these doublets were not spin doublets, Akito and his collaborators called these doublets pseudospin doublets. However they had no deep understanding of the origin of this symmetry. At the meeting celebrating Akito's 60th birthday in Santa Fe in 1990,[5] Ben Mottleson said in his talk that "at the present time there is no deeper understanding of these (approximate) degeneracies. They are just "facts" which, as we shall see, can be exploited."

About thirty years after the observation of these quasi degenerate pseudospin doublets it was discovered that this quasi symmetry was due to a relativistic symmetry.[6] Akito was enthralled by this discovery. In this paper in Akito's honor we shall discuss this relativistic symmetry and some of its consequences and possible deeper origins in quark and hadron physics. We will then discuuss the transformation from the conserved pseudospin wavefunctions to the spin wavefunctions. We will finish with some remembrances of Akito.

2. Symmetries of the Dirac Hamiltonian

The Dirac Hamiltonian with a Lorentz scalar potential, $V_S(\vec{r})$, and a potential which is the fourth component of a Lorentz vector potential, $V_V(\vec{r})$, is

$$H = \vec{\alpha} \cdot \vec{p} + \beta(V_S(\vec{r}) + M) + V_V(\vec{r}), \tag{1}$$

where $\vec{\alpha}$, β are the Dirac matrices, $\vec{p}$ is the momentum, M is the mass, $\vec{r}$ is the radial coordinate, and the velocity of light is set equal to unity, c = 1.

2.1. *Spin Symmetry: A Symmetry of the Dirac Hamiltonian*

The Dirac Hamiltonian has spin symmetry when the difference of the vector and scalar potentials in the Dirac Hamiltonian is a constant, $V_S(\vec{r}) - V_V(\vec{r}) = C_s$.[7] Spin symmetry is not usually thought of as a relativistic symmetry. However, hadrons[8] and anti-nucleons in a nuclear environment have spin symmetry.[9] These are relativistic systems and normally, in such systems, we would expect large spin-orbit splittings, but, in this limit, spin doublets are degenerate. The generators for this SU(2) spin algebra, $\vec{S}$, which commute with the Dirac Hamiltonian, $[H, \vec{S}] = 0$, for any potential, spherical or deformed, that satisfies $V_S(\vec{r}) - V_V(\vec{r}) = C_s$, are given by,[10]

$$\vec{S} = \begin{pmatrix} \vec{s} & 0 \\ 0 & U_p\, \vec{s}\, U_p \end{pmatrix}, \tag{2}$$

where $\vec{s} = \vec{\sigma}/2$ are the usual spin generators, $\vec{\sigma}$ the Pauli matrices, and $U_p = \boldsymbol{\vec{\sigma}} \cdot \boldsymbol{\hat{p}}$ is the helicity unitary operator introduced in[11] and $\hat{p} = \frac{\vec{p}}{p}$, the unit momentum. The generators are four by four matrices as is appropriate for the Dirac Hamiltonian.

This symmetry holds for deformed nuclei as well as spherical nuclei. For spherical nuclei, the orbital angular momentum is conserved as well and the generators are given by

$$\vec{L} = \begin{pmatrix} \vec{\ell} & 0 \\ 0 & U_p\, \vec{\ell}\, U_p \end{pmatrix}, \tag{3}$$

where $\hbar\, \vec{\ell} = \vec{r} \times \vec{p}$ is the non-relativistic angular momentum.

The relativistic harmonic oscillator for which $V_S(\vec{r}) = v_S r^2$ and $V_V(\vec{r}) = v_V r^2$ has, for $v_S(\vec{r}) - v_V(\vec{r}) = c_s$, an SU(3) symmetry, in addition to spin and orbital angular momentum symmetry. The generators of the SU(3) symmetry are the orbital angular momentum $\vec{L}$ and the quadrupole operator[12,13]

$$Q_m = \begin{pmatrix} q_m & 0 \\ 0 & U_p\, q_m\, U_p \end{pmatrix}, \tag{4}$$

where $\hbar\, q_m = \sqrt{\frac{3}{2}}([(\vec{r}\vec{r}]^2_m + [\vec{p}\vec{p}]^2_m)$ is the non-relativistic quadrupole SU(3) generator and the harmonic oscillator constants are set equal to unity.

2.2. *Pseudospin Symmetry: A Symmetry of the Dirac Hamiltonian*

Another SU(2) symmetry of the Dirac Hamiltonian occurs when the sum of the vector and scalar potentials in the Dirac Hamiltonian is a constant, $V_S(\vec{r}) + V_V(\vec{r}) = C_{ps}$.[7] The generators for this SU(2) algebra which commute with the

Dirac Hamiltonian, $[H, \tilde{\vec{S}}] = 0$, for any potentials, spherical or deformed, that satisfy $V_S(\vec{r}) + V_V(\vec{r}) = C_{ps}$, are given by[10]

$$\tilde{\vec{S}} = \begin{pmatrix} U_p\, \vec{s}\, U_p & 0 \\ 0 & \vec{s} \end{pmatrix}. \tag{5}$$

Pseudospin symmetry was shown to be a symmetry of this type.[6] The eigenfunctions of the Dirac Hamiltonian in this limit will have degenerate doublets of states, one of which has pseudospin aligned and the other with pseudospin unaligned. The "upper" matrix of the pseudospin generators in Eq (5), $U_p\, \vec{s}\, U_p$, have the spin intertwined with the momentum which enables the generators to connect the states in the doublet, which differ by two units of angular momentum. In finite nuclei both potentials go to zero at infinite distances. Hence $C_{ps} = 0$ and the scalar potential will be equal in magnitude but opposite in sign to the vector potential in the pseudospin limit. The approximate equality in the magnitude of the vector and scalar fields in nuclei and their opposite sign have been confirmed in relativistic mean field theories[13,14] and in QCD sum rules.[13,15]

This symmetry holds for deformed as well has spherical nuclei. For spherical nuclei, the pseudo orbital angular momentum is conserved as well and the generators are given by

$$\tilde{\vec{L}} = \begin{pmatrix} U_p\, \vec{\ell}\, U_p & 0 \\ 0 & \vec{\ell} \end{pmatrix}. \tag{6}$$

The relativistic harmonic oscillator has, for $v_S + v_V = c_{ps}$, a pseudo SU(3) symmetry, in addition to pseudospin and pseudo orbital angular momentum symmetry.[16,17] The generators are $\tilde{\vec{L}}$ along with the quadrupole operator,

$$\tilde{Q}_m = \begin{pmatrix} U_p\, q_m\, U_p & 0 \\ 0 & q_m \end{pmatrix}. \tag{7}$$

3. Consequences of Relativistic Pseudospin Symmetry

Akito has had a long history in studying magnetic moments and Gamow-Teller transitions in nuclei.[18–20] In this section we report on some of the predictions that pseudospin symmetry, particularly its relativistic nature, predicts about these observables.

One immediate consequence of pseudospin symmetry as a relativistic symmetry is that the "lower" matrix of the pseudospin generators in Eq (5), $\vec{s}$, does not change the radial wavefunction of the "lower" component of the Dirac eigenfunctions. Hence this symmetry predicts that the radial wavefunctions of the "lower" component are the same for the two states in the doublet. Previous to this discovery many relativistic mean field calculations of nuclear properties had been made. Hence this prediction was tested with existing calculations and, indeed, these wavefunctions are very similar for both spherical[10,14,21–24] and deformed nuclei.[25,26] Because of the momentum dependence of the "upper" matrix of the generators the relationship between the "upper" components involves a differential equation and these have also been tested in spherical[22] and deformed nuclei[26] with success.

Magnetic dipole and Gamow-Teller transitions between the two single-nucleon states in pseudospin doublets are forbidden non-relativistically because the states differ by two units of angular momentum. However, they are not forbidden relativistically which means that they are proportional to the lower component of the Dirac eigenfunction. This leads to a condition between the magnetic moments of the states and the magnetic dipole transtion between them because the radial amplitudes of the lower components of the two states in a pseudospin doublet are equal. Therefore the magnetic dipole transition between the two states in the doublet can be predicted if the magnetic moments of the states are known.[13,23] Likewise pseudospin symmetry also predicts Gamow-Teller transitions between a state in a parent nucleus to the partner state in the daughter nucleus if the Gamow-Teller transition to the same states in the parent and daughter nucleus is known. We do not have space to discuss these relationships in detail but one example occurs in the mirror nuclei ${}^{39}_{19}K_{20}$ and ${}^{39}_{20}Ca_{19}$. The ground state and first excited state of ${}^{39}_{19}K_{20}$ are interpreted as a $0d_{3/2}$ and $1s_{1/2}$ proton hole respectively, while the ground state and first excited state of ${}^{39}_{20}Ca_{19}$ are interpreted as a $0d_{3/2}$ and $1s_{1/2}$ neutron hole respectively. These states are members of the $\tilde{n}_r = 1, \tilde{\ell} = 1$ pseudospin doublet. Using the magnetic moment of ^{39}Ca a transition rate is calculated which is only about 37% larger than the measured value. However, the two states in the doublet are not pure single-particle states. A modification of these relations has been derived which take into account the fact that these states are not pure single particle states.[13,27] The modified relations give a transition rate that agrees with the measured value to within experimental error. Again using the mass 39 nuclei, the Gamow-Teller transitions from the ground state of ^{39}Ca to the ground and first excited state of ^{39}K are known, which is enough information to predict the transition from the ground state to the excited state. In the non-relativistic shell model an effective tensor term $g_{eff}\ [Y_2\sigma]^{(1)}$ is added to the magnetic dipole operator and the Gamow-Teller operator to produce a transition, where g_{eff} is a calculated effective coupling constant. However, the magnetic dipole transition calculated between the same states is an order of magnitude lower than the experimental transition[28] although the calculated Gamow-Teller agrees with the experimental value within the limits of experimental and theoretical uncertainty. This inconsistency has been a puzzle for the non-relativistic shell model. On the other hand the relativistic single-nucleon model gives a consistent description of both of these transitions. A global prediction of magnetic dipole transitions throughout the periodic table has had reasonable success as well.[13,27] However, a global prediction of Gamow-Teller transitions have not been done yet. Pseuodspin symmetry can also be used to relate quadrupole transitions between multiplets.[13]

4. Anti-nucleon in a Nuclear Environment

Charge conjugation changes a nucleon into an anti-nucleon. Under charge conjugation the scalar potential of a nucleon remains invariant while the its vector potential changes sign. Hence the pseudospin condition $V_S(\vec{r}) + V_V(\vec{r}) \approx C_{ps}$ for nucleons becomes the spin condition $\bar{V}_S(\vec{r}) - \bar{V}_V(\vec{r}) \approx C_{ps} = \bar{C}_s$ for anti-nucleons, where $\bar{V}_{S,V}(\vec{r})$ are the scalar and vector potentials of an anti-nucleon in a nuclear environment respectively. Hence approximate pseudospin for nucleons predicts approximate spin symmetry for anti-nucleons in a nuclear environment. Since the potentials are also very deep we would expect approximate $U(3)$ symmetry as well.[29]

Of course the anti-nucleons scattered from nuclei experience a strong annihilation potential while nucleons scattered from nuclei do not. However, since the nucleons do not feel an annihilation potential, this means that the sum of the scalar and vector annihilation potentials for nucleons must be zero which means that the difference of the anti-nucleon annihilation potentials is zero as well. Hence the anti-nucleon annihilation potentials conserve spin symmetry as well.

The sparse data on the scattering of polarized anti-protons on nuclei supports the prediction that spin symmetry is approximately conserved.[30] Perhaps more extensive data on the scattering of polarized anti-protons on nuclei will be forthcoming from GSI.

5. QCD Sum Rules

QCD sum rules have been used to show that $V_S \approx -V_V$ in nuclear matter.[15] Since the derviation is too complex to repeat here, we shall only attempt to justify the result.

The vector potential will be proportional to the nuclear matter density, which in nuclear matter is uniform, $V_V \approx \rho_N$. The value of ρ_N is taken to be the central matter density of nuclei which has been measured in electron scattering from nuclei.

In order to determine the scalar potential we need to know the nuclear scalar density. There is no direct measurement of the nuclear scalar density, but the scalar density of quarks in a nucleon can be measured experimentally. This scalar density of quarks is given in terms of what is called the sigma term

$$\sigma_N = 2m_q(\langle N|\bar{q}q|N\rangle - \langle 0|\bar{q}q|0\rangle), \tag{8}$$

where q is quark field operator, $\bar{q} = q^\dagger\gamma_0$, m_q is the quark mass, $|N\rangle$ is the nucleon state, and $|0\rangle$ is the vacuum state. This sigma term can be measured in pion-nucleon scattering[31,32] and, after a sophisticated analysis, is determined to be $\sigma_N \approx 45 \pm 8$ MeV. Averaging over all the nucleons in nuclear matter and ignoring nuclear interactions,

$$\sigma_N \rho_N = 2m_q(\langle \bar{q}q\rangle_{\rho_N} - \langle \bar{q}q\rangle_{\text{vac}}). \tag{9}$$

The scalar density of quarks in nuclear matter relative to the vacuum is then

$$\frac{\langle\bar{q}q\rangle_{\rho_N}}{\langle\bar{q}q\rangle_{\rm vac}} = 1 + \frac{\sigma_N \rho_N}{2\, m_q \langle\bar{q}q\rangle_{\rm vac}} = 1 - \frac{\sigma_N \rho_N}{m_\pi^2 f_\pi^2} \tag{10}$$

where the last term follows from the Gell-Mann-Oakes-Renner relation,[33]

$$2m_q \langle\bar{q}q\rangle_{\rm vac} = -m_\pi^2 f_\pi^2\,. \tag{11}$$

Both the nucleon mass and the scalar potential are Lorentz scalars. The effective mass will be proportional to the left-hand side of Eq. (10) with the first term on the right-hand side proportional to the mass of the nucleon and the second proportional to the scalar potential.

The detailed QCD sum rule gives

$$V_S = -\frac{4\pi^2 \sigma_N \rho_N}{M^2 m_q} \tag{16a}$$

$$V_V = \frac{32\pi^2 \rho_N}{M^2} \tag{16b}$$

Since all the quantities on the right-hand side of Eq. (16b) are positive, the scalar potential is attractive and the vector potential is repulsive, just as one finds in relativistic mean field models. Furthermore the ratio becomes

$$\frac{V_S}{V_V} = -\frac{\sigma_N}{8\, m_q} \approx -1.1, \tag{17}$$

using accepted values of the average quark mass in the proton ($\approx$5 MeV) and the value of σ_N ($\approx$45 MeV), which is uncannily close to the the ratio of the potentials at the center of the nucleus as determined in relativistic mean field models and indicative of pseudospin symmetry. Also the negative sign originates in the vacuum expectation of the quark scalar density given in Eq. (11). These features suggest that perhaps pseudospin has a more fundamental foundation in terms of QCD.

6. Quark Models with Spin and Pseudospin Symmetries

A successful phenomenological model of quarks uses the non-relativistic Schrodinger Hamiltonian with the Cornell potential

$$V(r) = -\frac{\alpha_s}{r} + \sigma r. \tag{18}$$

The first term arises from the single gluon exchange between quarks while the second term leads to confinement of the quarks inside the hadrons. Recently this Cornell potential was used in the Dirac Hamiltonian and studied under the conditions of spin symmetry and pseudospin symmetry.[34] When spin symmetry was imposed the effective energy-dependent potential derived in the non-relativistic limit was a Cornell type potential which is consistent with hadron phenomenology. However when pseudospin symmetry was imposed the effective energy-dependent potential derived in the non-relativistic limit was quite different and was bag-like. The size of the bag was comparable to the size of the nucleus. This is unrealistic because quarks in nuclei are confined to the size of the nucleon and do not traverse the size of the nucleus as a whole. Hence quarks do not have pseudospin symmetry but spin symmetry. The question is how does the quark spin symmetry imply pseudospin symmetry for nuclei. This is an interesting line of research which should be explored more. In particular the question of why hadrons have spin symmetry while nuclei have pseudospin symmetry needs to be answered.

7. Transformation from Pseudospin Symmetry to Spin Symmetry

The unitary operator $U_p = \boldsymbol{\vec{\sigma}} \cdot \boldsymbol{\hat{p}}$ transforms pseudospin states to spin states and visa- versa. In order to affect this transformation, we Fourier transform the pseudospin single partcle wavefunctions in coordinate space to momentum space. We assume pseudospin symmetry which means that the radial wavefunctions do not depend only on the pseudo-orbital angular momentum $\tilde{\ell}$ and not the total angular momentum j. The spherical wavefunctions in coordinate space

$$|\tilde{n}, \tilde{\ell}, j>= R_{ps}(\tilde{n}, \tilde{\ell}, r)[Y^{(\tilde{\ell})}(\theta_r, \phi_r)\chi]^{(j)} \tag{19}$$

becomes in momentum space

$$|\tilde{n}, \tilde{\ell}, j>= \frac{\sqrt{2}}{4\sqrt[3]{\pi}i^{\tilde{\ell}}} \int P_{ps}(\tilde{n}, \tilde{\ell}, p)[Y^{(\tilde{\ell})}(\theta_p, \phi_p)\chi]^{(j)} \exp(i\vec{r} \cdot \vec{p})\ p^2dp\ sin(\theta_p)d\theta_p d\phi_p \tag{20}$$

In the above $\tilde{n}, \tilde{\ell}$, are the pseudo radial node and pseudo angular momentum respectively, and j is the total angular momentum. $R_{ps}(\tilde{n}, \tilde{\ell}, r)$ and $Y^{(\tilde{\ell})}$ are the radial wavefunction and the spherical harmonic respectively, and χ is the spin function. $P_{ps}(\tilde{n}, \tilde{\ell}, p)$ is the wavefunction in momentum space. Doing the angular integration in Eq. (20), we have

$$R_{ps}(\tilde{n}, \tilde{\ell}, r) = \sqrt{\frac{2}{\pi}} \int P_{ps}(\tilde{n}, \tilde{\ell}, p) j_{\tilde{\ell}}(rp)\ p^2dp \tag{21}$$

where $j_{\tilde{\ell}}(rp)$ is the spherical Bessel function.

The unitary operator $U_p = \boldsymbol{\vec{\sigma}} \cdot \boldsymbol{\hat{p}}$ does not affect the magnitude of the momentum, $\vec{p}$, but changes the spherical harmonic in the integral in Eq. (20) while leaving the total angular momentum, j, unchanged;

$$U_p[Y^{(\tilde{\ell})}(\theta_p, \phi_p)\chi]^{(j)} = [Y^{(\ell)}(\theta_p, \phi_p)\chi]^{(j)} \tag{22}$$

where

$$\ell = \tilde{\ell} + 1, \ j = \tilde{\ell} + 1/2; \quad \ell = \tilde{\ell} - 1, \ j = \tilde{\ell} - 1/2. \tag{23}$$

The spin wavefunctions are

$$|n, \ell, j\rangle = U_p |\tilde{n}, \tilde{\ell}, j\rangle \tag{24}$$

where

$$n = \tilde{n}, \ j = \tilde{\ell} + 1/2; \quad n = \tilde{n} + 1, \ j = \tilde{\ell} - 1/2. \tag{25}$$

From Eq. (22), the radial wavefunction for the spin functions are

$$R_s(n, \ell, j, r) = \sqrt{\frac{2}{\pi}} \int P_{ps}(\tilde{n}, \tilde{\ell}, p) j_\ell(rp) \ p^2 dp. \tag{26}$$

where we note the Bessel function is of rank ℓ rather than $\tilde{\ell}$.

Because U_p is a orthogonal transformation, $U_p \times U_p = 1$, the spin wavefunctions will be orthogonal and normalized to unity if the pseudospin wavefunctions are orthogonal and normalized. For the same reason, the energies of the spin wavefunctions will be the same as the energies of the pseudospin wavefunctions. However the spin radial wavefunctions depend on j unlike the pseudospin radial wavefunctions which do not depend on j.

We take the pseudospin momentum space wavefunctions to be harmonic oscillator wavefunctions, $P_{ps}(\tilde{n}, \tilde{\ell}, p) = \sqrt{\frac{2\ \tilde{n}!}{(\tilde{n}+\tilde{\ell}+1/2)}} \exp(\frac{-p^2}{2}) L_{\tilde{n}}^{\tilde{\ell}+1/2}(p^2)$, where $L_{\tilde{n}}^{\tilde{\ell}+1/2}(p^2)$ is the Laguerre polynomial. We have set the constants in the harmonic oscillator Hamiltonian to be unity for simpicity. Then the pseudospin wavefunctions in radial space will also be harmonic oscillators, $R_{ps}(\tilde{n}, \tilde{\ell}, r) = \sqrt{\frac{2\ \tilde{n}!}{(\tilde{n}+\tilde{\ell}+1/2)}} \exp(\frac{-r^2}{2}) L_{\tilde{n}}^{\tilde{\ell}+1/2}(r^2)$. This follows from the fact that the harmonic oscillator Hamiltonian is invariant under the interchange of momentum and the radial coordinate.

However the spin wavefunctions will be more complicated. For example, the spin doublet $(1p_{1/2}, 1p_{3/2})$ wavefunctions are

$$R_s(1, 1, 1/2, r) = \frac{(\sqrt{2/3}(\sqrt{2}r(1 + 2r^2) + (-1 + r^2 - 2r^4) \exp(-r^2/2)\sqrt{\pi} Erfi(r/\sqrt{2}))}{(\pi^{3/4} r^2)} \tag{27}$$

$$R_s(1, 1, 3/2, r) = \frac{(4\sqrt{2/15}(-r + r^3 + \frac{(1+2r^2-r^4)}{\sqrt{2}} \exp(-r^2/2)\sqrt{\pi} Erfi(r/\sqrt{2}))}{(\pi^{3/4} r^2)} \tag{28}$$

where $Erfi$ is the imaginary error function. In Fig. 1 the spin doublet $(1p_{1/2}, 1p_{3/2})$ wavefunctions are plotted as a function of r.

The relativistic harmonic oscillator has an additional pseudo SU(3) symmetry which will mean that the spin doublets will have the same energy but nevertheless the spin radial wavefunctions will be different in shape.

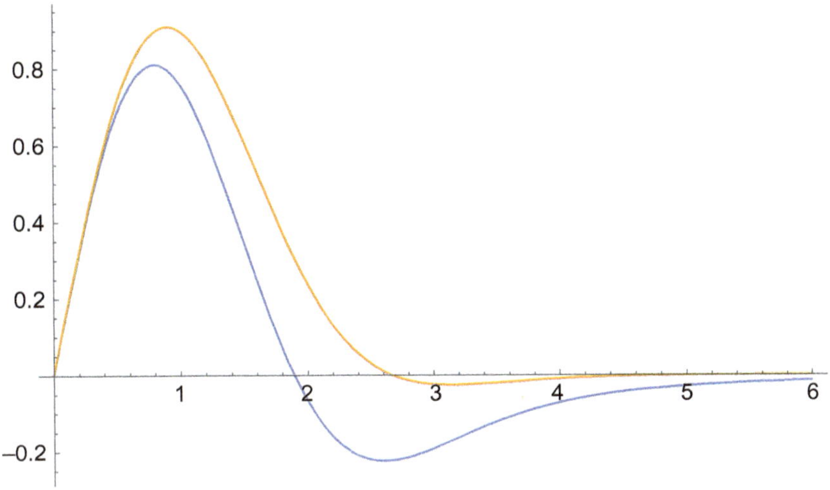

Fig. 1. $1p_{1/2}$ (Blue), $1p_{3/2}$ (Yellow) versus r.

8. Conclusions

Most features of nuclei can be understood within the non-relativistic shell model. Therefore it comes as a surprise that pseudospin symmetry in nuclei is a relativistic symmetry. However, it can never be an invariant symmetry since no nucleon bound states exist when the sum of the scalar and vector potentials is a constant.

We have seen that that magnetic and Gamow-Teller transitions that are forbidden non-relativistically can be explained due to the near equality of the "lower" components of the relativistic single-particle wavefunctions which is a result of pseudospin symmetry.

Approximate pseudospin symmetry in nuclei implies that spin symmetry will be approximately conserved in anti-nucleon scattering from nuclei. The new anti-proton facility at GSI could test this prediction by measuring the polarization of anti-protons scattered from nuclei. This prediction follows simply from charge conjugation and, if confirmed, will be another manifestation of the relativistic origin of pseudospin symmetry.

In addition, since QCD sum rules predict that the sum of the scalar and vector potentials is small in nuclear matter, perhaps there is a more fundamental rationale for pseudospin symmetry in terms of quark dynamics. With this understanding may come an answer to the question of why hadrons exhibit spin symmetry whereas nuclei exhibit pseudospin symmetry.

We have shown that a pseudospin wavefunction transforms into a spin wavefunction with the same energy but with a different radial wavefunction. In particular, with pseudospin symmetry, pseudospin doublets have identical radial wavefunctions, whereas spin doublets will have different radial wavefunctions. In the relativisitc harmonic oscillator, because of an additional pseudo SU(3), the spin doublets will have the same energy, but will still have different radial wavefunctions. Shell model calculations should make use of the spin functions derived from pseudospin symmetry.

9. Remembrances of Akito

I first met Akito Arima in 1964, 60 years ago. I had just received my Ph.D. and was a post-doc at Rutgers University. Akito was giving a seminar at nearby Princeton University. Akito and I talked to each other after his seminar and we hit it off right away. I saw Akito again in 1966 at MIT where I had my next post-doc. My wife, Anita, took him to the Boston Museum of Arts. She thought she would show him around the museum. But it turned out to be just the opposite. Akito gave her a tour explaining many of the paintings, particularly those in the Asian art section. When I joined the faculty at Yale University Akito visited me and we had him over for dinner. He had recently, along with his collaborators, published the paper on the observation of nearly degenerate pseudospin doublets. These doublets fascinated me at the time. Little did I know that about 30 years later I would come up with an explanation of this symmetry based on relativistic quantum mechanics that was discussed in this paper.

In the summer of 1970 my family and I visited Paris where I was a visiting Vincent Gillet at Saclay. By chance, Akito was visiting Vincent that summer as well. The three of us collaborated on the quartet model of light nuclei. Basically a quartet is a correlated state of two neutrons and two protons with spin and isospin zero. The model simplified the study of light nuclei in which alpha clustering is important. We published a paper together on quartets, the first paper in which I collaborated with Akito.[35] Vincent would regularly go swimming in a pool to improve his swimming skills. Neither Akito or I were good swimmers. So we joined Vincent and we would splash around in the pool together.

In 1975 our family moved to Santa Fe, New Mexico. I had accepted a position at Los Alamos National Laboratory (LANL). Akito visited us regularly. He enjoyed the vistas, visiting the museums, and visiting ancient dwellings of the indigenous people as seen in Fig. 2, Fig. 3.

One day I took Akito to Pecos National Monument. The park has ancient ruins of the original indigenous people of this area. Akito was taken by the quiet dignity of this place. Akito never took photos. He wrote haiku instead as seen in Fig. 4.

On one visit my two children, Chris and David, came with us to visit a Native American kiva, an underground room where the native people held meetings (Fig. 5, Fig. 6). They loved being with him. Chris said he was a physicist that they could relate to. Akito could make fun of himself as when he wore a cowboy hat in our garden (Fig. 7).

On one of his visits we had Akito over for dinner. One of our guests was Murray Gell Mann. Murray had recently moved to Santa Fe to join the Santa Fe Institute.

Fig. 2. Akito on a mesa overlooking a canyon with his haiku notebook in his hand.

Fig. 3. Akito before a wall of Native American petroglyphs.

Fig. 4. Akito writing haiku among the ancient dwellings of the indingenus people.

Fig. 5. Akito with David (yellow shirt) and Chris at the entrance to a kiva.

Fig. 6. Akito with David overlooking a canyon.

One of Murray's pursuits was the origins of languages. Murray would like to show off his proficiencies. I remember Murray correcting Akito's pronunciation of some Japanese words. Akito smiled and took the "corrections" gracefully.

One summer Akito brought his daughter Akiko along with him to Santa Fe (Fig. 8). Our youngest son, David, was in high school at the time. I asked him to show Akiko around Santa Fe while Akito and I went to LANL. David was delighted to escort this beautiful young woman around the town.

In 1990 I, D. H. Feng, T. Otsuka, and D. Strottman organized a conference in Santa Fe in honor of Akito to celebrate his 60th birthday.[5] Two hundred and thirty scientists from around the world came to honor Akito including two Nobel Prize winners. The talks included not only nuclear physics, but hadron physics, and talks on QCD and high energy physics. This diversity of scientists attest to the impact that Akito had on physics around the world.

Akito told me that after the War science in Japan was devastated. He told me the struggles he and others endured to build physics in Japan and his gratitude to the international community that helped them. Spurred by the loss he felt after the War, he, in turn, encouraged the development of physics all over the world in particular in smaller countries, but also in China, which had been isolated from international science.

At the banquet in Akito's honor he was awarded the Franklin Institute's Whetherill Medal "for his work in the advancement of the field of physics, and for his pivotal contributions to our understanding of atomic nuclei". Sir Denys Wilkinson then talked on "Akito and Transculturalism", which spoke to how Akito impacted science worldwide.

Fig. 7. Anita with Akito wearing a cowboy hat in our garden.

Akito wrote papers on the relativistic origins of pseudospin symmetry[24, 25, 36] and I wrote many papers on the relativistic origins of pseudospin symmetry, but none with Akito.

At the time that I arrived at LANL the interacting boson model (IBM) was being developed. Akito was puzzling over the physical meaning of the the O(6) symmetry limit of the model. A few years later Michael Kirson and I had developed an intrinsic

Fig. 8. Akito and Akiko in Santa Fe.

coherent state for the IBM and were able to demonstrate that the O(6) limit was a gamma soft rotor.[37,38] Also about that time I developed fermion models[39,40] that have the same symmetry limits as the IBM, but with one additional symmetry, O(7).[41,42] Akito was very interested in these models. We collaborated on a number

Fig. 9. Akito at a banquet in China with Anita.

Fig. 10. Akito with Igal Talmi at a conference.

Fig. 11. Akito in Erice overlooking the Mediterranean Sea.

of papers mostly with the aim of understanding the microscopic foundations of the IBM.[43–45]

I visited Akito many times in Japan and we saw each other often at International conferences (Fig. 9, Fig. 10, Fig. 11). We had a natural way with each other because we looked at physics in a similar way. We tried to understand physics by looking for patterns in the empirical data, rather than through large computer simulations. Both are necessary in understanding nuclei and they compliment each other.

I miss Akito. He was a big influence in my career. I regret not being able to see him in the twilight of his life. But he will be alive in my memories and thoughts and in my physics forever.

Acknowledgments

This research was supported by the US Department of Energy, Contract No. W-7405-ENG-36.

References

1. A. Arima, M. Harvey, and K. Shimizu, *Phys. Lett. B* **30**, 517 (1969).
2. K. T. Hecht and Adler, *Nucl. Phys. A* **137**, 129 (1969).
3. O. Sorlin and M. G. Porquet, *Prog. Part. Nucl. Phys.* **61**, (602).
4. A. Bohr, I. Hamamoto, and B. R. Mottelson, *Phys. Scr.* **26**, 267 (1982).
5. D. H. Feng, J. N. Ginocchio, T. Otsuka, and D. D. Strottman, *Nucl. Phys.* **A522** (1991).

6. J. N. Ginocchio, *Phys. Rev. Lett.* **78**, 436 (1997).
7. J. S. Bell and H. Ruegg, *Nucl. Phys. B* **98**, 151 (1997).
8. P. R. Page, T. Goldman, and J. N. Ginocchio, *Phys. Rev. Lett.* **86**, 204 (2001).
9. J. N. Ginocchio, *Phys. Rep.* **315**, 231 (1999).
10. J. N. Ginocchio and A. Leviatan, *Phys. Lett. B* **425**, 1 (1998).
11. A. L. Blokhin, C. Bahri, and J. P. Draayer, *Phys. Rev. Lett.* **74**, 4149 (1995).
12. Joseph N. Ginocchio, *Phys. Rev. C* **69**, 034318 (2004).
13. J. N. Ginocchio *Phys. Rep.* **414**, 165 (2005).
14. J. N. Ginocchio and D. G. Madland, *Phys. Rev. C* **57**, 1167 (1998).
15. T. D. Cohen *et al.*, *Prog. Part. Nucl. Phys.* **35**, 221 (1995).
16. Joseph N. Ginocchio, *Phys. Rev. Lett.* **95**, 252501 (2005).
17. Joseph N. Ginocchio, *Journal of Phys.* **267**, 012037 (2005).
18. A. Arima and H. Horie, *Progress of Theoretical Physics* **11**, 509 (1954).
19. A. Arima and H. Horie, *Progress of Theoretical Physics* **12**, 623 (1954).
20. Y. M. Zhao and A. Arima, *Phys. Rev C* **72**, 064333 (2005).
21. J. N. Ginocchio and A. Leviatan, *Phys. Rev. Lett.* **87**, 072502 (2001).
22. J. N. Ginocchio, *Phys. Rev. C* **66**, 064312 (2002).
23. J. N. Ginocchio, *Phys. Rev. C* **59**, 2487 (1999).
24. J. Meng, K. Sugawara-Tanabe, S. Yamaji, P. Ring, and A. Arima, *Phys. Rev. C* **58**, R628 (1998).
25. K. Sugawara-Tanabe, S. Yamaji, and A. Arima, *Phys. Rev. C* **65**, 054313 (2002).
26. J. N. Ginocchio A. Leviatan, J. Meng, and S. G. Zhou, *Phys. Rev. C* **69**, 03430 (2004).
27. P. von Neumann-Cosel and J. N. Ginocchio, *Phys. Rev. C* **62**, 014308 (2000).
28. E. Hagberg *et al.*, *Nucl. Phys. A* **571**, 555 (1994).
29. J. N. Ginocchio, *Phys. Rev. Lett.* **95**, 252501 (2005).
30. A. Martin *et al.*, *Nucl. Phys. A* **487**, 563 (1988).
31. J. Gasser, H. Leutwyler, and M. E. Saino, *Phys. Lett. B* **253**, 252 (1991).
32. M. E. Saino, *Proceedings of Ninth International Symposium on Meson-Nucleon Physics and the Structure of the Nucleon.* **hep-ph/0110413** (2001).
33. M. Gell-Mann, R. J. Oakes, and B. Renner, *Phys. Rev.* **175**, 2195 (1968).
34. B. P. Kosyakov, E. Y. Popov, and M. A. Vronskii, *Phys. Lett. B* **77**, 28 (2015).
35. A. Arima, V. Gillet, and J. N. Ginocchio, *Phys. Rev. Lett.* **25**, 1043 (1970).
36. K. Sugawara-Tanabe, J. Meng, S. Yamaji, and A. Arima, *Jounral of Phys. G.* **25**, 811 (1999).
37. J. N. Ginocchio and M. Kirson, *Phys. Rev. Lett.* **44**, 1744 (1980).
38. J. N. Ginocchio and M. Kirson, *Nucl. Phys.* **A350**, 31 (1980).
39. J. N. Ginocchio, *Phys. Lett. B* **85**, 9 (1979).
40. J. N. Ginocchio, *Ann. of Physics* **126**, 234 (1980).
41. R. F. Casten, C. L. Wu, D. H. Feng, J. N. Ginocchio, and Xi Han, *Phys. Rev. Lett.* **56**, 2578 (1986).
42. W. M. Zhang D. H. Feng, J. N. Ginocchio, and Xi Han, *Phys. Rev. Lett.* **59**, 2032 (1987).
43. A. Arima, J. N. Ginocchio, and N. Yoshida, *Phys. Lett. B* **101**, 112209 (1981).
44. A. Arima, J. N. Ginocchio, and N. Yoshida, *Nucl. Phys.* **A384**, 112 (1982).
45. Y. M. Zhao, A. Arima, J. N. Ginocchio, and N. Yoshinaga, *Phys. Rev.* **A384**, 112 (1982).

https://doi.org/10.1142/9789819813353_0006

Memories of Argonne National Laboratory in 1963

A. M. Green (retired)
Department of Physics, University of Helsinki, Finland
greenanthony38@gmail.com

Once upon a time, Argonne National Laboratory (Illinois, USA) arranged summer activities for Nuclear-Physics-Leaning theoreticians. Some of the 1963 participants are seen in this group photograph.

Fig. 1. Photograph for participants in summer activities in Argonne National Laboratory (Illinois, USA), 1963.

I shall not attempt to name all those seen here — but only those with whom I interacted or are familiar to me. I will leave it to the reader to identify in detail faces with names. From the left: Harry Lipkin, Murray Peshkin, X, Fritz Coester, X, X, Tony Green, Gerry Brown, Akito Arima, X, Dieter Kurath, 7X, Richard Ferrell, David Inglis, X. I give a collective apology to the 12 Xs for not remembering their names — it's 60 years ago, Below I will concentrate on two of the above participants — Akito Arima and Gerry Brown, since they were present at a 'game-changing-time' in my life. In addition I will mention Ammon Katz (who is not in the above photograph) since my association with him could have been even more 'game-changing'.

1. Akito Arima

It was as summer visitors to Argonne National Laboratory in 1963 that I first met Akito Arima. Here he is seen in the photograph with his two colourful bodyguards. If I Remember Correctly (IIRC) — after a time gap of 60 years — we shared an office, which naturally led to discussions on a variety of topics. He showed a genuine interest in what I was doing — something involving the messy world of 3j- and 6j-symbols. But he soon introduced me to the wonderful world of 9j-symbols with their lovely symmetries. I was impressed by this and also by the fact that he knew by heart the actual values of several such symbols. IIRC, he then took this one stage further by replacing the 9 numbers in the 9j-symbol by 9 Japanese characters. Apparently, this can become two different poems depending on whether you read the characters horizontally or vertically. I was told that these combinations were not at all easy to construct — even if you knew Japanese. Here I also learnt that he practiced the martial art of Kendo with its bamboo swords. But he never invited me to join him in that activity. Actually, I preferred to play tennis with his other

Fig. 2. Akito Arima, Tony Green, and Gerry Brown in Argonne National Laboratory (Illinois, USA), 1963.

'body guard' — a tennis racket sounded less dangerous. However, I put my life in danger another way as explained later.

After that summer, we did not meet again for 2 or 3 years, when we were both in Princeton collaborating with Gerry Brown. One of Gerry's 'specialities' was (IIRC) the Tuesday evening seminars. These took place in the Palmer Laboratory coffee room at about 6 or 7pm. The time was chosen so that we could have dinner beforehand and then return for the seminar — not very convenient for family life. These seminars were meant to be very informal — only chalk and blackboard — no transparencies — way before powerpoint. One particular occasion comes to mind. We had a special guest (Akito) and so the preseminar dinner was also special. This was probably the Chinese restaurant on Witherspoon Street, since Gerry liked the cheeks in their fish speciality. Naturally, being a special occasion, we had a beer — or two — before making our way — somewhat red faced — back to Palmer Lab. There our joviality rapidly ended — Eugene Wigner was waiting for the seminar to begin. That was the only time he ever appeared at these seminars. But it was sufficient to curb our Tuesday night eating habits. I think that Wigner — the 'father of 9j-symbols' — came to see how his 1937 invention was being used and developed after 30 years by the young 'Akito'. With Akito in mind, for many years my lectures on theoretical nuclear physics always contained a lengthy section on 3,6 and 9j-symbols and their many manipulations. However, I suspect such a section would now be considered out-of-date with students thinking the symbols were some sort of Emoji. Hopefully Akito's 9j-poems do not suffer the same fate. It was several years after my retirement that I finally disposed of my tables of 3,6 and 9j-symbols. So their carbon footprint, started by Akito, was finally ended.

Akito and I never met again for many years. In the meantime, he had moved away from 9j-symbols to become, amongst other posts, the President of The Institute of Physical and Chemical Research (RIKEN) and also holding several ministerial portfolios. The latter eventually brought him to Helsinki. Some days before his visit, Dan Riska and I received notes to say we were invited by Akito to tea at the Japanese Embassy. So at the appointed time we arrived in our best clothes at the Embassy. He greeted us at the door and behaved as I remembered him — very friendly — in spite of his ministerial rank. During the tea drinking we small-talked and reminisced. Luckily, the tea was served 'western style', since an alternative could have been 'japanese style' — sitting cross legged on the floor. On the hour, a secretary entered the room to remind him of his ministerial duties. He politely walked with us to the door. The final words he said to me — with a smile on his face — were "Tony, there was no need to have worn a suit" — he still had his good sense of humour. We never met again.

2. Gerry Brown

Gerry's connection with Akito I believe began that summer in Argonne leading to Akito's time with Gerry in Princeton — as mentioned above. However, Gerry and I started to collaborate earlier in 1958 in Birmingham UK, when he became my

Master's Thesis advisor. The topic was the P-2P reaction on closed shell nuclei. I soon learnt one of Gerry's methods for doing research — namely — "Tony, leave something for someone else to do". This was his answer to a question of detail. He preferred to concentrate on the broad picture using simple models. A good example of this was the Brown-Bolsterli model for Dipole resonances. On the non-physics side, he liked to play tricks on people. The first one he played on me was at the Easter Physical Society Meeting in Oxford 1959, where I was due to give a short contribution on the P-2P reaction — my first of many verbal contributions over the next 40+ years. However, just before I was due to speak, Gerry who was sitting next to me — to my horror — got up and left the room. I had expected him to defend me. On leaving he said "I'll be back" or words to that effect. I should point out that — years later — this phrase is often attributed to Arnold Schwarzenegger. But Gerry beat him to it. I should add that Gerry did return before I started my contribution. However, Gerry left Birmingham for a time so I was transferred to Professor Rudolf Peierls for my PhD Thesis on the Momentum Dependence of the NN Interaction. So for the rest of my time in B'ham I had little to do with Gerry — except on the 'sports field'. There I occasionally joined a volleyball game. I forget whether this was voluntary or 'strongly encouraged' by Gerry, who was the organiser. This was understandable, since Gerry was by far the tallest person in the Department — making him the best player. Later, when we were both in Princeton we frequently played tennis 'on a more even playing field', where height was much less of an advantage. In relation to Gerry, "Fate was kind to me". After leaving B'ham in 1961 with my PhD, I went to work with Keith Brueckner in La Jolla. Unfortunately, before I got to La Jolla, Brueckner moved to Washington DC. Eventhough La Jolla was a marvelous place to be naturewise — especially after spending 6 years in B'ham — it was clear that I could not stay there after 1 year. Luckily the department sent several of us to the Annual New Year's Physical Society Meeting in New York. There I met Arnold Glick from the University of Maryland(UM), whom I had got to know a few years earlier at a Spring School in Naples. I told him about my problem and soon I was offered a post-doc position in Maryland — Lucky Break number One. In the Physics Department at UM the main theoretical nuclear physicist was Dick Ferrell — seen in the above group photo. Then came Lucky Break number Two. Dick invited Gerry, who was travelling around the USA, to give a seminar. So Gerry and I were 'reunited' — ending in Gerry asking me to be with him in Copenhagen the following year. On arrival at the Niels Bohr Institute in Copenhagen, Gerry's first words were "Tony — don't bother to unpack. You are going to Princeton next year". This was another of Gerry's modus operandi — collaborate with unmarried coworkers — they are easier to move around, attend evening seminars and play tennis frequently. On reflection, Gerry encouraged my unattached status by rarely 'arranging' the alternative. During my next four years in Princeton, under the ever watchful eye of Gerry, the topics of interest to Gerry were varied: 1) The most basic models actually considered nuclei as made up of nucleons interacting through the strong NN interaction. This first required a knowledge of the interaction between nucleons — taking us also into a study of 3- and 4-nucleon interactions mediated by mesons (Ebbe Nyman and A.M.G). With standard NN interactions matrix elements between a variety of shell model states were generated and used in the calculation of nuclear levels(Tom Kuo). Larry Zamick was also engaged in these types of microscopic methods for studying nuclei. In addition, the rotational bands near closed shell nuclei were studied by treating them as multi-Particle-multi-Hole (nP-nH) states in a deformed oscillator well. This started off with treating the famous O+ state at 6.06MeV in O16 as 4P-4H and ended up using 8P-8H in Ca40 (Bill Gerace and A.M.G). Gerry played an active role in each of these projects. 2) A step away from the above was to first couple the nucleons into pairs and consider these pairs as the basic nuclear building blocks. Akito was involved in this type of approach with Francesco Iachello. Later this was developed into the so-called IBM (Interacting Boson Model and not as some think IBM means — Iachello Believes in Miracles or Improvement on Bohr and Mottelson). 3) At this time there was activity treating simple nuclei in terms of interacting Alpha particles. Akito was interested in this topic using methods based heavily on symmetry arguments in order to understand, for example, the lowest lying 0+ in O16 mentioned above. 4) Some years earlier Igal Talmi had wriitten down a formula relating the masses of neighbouring nuclei. In this way it was possible to predict the masses of many unknown nuclei. This idea was now extended to the Garvey-Kelson Mass Formula, which was more general and applicable in situations where the original Talmi formula could not be used.

In the above, I mention some of the topics and the associated physicists in which the Princeton Physics Department played a role. Others that come to mind are: George Bertsch, Jerzy Borysowics, Aldo Cavello, Andy Jackson, Alex Lande, Eivand Ostgaard, Harvey Picker, George Ripka, Giovani Sartoris and Chun Wa Wong I apologise to those I have not included in this list. Also to those in the list, I apologise for forgetting — after almost 60 years — their research topics. Here I have also included names that were mainly based in Rutgers University, since Gerry considered them as part of his Extended Family on account of their regular visits to Princeton. I should add that all of the people mentioned above spent an extended period of time in Princeton. Of course, in addition, there were numerous others that visited just for a few days to give a seminar. Perhaps the most famous of these was Gregory Breit — Gerry's PhD supervisor. I believe that his presence made Gerry nervous. This was compounded when Breit's raincoat was stolen from a coat rack in the local French restaurant. Gerry then had to run home to fetch one of his own raincoats. The problem was that Breit stood at about 1/2 m shorter than Gerry. The subject of Breit brings up the question as to How did Gerry learn physics?. Some simply take a book and read it from cover to cover. Gerry's approach was different. In some cases he would give a lecture course on the topic he was learning. Here I have in mind a course on SU(3) when he was working on the rotational bands in O16. Another example was his learning of basic Many Body Techniques in Copenhagen. He drafted David Thouless to give a course and sat in the front row taking copious notes and asking many questions. In the Physics Department, Gerry's domain was restricted to two large rooms — the Ground Floor room was his office and the First Floor room for some of his coworkers. Actually his office is a misnomer, since he only reserved one corner for himself — the rest being used to house other coworkers in the 4 or 5 desks spread around the room. Over the 4 years I occupied one of these desks, I observed several visitors that did not really fit into the rest of Gerry's Bubble. In particular I remember Leonardo Castillejo and Harry Lipkin. Leonardo, as always, was interested in other people's physics problems to which he often made positive contributions. I heard that, when Gerry was proposing to the Physics Faculty for Leonardo to be a visiting Lecturer, a complaint was raised that the students should only be lectured to in well spoken English and without a corrupted accent. I believe that Gerry then pointed out that Leonardo's English and accent were superior to any one else in the room in spite of his Spanish sounding name. Leonardo, at the time, was interested in solving 3-body equations numerically. This was before we each had a terminal on our desks and at home. So Leonardo would have to trudge his way to the Computer Centre sitting in another building and return with a stack of computer output. I had a feeling that he was a burden on the Computer paper budget. Harry Lipkin was somewhat out of place in Gerry's Bubble, since he was more interested in the goings on of Sam Treiman and Marvin Goldberger. My main memory of Harry was his technique for making phone calls — (IIRC) if you get an engaged signal, hang up and immediately dial again. When he finally succeeded, he would proceed to have a discussion in a loud voice — not ideal for the other 5 people trying to work in the same room. On the non-physics side, several events/occasions come to mind. Gerry believed that a healthy body led to a healthy mind. So he was sympathetic to the Wednesday afternoon soccer games in which Tom Kuo was the backbone. These gatherings could attract a whole variety of people — sometimes upto 30 arriving with their differing degrees of soccer competence and not all physicists. So we sometimes played 15 a side. But Gerry always kept away from this rough sport. Over the years several physicists asked me "How can you work with that guy? He's so unpleasant." But this was not the case for those inside 'his bubble'. However, even with insiders he had a habit of sometimes being less than complimentary. Once he suggested that I should write the first draft of one of our joint papers. The result was for Gerry to say "Tony — this reads like Uncle Remus telling about Brer Rabbit". So he made some changes. On another occasion he came back from some tour which included a place where I had earlier given a seminar — with the result "Tony — I had to spend time clearing up the wrong impressions you had given about our work". On the positive side — once he was explaining to me some work he had 'in progress' with Hans Bethe. I asked a question which greatly simplified their work — with the result "Tony — you're talking poetry". Gerry and I were once invited to spend Thanksgiving at Norman Austern's home in Pittsburgh. This meant flying together. For some reason the airhostess asked for our surnames. We truthfully replied "Brown and Green" — she wrote these down with a strange disbelieving look on her face. Norman had only recently got married and this was his first Thanksgiving dinner as the host. He tried to hide this fact of being a 'beginner' by starting to cut up the turkey in their kitchen. At this point, Gerry took over by insisting Norman 'performed the operation' in front of their guests. This illustrated another of Gerry's traits — not being shy in coming forward. Occasionally, this somewhat brash approach would end him in trouble. At one Gordon conference he had been particularly aggressive towards the large Israeli contingent. They retaliated at one of the evening 'liquid-discussion-sessions' by pinning a notice on Gerry's back saying "Arab Lover". It took him some time before he realised why people were laughing. At another Gordon Conference Gerry cut his foot while swimming. This meant a visit to the local medical centre, where — being summertime — the only doctor on duty was a gynaecologist. For the rest of the time at the conference, people were asking about the next visit to his gynaecologist. This was not the first time I had seen Gerry needing medical attention. As said earlier, I changed supervisors for my PhD because Gerry went to the USA for an extended visit. On his last day in Birmingham, the Department had a farewell party for him and, for an inner circle, this continued at his home. However, he unexpectedly turned up again the next morning at the Department using a crutch and being supported by Leonardo Castillejo. Apparently he had failed to jump over a bush in his garden. The broken leg delayed his visit to the USA. It was during my years in Princeton that we became real friends — compared with Copenhagen where we were mainly collaborators. We would now play tennis frequently — a sport where we were basically equals — unlike in our earlier sports activity (volleyball) in Birmingham. Occasionally Gerry's 'will-to-win' would show through by accusing me of playing like a squirrel as I ran around waiting for him to make a mistake. Often the game was followed by a dinner/lunch invitation at his nearby house. Here I played another role as Tony the Peacemaker between Gerry and his first wife Traudl, since they frequently did not see eye to eye — to put it mildly. Another way by which Gerry tried to alleviate this problem was to import and employ young Danish girls. If I understand correctly these ladies played a multiple role — as secretaries for Gerry, companions for Traudl and, last but not least, to brighten up the life of the all male University. They succeeded in each of these roles. I believe that Gerry was never truly happy in Princeton, since Theoretical Nuclear Physics was not appreciated by some of the other Professors. Before his arrival, this discipline was essentially in the hands of an Associate Professor — not the best way to promote the subject in Professorial meetings. I personally faced this lack of respect for Nuclear Physics, while on a committee preparing examination questions. When it came around to discussing the questions submitted by me, the chairman, who will remain anonymous, said Looking at Nuclear Physics questions is like looking at someone elses dirty washing — or words to that effect — not encouraging. Actually, Gerry probably alienated other professors by referring to (for example) Marvin Goldberger as Murph the Surf — at the time a popular Olympic windsurfer. Possibly, Gerry was expected to be in the mold of Eugene Wigner — a onetime Nuclear Physicist. However, I once heard Gerry say that he could never create Nuclear Physics like

Wigner — but perhaps like Hans Bethe, with whom he collaborated a lot. Another indication that Gerry was not happy in Princeton was his moving to Stonybrook, even though the latter Institute had earlier been considered to have a much lower status than Princeton. However, this had changed rapidly in the mid 60's when John Toll became President of Stonybrook. He rapidly went 'head-hunting' and returned with Frank Yang (Nobel Prize 1957) and later Gerry.

3. Ammon Katz

I realise that this issue is meant to be mainly about Akito Arima. However, I feel as the author of this section, that I can add a few words of a personal nature — poetic licence. For me the most memorable event in my Argonne visit was with Ammon Katz — another of the summer visitors — an Israeli physicist. I had met him at the same Spring School in Naples — where I had met Arnold Glick. One day he asked me if I was interested in going flying. Naturally, as a young 25 year old I said "Yes". Possibly one reason for the invitation was that I had a car — a convenient way to get to the airfield. However, on the way to the airport I began to have second thoughts when I realised that he had only just received his flying license while on a sabbatical in California. The reason for him to be flying while at Argonne was that he needed to fly a certain number of hours each month for the first 6 months of getting his pilot's license. The plane was a 2 seater Piper Colt. Being a new pilot, he was very eager to explain flying in some detail — such as "If the right wing goes 60km/hr slower, we will go into a tailspin which I couldn't correct at this altitude". We were flying for about 1/2hr during which we flew over Argonne Laboratory. Needless to say, I was very glad to get back to Terra Firma. I was not invited to go flying a second time — thank goodness,

So we see that the Spring School in Naples in about 1960 featured in two ways in my life: 1) Through Arnold Glick and Dick Ferrell I came into contact with Gerry Brown again. This was without doubt a 'Life — Changer' for me. 2) Through Ammon Katz I learnt to never again go flying with novices. This shows that, when going to conferences etc, it's the contacts you make in the corridors and swimming pools that are longer lasting than the information gleaned in the lectures.

https://doi.org/10.1142/9789819813353_0007

The Role of Dynamic Symmetries in Nuclei

Francesco Iachello
Center for Theoretical Physics, Sloane Physics Laboratory,
Yale University, New Haven, CT 06520-8120, USA
francesco.iachello@yale.edu

The concepts of dynamic symmetry, dynamic supersymmetry and critical symmetry are briefly reviewed. These concepts have provided benchmarks for the classification of collective states in nuclei. Examples of dynamic symmetries, dynamic supersymmetries and critical symmetries are presented. The contributions of Akito Arima to the development of these concepts are described.

1. Introduction

In the last 40 years, the concepts of dynamic symmetry and its generalizations dynamic supersymmetry and critical symmetry have played a major role in Nuclear Physics. Akito Arima made major contributions to this subject through the study of the dynamic symmetries of the Interacting Boson Model. In this article, a review of applications of symmetry methods in Nuclear Physics is given, with emphasis to the dynamic symmetries of the Interacting Boson Model (IBM) to which Akito Arima greatly contributed.

2. Dynamic Symmetry

A dynamic symmetry is that situation in which the Hamiltonian H describing a physical system contains only invariant Casimir operators of a chain of algebras originating from an algebra G,

$$G \supset G' \supset G'' \supset \cdots . \qquad (1)$$

It is called dynamic because it is a symmetry of the Hamiltonian which produces the dynamics of the system.

Dynamic symmetries in nuclei have been mostly studied within the framework of the Interacting Boson Model.[1,2] In this model, even-even nuclei are described as a collection of correlated pairs of nucleons treated as bosons. The bosons have angular momentum $J = 0$ and $J = 2$ (s and d bosons). The model has algebraic structure $G \equiv U(6)$, since all operators can be constructed in terms of elements of $U(6)$,

$G_{\alpha\beta} = b^\dagger_\alpha b_\beta$, where $b^\dagger_\alpha\,(\alpha = 1,\ldots,6) \equiv s^\dagger, d^\dagger_\mu(\mu = 0,\pm1,\pm2)$ and $b_\alpha\,(1,\ldots,6) \equiv s, d_\mu\,(\mu = 0,\pm1,\pm2)$, with generic form

$$H = E_0 + \sum_{\alpha\beta} \varepsilon_{\alpha\beta} G_{\alpha\beta} + \frac{1}{2} \sum_{\alpha\beta\gamma\delta} u_{\alpha\beta\gamma\delta} G_{\alpha\gamma} G_{\beta\delta} \tag{2}$$

to be diagonalized within the totally symmetric representations

$$[N] = \square\square\cdots\square \tag{3}$$

of $U(6)$. Here N represents the total number of valence pairs (particles or holes) and, in first approximation, no difference is made between proton pairs and neutron pairs.

The dynamic symmetries of the Interacting Boson Model can be studied by considering all possible subalgebra chains that originate from $U(6)$ with the constraint that rotational invariance be maintained. This study was done by Akito Arima and I in the 1970's. It was found that there are three and only three possibilities corresponding to the chains[3–5]

$$\begin{array}{lll} & U(5) \supset SO(5) \supset SO(3) \supset SO(2) & (I) \\ & \nearrow & \\ U(6) \longrightarrow & SU(3) \supset SO(3) \supset SO(2) & (II) \\ & \searrow & \\ & SO(6) \supset SO(5) \supset SO(3) \supset SO(2) & (III). \end{array} \tag{4}$$

In these cases, all properties can be calculated in explicit analytic form. In particular, the energies of the states are given in terms of quantum numbers

$$\begin{aligned} E^{(I)}\,(N, n_d, v, n_\Delta, L, M_L) &= E_0 + \varepsilon n_d + \alpha n_d(n_d+4) + \beta v(v+3) \\ &\quad + \gamma L(L+1), \\ E^{(II)}\,(N, \lambda, \mu, K, L, M_L) &= E_0 + \kappa\left(\lambda^2 + \mu^2 + \lambda\mu + 3\lambda + 3\mu\right) \\ &\quad + \kappa' L(L+1), \\ E^{(III)}\,(N, \sigma, \tau, \nu_\Delta, L, M_L) &= E_0 + A\sigma\,(\sigma+4) + B\tau\,(\tau+3) \\ &\quad + CL(L+1), \end{aligned} \tag{5}$$

where the various terms are the eigenvalues of the invariant Casimir operators in the appropriate irreducible representations. These energy formulas have provided benchmarks for the analysis of experimental data. In the last 40 years, several examples of nuclei described well by the dynamic symmetries of the IBM have been found, three of which are given in Figs. 1–3. Particularly relevant was the discovery by Cizewski, Casten *et al.* of the third symmetry $SO(6)$[6] unknown at the time. The analysis of spectra in terms of symmetries is not limited to energies, but it includes intensities of transitions and other measurable quantities.[2]

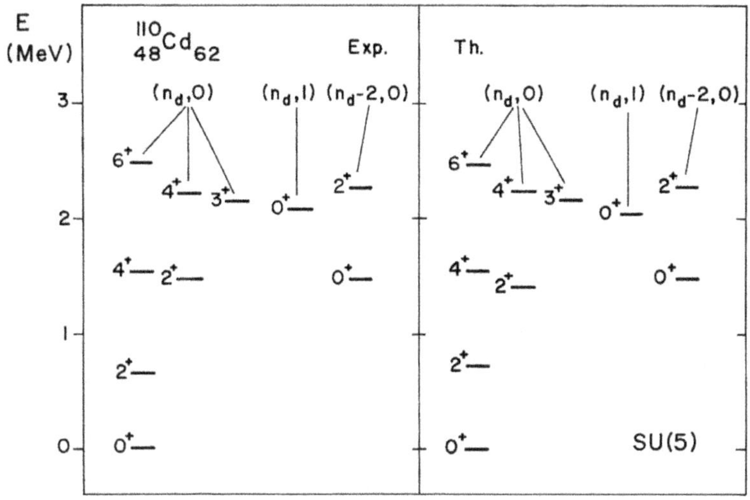

Fig. 1. An example of $U(5)$ dynamic symmetry in nuclei: ${}^{110}_{48}\mathrm{Cd}_{62}$. Adapted from 3.

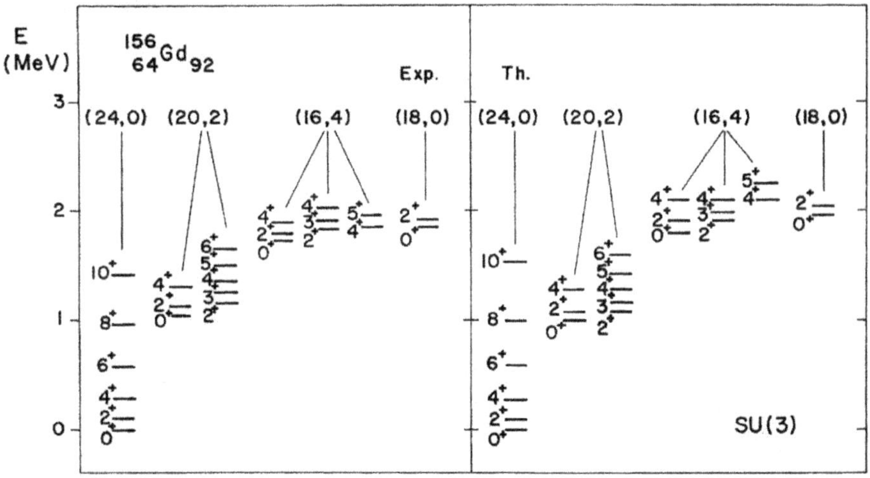

Fig. 2. An example of $SU(3)$ dynamic symmetry in nuclei: ${}^{156}_{64}\mathrm{Gd}_{92}$. Adapted from 4.

The dynamic symmetries of the Interacting Boson Model have been the subject of numerous investigations to determine what are the limits of their applicability. Maps of the periodic table have been developed to determine the regions where they are found. An example of the map of nuclei in the region of proton number 82-126, is shown in Fig. 4.[7] Also the extent to which symmetries persist as the excitation energy is increased has been the subject of many studies. It has been found that symmetry persists to high excitation energy, although states belonging to a symmetry, for example $U(5)$, may coexists, at high excitation energy, with

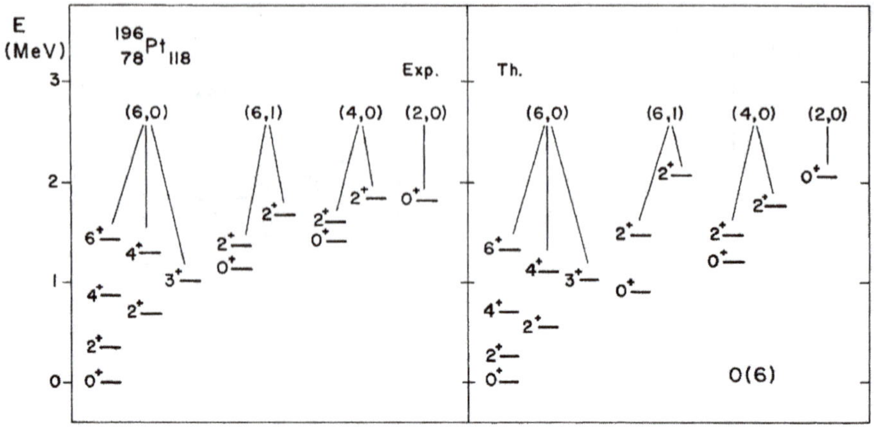

Fig. 3. An example of $SO(6)$ dynamic symmetry in nuclei: ${}^{196}_{78}\mathrm{Pt}_{118}$. Adapted from 5.

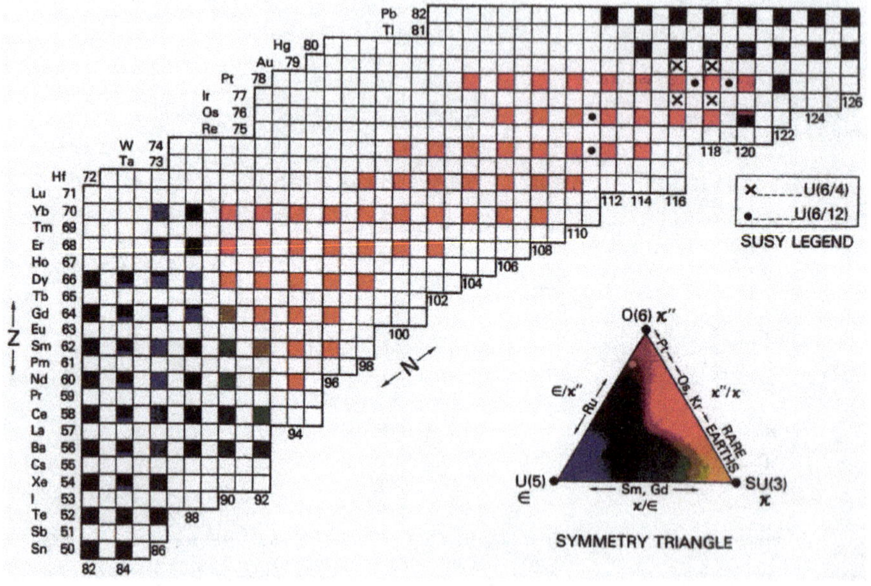

Fig. 4. A representation of the nuclear chart with color coded symmetries. Adapted from 7.

states of a different symmetry, for example $SO(6)$. An example is shown in Fig. 5,[8] where states of $U(5)$ symmetry are compared with Eq. (5).

Associated with symmetry there is a geometry. The geometry associated with an algebraic structure is called a coset space. In the case of the Interacting Boson Model, the geometry is that of quadrupole shapes. The geometric description of collective quadrupole states in nuclei has been the cornerstone of Nuclear Structure for decades.[9] The nucleus is seen as a liquid drop with radius

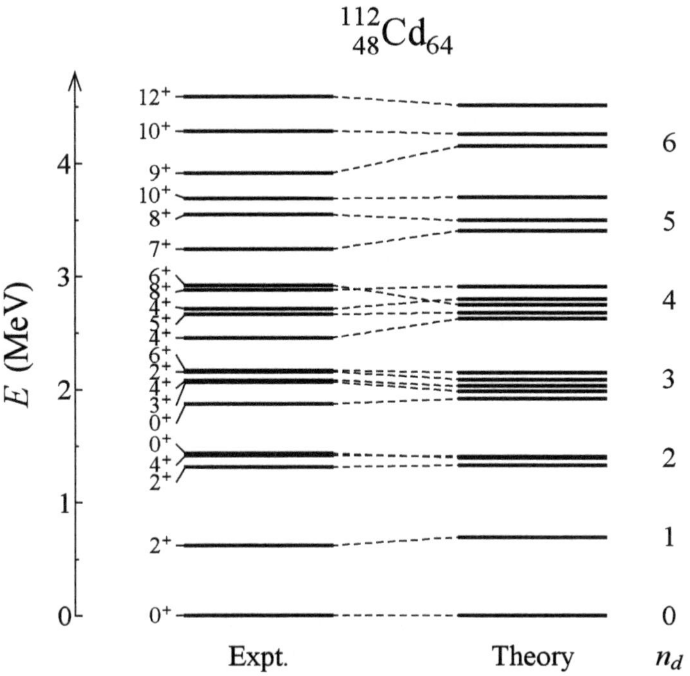

Fig. 5. An example of $U(5)$ dynamic symmetry at high excitation energy: $^{112}_{48}\mathrm{Cd}_{64}$. Adapted from 8.

$R = R_0\left(1 + \sum_\mu \alpha_\mu Y_{2\mu}(\theta, \varphi)\right)$. By the method of intrinsic or coherent states[10–12] one can find that the shapes associated with the three symmetries are:

$$
\begin{array}{llll}
 & U(5) & \text{Spherical} & (I) \\
 & \nearrow & & \\
U(6) & \longrightarrow SU(3) & \text{Axially deformed} & (II) \\
 & \searrow & & \\
 & SO(6) & \gamma\text{-unstable} & (III).
\end{array}
\tag{6}
$$

Soon after the introduction of the Interacting Boson Model, it was realized that an even more accurate description of nuclei could be obtained by distinguishing between proton pairs and neutron pairs treated as bosons (Interacting Boson Model-2).[13,14] Akito Arima, together with Takaharu Otsuka, made major contributions to the development of the proton-neutron Interacting Boson Model (IBM-2) by deriving a microscopic connection between the IBM and the shell model.[15] Proton bosons and neutron bosons can be labeled by a quantity called F-spin, similar but not identical to isospin. Proton bosons were assigned z-projection $F_z = +\frac{1}{2}$ and neutron bosons $F_z = -\frac{1}{2}$. The algebraic structure of IBM-2 is $G \equiv U_\pi(6) \oplus U_\nu(6)$, since all operators can be constructed in terms of elements of $U_\pi(6)$ and $U_\nu(6)$, which in turn are constructed from boson creation operators, $s^\dagger_\pi, d^\dagger_{\pi,\mu}, s^\dagger_\nu, d^\dagger_{\nu,\mu}$ and annihilation operators, $s_\pi, d_{\pi,\mu}, s_\nu, d_{\nu,\mu}$. The Hamiltonian is now

$$H = H_\pi + H_\nu + V_{\pi\nu} \tag{7}$$

where H_π, H_ν are as in Eq. (1) with an index π, ν and

$$V_{\pi\nu} = \sum_{\alpha\beta\gamma\delta} w_{\alpha\beta\gamma\delta} G_{\pi,\alpha\beta} G_{\nu,\gamma\delta} \tag{8}$$

to be diagonalized within the representations

$$[N_\pi] \otimes [N_\nu] = \oplus \sum_{k=0}^{N_\nu} [N_\pi + N_\nu - k, k], \quad \text{if } N_\pi \geq N_\nu. \tag{9}$$

The sum extends to N_π if $N_\pi \leq N_\nu$. Here N_π and N_ν are the numbers of proton and neutron bosons. One can see that one has in IBM-2 extra states corresponding to mixed symmetry representations $[N_1, N_2]$ with $N_2 \leq N_1 \neq 0$, as for example

$$\begin{array}{l} [N_1, 1] = \square\square\cdots\square \\ \qquad\qquad\;\; \square \end{array} \tag{10}$$

The Interacting Boson Model-2 has many possible dynamic symmetries.[2] Three of them correspond to the three symmetries of the IBM-1,

$$\begin{aligned} U_\pi(6) \oplus U_\nu(6) &\supset U_\pi(5) \oplus U_\nu(5) \supset U_{\pi\nu}(5) \supset O_{\pi\nu}(5) \\ &\supset O_{\pi\nu}(3) \supset O_{\pi\nu}(2) \quad (I_2) \\ U_\pi(6) \oplus U_\nu(6) &\supset SU_\pi(3) \oplus SU_\nu(3) \supset SU_{\pi\nu}(3) \\ &\supset O_{\pi\nu}(3) \supset O_{\pi\nu}(2) \quad (II_2) \\ U_\pi(6) \oplus U_\nu(6) &\supset O_\pi(6) \oplus O_\nu(6) \supset O_{\pi\nu}(6) \supset O_{\pi\nu}(5) \\ &\supset O_{\pi\nu}(3) \supset O_{\pi\nu}(2) \; (III_2) \end{aligned} \tag{11}$$

but there is now a fourth symmetry, called $SU(3)^*$, which arises when combining the algebra of $SU_\pi(3)$ with that of $\overline{SU_\nu(3)}$ obtained as an inner automorphism of $SU_\nu(3)$,

$$U_\pi(6) \oplus U_\nu(6) \supset SU_\pi(3) \oplus \overline{SU_\nu(3)} \supset SU_{\pi\nu}(3)^* \supset O_{\pi\nu}(3) \supset O_{\pi\nu}(2) \; (IV). \tag{12}$$

In Eqs. (10)–(11), the index $\pi\nu$ denotes algebras obtained by the sum of algebras with index π and ν. The algebra $\overline{SU(3)}$ is obtained from $SU(3)$ by changing the sign of the quadrupole moment operator.

Similarly, when considering the geometric structure of IBM-2, one has one more shape in addition to the three shapes of Eq. (4),

$$U_\pi(6) \oplus U_\nu(6) \to SU(3)^* \quad \text{Triaxial} \quad (IV). \tag{13}$$

The triaxiality here is obtained by the combination of a prolate shape for protons with an oblate shape for neutrons or viceversa.

The introduction of the Interacting Boson Model-2 opened the way to the study of dynamic symmetries in two-component (proton-neutron) systems. States with $F = F_{\text{max}} - 1$ (the so-called scissor modes) were found by Richter and collaborators in 1984.[16] In later years, this subject was taken up again and new examples of F-spin symmetry were found by Pietralla and collaborators.[17]

A further extension of the Interacting Boson Model to include proton-neutron pairs called Interacting Boson Model-3 and -4 was introduced in the 1980's.[18,19] Symmetries of these models have been the subject of several investigations.[20]

The concept of dynamic symmetry was also expanded to situations in which only part of the spectrum is given in closed analytic form and part is not, called partial dynamic symmetry.[21,22] Symmetries of IBM-3 and IBM-4 and partial dynamical symmetries will not be discussed here.

3. Dynamic Supersymmetry

A dynanic supersymmetry is that situation in which the Hamiltonian H for a mixed system of bosons and fermions contains only invariant Casimir operators of a chain of algebras graded or not originating from a superalgebra G^*,

$$G^* \supset G'^* \supset G''^* \supset \cdots . \tag{14}$$

As in the previous case of dynamic symmetry, the supersymmetry is called dynamic because it is the symmetry of the Hamiltonian which produces the dynamics of the system. The difference between a dynamic symmetry and a dynamic supersymmetry is that the former describes a system of bosons *or* fermions, while the latter describes a system of bosons *and* fermions.

Dynamic supersymmetries were introduced in Nuclear Physics in 1980[23] within the framework of the Interacting Boson-Fermion Model.[24,25] Akito Arima made an important contribution to the Interacting Boson-Fermion Model with the extension of the Interacting Boson Model to odd-even nuclei.[26] In nuclei with an odd number of particles, at least one particle is unpaired. In addition, in even-even nuclei, at higher excitation energies, $E_x \gtrsim 1.5$ MeV, some of the pairs break and one has states with broken pairs. The basic ingredients of the Interacting Boson-Fermion Model are thus correlated pairs, treated as bosons, and individual unpaired nucleons. The building blocks are boson creation and annihilation operators $s^\dagger, d^\dagger_\mu, s, d_\mu$, and fermion creation and annihilation operators, $a^\dagger_{j,m}, a_{j,m}$ where j, m denote the total angular momentum and its z-projection. The model Hamiltonian is

$$H = H_B + H_F + V_{BF}. \tag{15}$$

The Hamiltonian (15) can be written in terms of elements of a Lie superalgebra, $G^* \equiv U(6/\Omega)$, where $\Omega = \sum_j (2j+1)$. The elements of the superalgebra can be conveniently written as a matrix

$$\begin{pmatrix} G_{\alpha\beta} & F^\dagger_{\alpha i} \\ F_{i\alpha} & G_{ij} \end{pmatrix} \tag{16}$$

with $G_{\alpha\beta} = b_\alpha^\dagger b_\beta, G_{ij} = a_i^\dagger a_j, F_{\alpha i}^\dagger = b_\alpha^\dagger a_i, F_{i\alpha} = a_i^\dagger b_\alpha$. A generic form of the Hamiltonian is

$$H_B = E_0 + \sum_{\alpha\beta} \varepsilon_{\alpha\beta} G_{\alpha\beta} + \frac{1}{2} \sum_{\alpha\alpha'\beta\beta'} v_{\alpha\alpha'\beta\beta'} G_{\alpha\beta} G_{\alpha'\beta'}$$

$$H_F = E_0' + \sum_{ik} \eta_{ik} G_{ik} + \frac{1}{2} \sum_{ii'kk'} u_{ii'kk'} G_{ik} G_{i'k'} \tag{17}$$

$$V_{BF} = \sum_{\alpha\beta ik} w_{\alpha\beta ik} G_{\alpha\beta} G_{ik}$$

A property of supersymmetry is that in the Hamiltonian Eq. (15) all couplings are related by supersymmetry transformations. One of the consequences of supersymmetry is that, if bosonic states are known, one can predict fermionic states. Both are given by the same energy formula. It is thus a most intricate type of symmetry. In nuclei, supersymmetry implies a relation between spectra of even-even nuclei and those of odd-even nuclei (odd proton or odd neutron).

In view of the fact that nucleons can occupy several different j states and bosons have three possible types of symmetry, $U(5), SU(3), SO(6)$, the number of possible supersymmetries is quite large. One possible supersymmetry is that in which bosons have $SO(6)$ symmetry and fermions occupy a single particle level with $j = 3/2$. The corresponding superalgebra is $U(6/4)$[23] with chain

$$\begin{aligned} U(6/4) &\supset U^B(6) \oplus U^F(4) \supset SO^B(6) \oplus SU^F(4) \\ &\supset Spin^{BF}(6) \supset Spin^{BF}(5) \supset Spin^{BF}(3) \supset Spin^{BF}(2) \end{aligned} \tag{18}$$

and energy formula

$$\begin{aligned} &E\left(\mathcal{N}, N_B, N_F, \Sigma, (\sigma_1, \sigma_2, \sigma_3), (\tau_1, \tau_2), \nu_\Delta, J, M\right) \\ &= E_0 + A'\Sigma(\Sigma+4) + A\left[\sigma_1(\sigma_1+4) + \sigma_2(\sigma_2+2) + \sigma_3^2\right] \\ &\quad + B\left[\tau_1(\tau_1+3) + \tau_2(\tau_2+1)\right] + CJ(J+1), \end{aligned} \tag{19}$$

where $\mathcal{N} = N_B + N_F$ is the total number of bosons plus fermions. This type of supersymmetry was found in 1981[23,27] in the spectra of the Pt-Ir nuclei, as shown in Fig. 6. It was concluded then that supersymmetry in nuclei had been found.[28,29]

The study of supersymmetry in nuclei was not much pursued in subsequent years. In the late 1990's, as a result of improved experimental techniques, it was taken up again. These developments were based on a more accurate description of nuclei called Interacting Bobon-Fermion Model-2 in which the basic constituents are proton bosons and neutron bosons (Interacting Boson Model-2) and unpaired protons and neutrons.[30,31] The algebraic structure of this model is $G^* \equiv U_\pi(6/\Omega_\pi) \oplus U_\nu(6/\Omega_\nu)$, where the index π, ν distinguishes protons from neutrons.[25] If supersymmetry occurs for this very complex system one expects to have supersymmetric partners composed of a quartet of nuclei, even-even, even-odd,

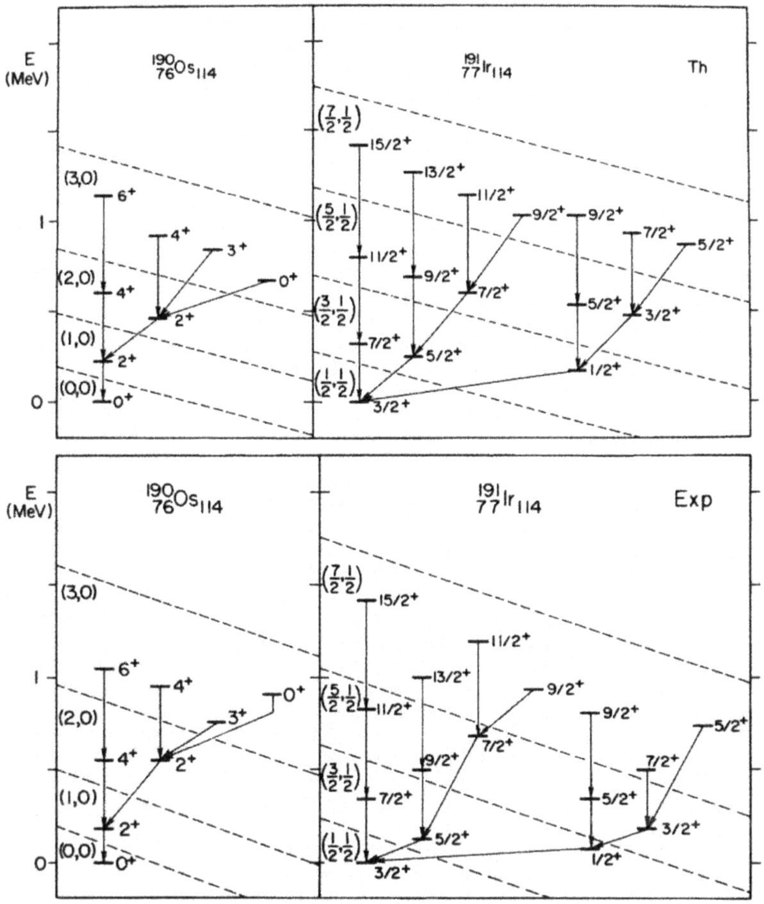

Fig. 6. An example of $U(6/4)$ dynamic supersymmetry in nuclei: the pair of nuclei $^{190}_{76}Os_{114}$-$^{191}_{77}Ir_{114}$. Adapted from 27.

odd-even and odd-odd. Spectra of even-even, even-odd and odd-even nuclei had been known for some time. However, spectra of odd-odd nuclei were very difficult to measure, since the density of states in these nuclei is very high and the energy resolution of most detectors was not sufficiently good. In a major effort led by Jolie and Graw that involved several laboratories for several years it was possible to measure spectra of odd-odd nuclei to the degree of accuracy needed to test supersymmetry. In particular it was possible to measure the spectrum of ^{196}Au, the missing supersymmetric partner of ^{194}Pt, ^{195}Pt and ^{195}Au.[34–36] The comparison between the observed spectrum of ^{196}Au and that expected by the supersymetry $U_\pi(6/4) \oplus U_\nu(6/12)$ is shown in Fig. 7. An analysis of electromagnetic transition rates and of intensitities of transfer reactions that measure explicitly the fermionic sector of supersymmetry was also done[37] and it was concluded that supersymmetry in nuclei was confirmed.

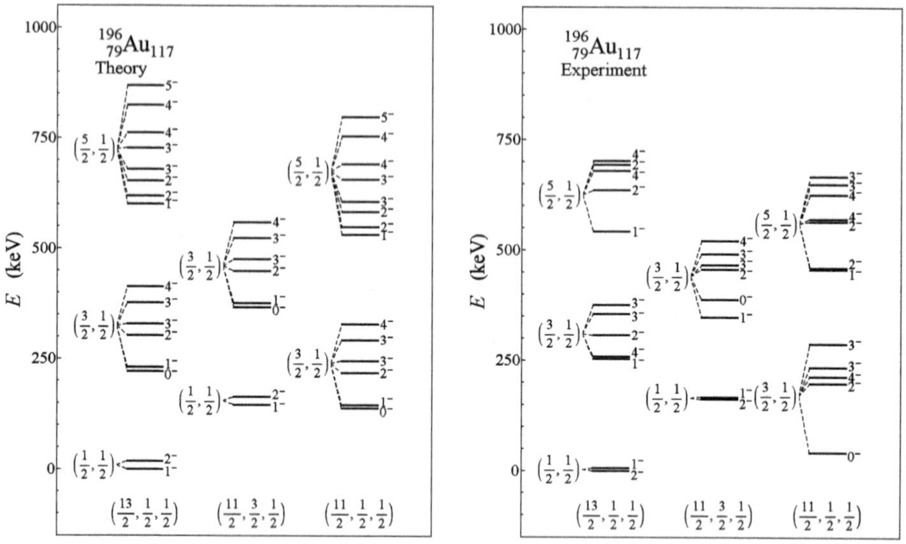

Fig. 7. An example of $U_\pi(6/4) \oplus U_\nu(6/12)$ dynamic supersymmetry in nuclei: ${}^{196}_{79}\mathrm{Au}_{117}$. Adapted from 34.

4. Critical Symmetry

In the 2000's, a new concept was introduced.[38–40] Because of the connection with the theory of phase transitions it was called "critical symmetry". This concept provides spectral signatures of finite quantal systems at the critical value of quantum phase transitions. It is similar to the concept of dynamic symmetry, since it provides a complete solution to the eigenvalue spectrum and intensites of electromagnetic transitions, but different from it in that it is formulated in a differential rather than algebraic language, and, in the form used up to now, it corresponds to situations in which the potential in the Schödinger formulation is discontinuous.

One of the best studied cases of critical symmetry is that of nuclei. The differential framework here is the Bohr equation,[41] with Hamiltonian

$$H = -\frac{\hbar}{2B}\left[\frac{1}{\beta^4}\frac{\partial}{\partial\beta}\beta^4\frac{\partial}{\partial\beta} + \frac{1}{\beta^2 \sin 3\gamma}\frac{\partial}{\partial\gamma} - \frac{1}{4\beta^2}\sum_\kappa \frac{Q_\kappa^2}{\sin^2\left(\gamma - \frac{2}{3}\pi\kappa\right)}\right] + V(\beta,\gamma), \tag{20}$$

where β, γ are the intrinsic Bohr variables and θ_i the three Euler angles characterizing the orientation in space of the ellipsoid. The main properties of these symmetries is that the spectrum is given in terms of zeros of special functions, in particular of Bessel functions. Three solutions were worked out.[38–40] The first solution, called $E(5)$, describes nuclei at the critical value of the quantum phase transition between the first and third dynamic symmetry of the Interacting Boson Model ($U(5) - SO(6)$ spherical to γ-unstable transition). The spectrum is given by a closed formula, similar to Eq. (5),

$$E^{(I_c)}(s,\tau,\nu_\Delta,L,M) = E_0 + A\,(x_{s,\tau})^2 + CL(L+1)$$
$$x_{s,\tau} = \text{s-th zero of } J_\nu(z); \quad \nu = \tau + 3/2. \tag{21}$$

Experimental examples of this symmetry were found.[42] The second solution, called $X(5)$, describes nuclei at the critical point of the quantum phase transition between the first and second dynamic symmetry of the Interacting Boson Model ($U(5) - SU(3)$ spherical to axially deformed transition). The spectrum is given by

$$E^{(II_c)}(s,L,s',K,M) = E_0 + A\,(x_{s,L})^2 + B\,(x_{s',L})^2$$
$$x_{s,L} = s\text{-th zero of } J_\nu(z); \quad \nu = \left(\frac{L(L+1)}{3} + \frac{9}{4}\right)^{1/2} \tag{22}$$
$$x_{s',K} = s'\text{-th zero of } J_{\nu'}(z); \quad \nu' = \frac{K}{2}.$$

(This second solution is only approximate and it should be properly called a quasisymmetry.) Experimental examples of this symmetry were found. One of these example is shown in Fig. 8.[43] After the original discovery of $E(5)$ and $X(5)$ by Casten and Zamfir, several other examples of these symmetries were found.[44] The third solution, called $Y(5)$, describes nuclei at the critical value of the quantum

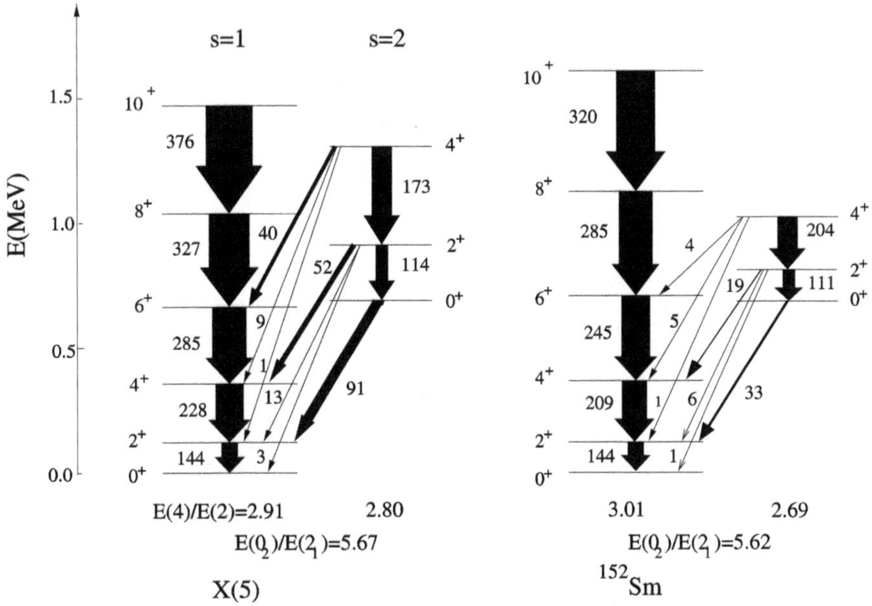

Fig. 8. An example of critical symmetry $X(5)$: ${}^{152}_{62}\text{Sm}_{90}$. Adapted from 43.

phase transition between the second and fourth dynamic symmetry of the Interacting Boson Model-2 ($SU(3) - SU(3)^*$ axial to triaxial transition). The spectrum is given by

$$E^{(III_c)}(n_\beta, L, s', K, M) = E_0 + A(x_{s',M'})^2 + BL(L+1) + B'n_\beta$$
$$x_{s',M'} = s'\text{-th zero of } J_\nu(z); \quad M' = \frac{K}{2}. \tag{23}$$

(This solution is also approximate and should be called a quasisymmetry.) The situation here is not as rich as that of the $E(5)$ and $X(5)$ symmetries and only few examples, $^{166-168}$Er, have been found.

The study of "critical" symmetries became in the 2000's one of the most active areas of research in Nuclear Physics. The role of critical symmetries within the context of quantum phase transitions in nuclei was reviewed in several articles in 2010–2011.[45–48]

5. Conclusions

Dynamic symmetries, dynamic supersymmetries and critical symmetries provide a classification scheme for finite quantal systems. These concepts have been extensively applied to Nuclear Physics where they have provided benchmarks for the analysis of experimental data. Several examples of all three types of symmetry of the Interacting Boson Model, of supersymmetries of the Interacting Boson-Fermion Model and of critical symmetries have been found. When the symmetry is broken, diagonalization of the Hamiltonian in a symmetry adapted basis, has provided an invaluable tool for nuclear stucture studies. Nuclear physics is presently engaged in an expansion into uncharted regions of the periodic table. Symmetries provide a roadmap to explore these regions as they will become available at radioactive beam facilities (GSI/FAIR, GANIL/SPIRAL2, RIKEN, FRIB). The roadmap provided by the symmetries of the Interacting Boson Model-1 is schematically shown in Fig. 9,

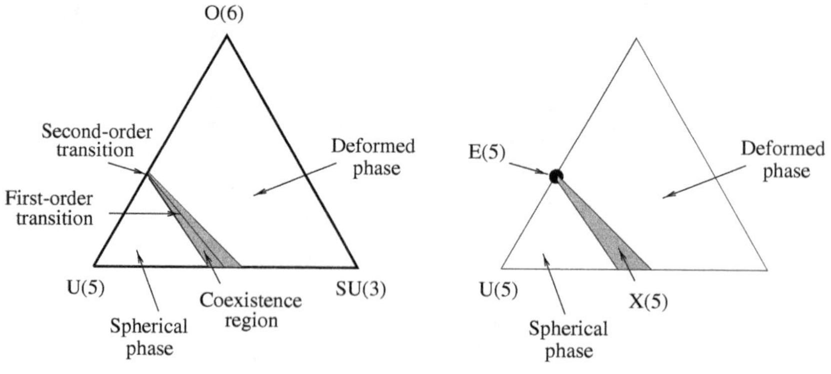

Fig. 9. Phase diagram of nuclei as one fluid system (no distinction between protons and neutrons). Left panel: Symmetries of the algebraic formulation. Right panel: Symmetries and Quasisymmetries of the differential formulation.

where the symmetries of the algebraic formulation are shown on the left and those of the differential formulation on the right. Akito Arima made crucial contributions to the development of the concept of dynamic symmetries in the Interacting Boson Model. His work will be remembered for many years to come.

Acknowledgements

I wish to thank Noam Gavrielov for helping me formatting this manuscript. This paper is dedicated to the memory of Akito Arima, to his friendship to me and to the invaluable contributions he made to the field of Nuclear Physics.

References

1. A. Arima and F. Iachello, *Phys. Rev. Lett.* **35**, 1069 (1975).
2. For a review, see F. Iachello and A. Arima, *The Interacting Boson Model*, Cambridge University Press, Cambridge (1987).
3. A. Arima and F. Iachello, *Ann. Phys. (NY)* **99**, 253 (1976).
4. A. Arima and F. Iachello, *Ann. Phys. (NY)* **111**, 201 (1978).
5. A. Arima and F. Iachello, *Ann. Phys. (NY)* **123**, 468 (1979).
6. J. A. Cizewski, R. F. Casten, G. J. Smith *et al.*, *Phys. Rev. Lett.* **40**, 167 (1978).
7. R. F. Casten and D. H. Feng, *Physics Today* **37**(11), 26 (1984).
8. M. Délèze, S. Drissi, J. Jolie *et al.*, *Nucl. Phys. A* **554**, 1 (1993).
9. A. Bohr and B. R. Mottelson, *Nuclear Structure*, Vol. 2, Benjamin (1975).
10. A. E. L. Dieperink, O. Scholten, and F. Iachello, *Phys. Rev. Lett.* **44**, 1747 (1980).
11. J. Ginocchio and M. W. Kirson, *Phys. Rev. Lett.* **44**, 1744 (1980).
12. A. Bohr and B. R. Mottelson, *Physica Scripta* **22**, 468 (1980).
13. A. Arima, T. Otsuka, F. Iachello, and I. Talmi, *Phys. Lett. B* **66**, 205 (1977).
14. T. Otsuka, A. Arima, F. Iachello, and I. Talmi, *Phys. Lett. B* **76**, 139 (1978).
15. T. Otsuka, A. Arima, and F. Iachello, *Nucl. Phys. A* **309**, 1 (1978).
16. D. Böhle, A. Richter, W. Steffen *et al.*, *Phys. Lett. B* **137**, 27 (1984).
17. N. Pietralla, C. Fransen, D. Belic *et al.*, *Phys. Rev. Lett.* **83**, 1303 (1999).
18. J. P. Elliott and A. P. White, *Phsy. Lett. B* **97**, 169 (1980).
19. J. P. Elliott and J. A. Evans, *Phys. Lett. B* **101**, 216 (1981).
20. P. van Isacker and D. Juillet, *Eur. Phsy. J. A* **15**, 233 (2002), and references therein.
21. A. Leviatan and P. van Isacker, *Phys. Rev. Lett.* **89**, 222501 (2002).
22. A. Leviatan, *Prog. Part. Nucl. Phys.* **66**, 93 (2011), and references therein.
23. F. Iachello, *Phys. Rev. Lett.* **44**, 772 (1980).
24. F. Iachello and O. Scholten, *Phys. Rev. Lett.* **43**, 679 (1978).
25. For a review, see F. Iachello and P. van Isacker, *The Interacting Boson-Fermion Model*, Cambridge University Press, Cambridge (1991).
26. A. Arima and F. Iachello, *Phys. Rev. C* **14**, 761 (1976).
27. A. B. Balantekin, I. Bars, and F. Iachello, *Nucl. Phys. A* **370**, 284 (1981).
28. For a review, see J. Vervier, *Rivista Nuovo Cimento* **10**(9) (1987).
29. F. Iachello, *Rivista Nuovo Cimento* **19**(7) (1996).
30. C. E. Alonso, J. M. Arias, R. Bijker *et al.*, *Phys. Lett. B* **144**, 141 (1984).
31. R. Bijker, *Dynamical Boson-Fermion Symmetries in Nuclei*, Ph.D. Thesis, University of Groningen (1984).
32. P. van Isacker, J. Jolie, K. Heyde, and A. Frank, *Phys. Rev. Lett.* **54**, 653 (1985).
33. T. Hübsch, V. Paar, and D. Vretenar, *Phys. Lett. B* **151**, 320 (1985).

34. A. Metz, J. Jolie, G. Graw *et al.*, *Phys. Rev. Lett.* **83**, 1542 (1999).
35. A. Metz, Y. Eisermann, A. S. Gollwitzer *et al.*, *Phys. Rev. C* **61**, 064313 (2000).
36. J. Gröger, J. Jolie, R. Krueken *et al.*, *Phys. Rev. C* **62**, 064304 (2000).
37. J. Barea, R. Bijker, and A. Frank, *Phys. Rev. Lett.* **94**, 152501 (2005).
38. F. Iachello, *Phys. Rev. Lett.* **85**, 3580 (2000).
39. F. Iachello, *Phys. Rev. Lett.* **87**, 052502 (2001).
40. F. Iachello, *Phys. Rev. Lett.* **91**, 132502 (2003).
41. A. Bohr, *Mat. Fys. Medd. Dan. Vid. Selsk.* **26**(14) (1952).
42. R. F. Casten and N. V. Zamfir, *Phys. Rev. Lett.* **85**, 3584 (2000).
43. R. F. Casten and N. V. Zamfir, *Phys. Rev. Lett.* **87**, 052503 (2001).
44. D. Tonev, A. Dewald, T. Klug *et al.*, *Phys. Rev. C* **69**, 034334 (2004).
45. P. Cejnar and J. Jolie, *Prog. Part. Nucl. Phys.* **62**, 210 (2009).
46. P. Cejnar, J. Jolie, and R. F. Casten, *Rev. Mod. Phys.* **82**, 2155 (2010).
47. F. Iachello and M. A. Caprio, in *Understanding Quantum Phase Transitions*, L. Carr, ed., CRC Press, Boca Raton, Fl (2011).
48. F. Iachello, *Rivista Nuovo Cimento* **34**, 617 (2011).

https://doi.org/10.1142/9789819813353_0008

The Spin–Orbit Interaction in the Interacting Boson Model

P. Van Isacker
Grand Accélérateur National d'Ions Lourds, CEA/DRF-CNRS/IN2P3,
Bvd Henri Becquerel, BP 55027, F-14076 Caen, France

Wigner's SU(4) symmetry assumes nuclear forces to be independent of the spin and isospin orientation of the nucleons. It is strongly broken in most atomic nuclei, mainly as a result of the spin–orbit interaction. Yet SU(4) symmetry favours pairs of nucleons in spatially symmetric configurations and is at the basis of models that describe the nucleus in terms of bosonic pairs of nucleons. It is shown that the spin–orbit interaction introduces bosons that correspond to nucleon pairs in spatially unfavoured configurations and construct the most general SU(4)-invariant version of the interacting boson model, which includes spatially favoured as well as unfavoured bosons. The model is applied to charge-exchange reactions towards self-conjugate ($N = Z$) nuclei.

1. Introduction

A standard approach in quantum many-body physics is to treat a system of interacting fermions in terms of composite particles, which in turn can behave either as fermions or bosons. An example of the former is provided by the fractional quantum Hall effect,[1] which has found an explanation in terms of electrons associated with (an even number of) magnetic flux quanta.[2] The fractional quantum Hall effect can then be understood as its integer version applied to composite fermions. If the composite particles are bosonic in nature, and in particular if they correspond to pairs of fermions, they are generally associated with collective excitations, and the original fermionic many-body problem can then be mapped onto a simpler bosonic one while the violation of the Pauli principle is accounted for by taking appropriate interactions between the bosons. Mappings from fermion to boson systems can be divided into two main classes, either based on an expansion that preserves commutation rules[3] or on a correspondence of states.[4,5] An extensive literature exists on diverse aspects of the two classes of mappings and their many subclasses, as reviewed, for example, in Ref. 6.

In the study of atomic nuclei, the passage from fermions to bosons has received much attention precisely because of its connection with collective behaviour of the nucleons. Phonon models of various aspects of the nucleus are ubiquitous.[7] One of the most successful approaches of this kind is the interacting boson model (IBM) of Arima and Iachello,[8] which assumes that collective excitations of the nucleus can be described in terms of s and d bosons. Although it started out as a phenomenological model,[9] it was realised early on that a microscopic justification can be formulated by treating pairs of valence nucleons as bosons with angular momentum $J = 0$ or $J = 2$.[10,11] The shell-model underpinning of the IBM proved to be highly fruitful because it showed the direction of various extensions, among which the neutron-proton version IBM-2.[12] As emphasised by Elliott and co-workers,[13,14] an important aspect of the shell-model justification of the IBM is the existence of corresponding symmetries in the fermion and boson spaces. This 'symmetry' approach culminated with the formulation of the so-called IBM-4,[14] in which the SU(4) labels, the orbital angular momentum L, spin S and isospin T are mapped onto the boson space.

The IBM-4 is the most sophisticated version of the boson model to date and incorporates Wigner's SU(4) model[15] with its full panoply of quantum numbers. Yet it is known that SU(4) symmetry is strongly broken in most nuclei, predominantly by the strong spin–orbit interaction, and that L and S are not conserved quantum numbers. It is the purpose of this contribution to show that the breaking of SU(4) symmetry can be incorporated in an algebraic framework, leading to the most general formulation of the IBM. An unavoidable consequence of the existence of the spin–orbit interaction is the introduction of bosons that correspond to fermion pairs in spatially unfavoured configurations.

2. The Algebraic Many-Fermion Problem

The many-nucleon problem can be algebraically formulated in terms of the bilinear fermion operators

$$(a^\dagger_{l^1/_2{}^1/_2} \times \tilde{a}_{l^1/_2{}^1/_2})^{(LST)}_{M_L M_S M_T}, \tag{1}$$

where $a^\dagger_{lm_l{}^1/_2m_s{}^1/_2m_t}$ creates a nucleon with orbital angular momentum l, spin $^1/_2$ and isospin $^1/_2$, with projections m_l, m_s and m_t, respectively. The corresponding annihilation operators are $a_{lm_l{}^1/_2m_s{}^1/_2m_t}$, with furthermore $\tilde{a}_{lm_l{}^1/_2m_s{}^1/_2m_t} \equiv -(-)^{l+m_l+m_s+m_t} a_{l-m_l{}^1/_2-m_s{}^1/_2-m_t}$. The superscripts in Eq. (1) denote the coupling in orbital angular momentum L, spin S and isospin T, and the subscripts refer to their projections M_L, M_S and M_T. The case of fermions occupying a single-l orbital, considered here, will be generalised later to that of several orbitals. The operators (1) close under commutation and generate the algebra $\mathrm{U}[4(2l+1)]$, where $(2l+1)$ is the orbital dimension of the single-particle space and the factor 4 accounts for the spin–isospin degrees of freedom.

The dynamical algebra $\mathrm{U}[4(2l+1)]$ has a possible substructure of the form

$$\mathrm{U}[4(2l+1)] \supset \big(\mathrm{U}_L(2l+1) \supset \cdots \supset \mathrm{SO}_L(3)\big) \otimes \big(\mathrm{SU}_{ST}(4) \supset \mathrm{SU}_S(2) \otimes \mathrm{SU}_T(2)\big), \tag{2}$$

which exists for any orbital angular momentum l. The dots in the first reduction in Eq. (2) indicate some orbital classification [*e.g.*, Elliott's SU(3)[16]], which ends in $\mathrm{SO}_L(3)$, the algebra associated with the orbital angular momentum L. The second reduction in Eq. (2) starts from Wigner's algebra $\mathrm{SU}_{ST}(4)$[15] and ends in the spin and isospin algebras $\mathrm{SU}_S(2)$ and $\mathrm{SU}_T(2)$. The algebra $\mathrm{SU}_{ST}(4)$ has the generators

$$\begin{aligned}\hat{S}_\mu &= \sqrt{2l+1}(a^\dagger_{l^1/_2{}^1/_2} \times \tilde{a}_{l^1/_2{}^1/_2})^{(010)}_{0\mu 0},\\ \hat{T}_\nu &= \sqrt{2l+1}(a^\dagger_{l^1/_2{}^1/_2} \times \tilde{a}_{l^1/_2{}^1/_2})^{(001)}_{00\nu},\\ \hat{Y}_{\mu\nu} &= \sqrt{2l+1}(a^\dagger_{l^1/_2{}^1/_2} \times \tilde{a}_{l^1/_2{}^1/_2})^{(011)}_{0\mu\nu},\end{aligned} \tag{3}$$

which are scalar in orbital space.

The short-range attractive nature of the nuclear interaction favours spatially symmetric states, for example, two-particle states with even orbital angular momentum L. Many-body states with different orbital symmetry are separated in energy by means of the 'Majorana' interaction

$$\hat{C}_2[\mathrm{SU}_{ST}(4)] = \hat{S}^2 + \hat{T}^2 + \hat{Y}^2, \tag{4}$$

that is, the quadratic Casimir operator of SU(4), with eigenstates characterised by the labels $(\lambda\mu\nu)$ associated with the irreducible representations of this algebra. Eigenstates of $\hat{C}_2[\mathrm{SU}_{ST}(4)]$ with the same energy are known as supermultiplets,[15] which have a high degree of degeneracy since the only criterion for their separation in energy is orbital symmetry.

As the atomic mass number A grows, SU(4) symmetry is increasingly broken by spin- and isospin-dependent components of the nuclear interaction. The effect can be represented by a one-body spin–orbit Hamiltonian, which for a single-l orbital is

$$\hat{H}_{\mathrm{so}} = -g\sum_{i=1}^{A} \hat{\mathbf{l}}_i \cdot \hat{\mathbf{s}}_i = \frac{1}{2}(l+1)g\,\hat{n}_- - \frac{1}{2}lg\,\hat{n}_+, \tag{5}$$

where $\hat{n}_\pm$ are the nucleon-number operators for the two orbitals $j = l \pm {}^1/_2$.

As they define two basic structural aspects of nuclei, the operators (4) and (5) set the scene for the results presented in this contribution. The goal is to construct a generic boson model that represents both operators.

To obtain a boson representation of a bosonic fermion system (*i.e.*, with an even number of fermions), one considers fermion pairs that are mapped onto bosons,

$$(a^\dagger_{l^1/_2{}^1/_2} \times a^\dagger_{l^1/_2{}^1/_2})^{(LST)}_{M_L M_S M_T} \mapsto b^\dagger_{LM_L SM_S TM_T}, \tag{6}$$

where $L+S+T$ must be odd because of anti-symmetry. The bosons are henceforth considered as elementary and their quantum numbers L, S and T will be written with small letters ℓ, s and t. The origin of the bosons $b^\dagger_{\ell st}$ should not be forgotten, however, which implies that ℓ, s and t should be integer and $\ell + s + t$ must be odd. The bosons are assumed to satisfy ideal commutation relations, which are only approximately satisfied by the fermion pairs. Violations of the Pauli principle must be corrected by considering appropriate interactions between the bosons.

Once a set of bosons $\{b^\dagger_{\ell st}\}$ has been chosen, the many-body problem can be formulated algebraically in terms of the bilinear boson operators

$$(b^\dagger_{\ell st} \times \tilde{b}_{\ell' s' t'})^{(\lambda\sigma\tau)}_{m_\lambda m_\sigma m_\tau}, \tag{7}$$

which close under commutation.

3. Boson Realisations of Wigner's SU(4) Algebra

Can an SU(4) algebra with the same interpretation as in the fermion case be constructed in terms of boson operators? This correspondence must be established despite the differing realisations. In the fermion realisation l is the single-particle orbital angular momentum, and the spin and isospin of the nucleon is $1/2$. In a boson realisation ℓ, s and t should be integer and $\ell + s + t$ must be odd. Two cases must be distinguished: ℓ is even or odd.

Bosons with even ℓ. If ℓ is even, $(st) = (01)$ or (10). The number of spin–isospin states is six and the dynamical algebra $\mathrm{U}[6(2\ell+1)]$ allows the reduction

$$\begin{aligned}\mathrm{U}[6(2\ell+1)] \supset\ & (\mathrm{U}_L[(2\ell+1)] \supset \cdots \supset \mathrm{SO}_L(3))\\ & \otimes (\mathrm{U}_{ST}(6) \supset \mathrm{SU}_{ST}(4) \supset \mathrm{SU}_S(2) \otimes \mathrm{SU}_T(2)).\end{aligned} \tag{8}$$

The generators of $\mathrm{SU}_{ST}(4)$ are

$$\begin{aligned}\hat{S}^{\mathrm{e}}_\mu &= \sqrt{2(2\ell+1)}(b^\dagger_{\ell 10} \times \tilde{b}_{\ell 10})^{(010)}_{0\mu 0},\\ \hat{T}^{\mathrm{e}}_\nu &= \sqrt{2(2\ell+1)}(b^\dagger_{\ell 01} \times \tilde{b}_{\ell 01})^{(001)}_{00\nu},\\ \hat{Y}^{\mathrm{e}}_{\mu\nu} &= \sqrt{2\ell+1}\big((b^\dagger_{\ell 10} \times \tilde{b}_{\ell 01})^{(011)}_{0\mu\nu} \pm (b^\dagger_{\ell 01} \times \tilde{b}_{\ell 10})^{(011)}_{0\mu\nu}\big),\end{aligned} \tag{9}$$

of which it can be shown that they close under commutation and satisfy the same SU(4) commutation relations as the fermion operators (3).

Bosons with odd ℓ. If ℓ is odd, $(st) = (00)$ or (11). The number of spin–isospin states is ten and the dynamical algebra $\mathrm{U}(10\Lambda_\ell)$ allows the reduction

$$\begin{aligned}\mathrm{U}[10(2\ell+1)] \supset\ & (\mathrm{U}_L(2\ell+1) \supset \cdots \supset \mathrm{SO}_L(3))\\ & \otimes (\mathrm{U}_{ST}(10) \supset \mathrm{SU}_{ST}(4) \supset \mathrm{SU}_S(2) \otimes \mathrm{SU}_T(2)).\end{aligned} \tag{10}$$

In this case $\mathrm{SU}_{ST}(4)$ is generated by

$$\begin{aligned}\hat{S}^{\mathrm{o}}_\mu &= \sqrt{6(2\ell+1)}(b^\dagger_{\ell 11} \times \tilde{b}_{\ell 11})^{(010)}_{0\mu 0},\\ \hat{T}^{\mathrm{o}}_\nu &= \sqrt{6(2\ell+1)}(b^\dagger_{\ell 11} \times \tilde{b}_{\ell 11})^{(001)}_{00\nu},\\ \hat{Y}^{\mathrm{o}}_{\mu\nu} &= \sqrt{2\ell+1}\big((b^\dagger_{\ell 00} \times \tilde{b}_{\ell 11})^{(011)}_{0\mu\nu} + (b^\dagger_{\ell 11} \times \tilde{b}_{\ell 00})^{(011)}_{0\mu\nu} \pm 2(b^\dagger_{\ell 11} \times \tilde{b}_{\ell 11})^{(011)}_{0\mu\nu}\big).\end{aligned} \tag{11}$$

Again, these operators close under commutation and satisfy the same SU(4) commutation relations as the fermion operators (3).

In the first case the two-nucleon pairs are in a spatially symmetric [favoured SU(4)] state, while in the second case they are in a spatially anti-symmetric (unfavoured) state. It is therefore natural to choose dominant pairs with even ℓ, as considered by Elliott and Evans in their elaboration of IBM-4.[14] However, in heavier nuclei, where SU(4) symmetry is increasingly broken mainly by the spin–orbit interaction, mixing with unfavoured SU(4) representations occurs, and it is therefore of interest to explore mappings that include spatially unfavoured bosons with odd ℓ.

With the help of the preceding results a generic boson model can now be constructed with favoured bosons (ℓ01) and (ℓ10) with even ℓ, as well as unfavoured bosons (ℓ00) and (ℓ11) with odd ℓ. The dynamical algebra is $\mathrm{U}(6\Lambda_\mathrm{e}+10\Lambda_\mathrm{o})$, where Λ_e is the orbital degeneracy of the even-ℓ bosons, $\Lambda_\mathrm{e}=\sum_{\ell\,\mathrm{even}}(2\ell+1)$, and Λ_o is that of the odd-ℓ bosons, $\Lambda_\mathrm{o}=\sum_{\ell\,\mathrm{odd}}(2\ell+1)$. A generic scheme, which includes Wigner's supermultiplet classification, reads

$$\begin{aligned}\mathrm{U}(6\Lambda_\mathrm{e}+10\Lambda_\mathrm{o}) &\supset \mathrm{U}(6\Lambda_\mathrm{e})\otimes\mathrm{U}(10\Lambda_\mathrm{o})\\ &\supset \big(\mathrm{U}_L(\Lambda_\mathrm{e})\otimes\mathrm{U}_L(\Lambda_\mathrm{o})\supset\cdots\supset\mathrm{SO}_L(3)\big)\\ &\quad\otimes\big(\mathrm{U}^\mathrm{e}_{ST}(6)\otimes\mathrm{U}^\mathrm{o}_{ST}(10)\supset\mathrm{SU}^\mathrm{e}_{ST}(4)\otimes\mathrm{SU}^\mathrm{o}_{ST}(4)\supset\mathrm{SU}_{ST}(4)\big).\end{aligned}\tag{12}$$

The dots indicate some orbital classification, the details of which dependent on Λ_e and Λ_o, and which ends in the algebra $\mathrm{SO}_L(3)$. The quantum numbers L, S and T, as well as the supermultiplet labels $(\lambda\mu\nu)$, have an interpretation identical to that in the shell model, allowing therefore to establish a mapping to the boson model. The classification (12) defines in fact the most general boson model, in which the full panoply of shell-model labels L, S, J, T and $(\lambda\mu\nu)$ can be represented.

Particular models of the type (12) have been studied in the past. The most elementary version adopts bosons with $\ell=0$, implying $\Lambda_\mathrm{e}=1$ and $\Lambda_\mathrm{o}=0$, resulting in the dynamical algebra U(6).[17] The IBM-4 of Elliott and Evans[14] considers bosons with $\ell=0$ and $\ell=2$, implying $\Lambda_\mathrm{e}=6$ and $\Lambda_\mathrm{o}=0$, resulting in the dynamical algebra U(36).

In line with the preceding results it is therefore possible to consider isovector ($t=1$) bosons with $\ell=0,1,2,\ldots$,

$$b^\dagger_{00,00,1m_t},\; b^\dagger_{1m_1,1m_s,1m_t},\; b^\dagger_{2m_2,00,1m_t},\ldots\tag{13}$$

as well as isoscalar ($t=0$) bosons also with $\ell=0,1,2,\ldots$,

$$b^\dagger_{00,1m_s,00},\; b^\dagger_{1m_1,00,00},\; b^\dagger_{2m_2,1m_s,00},\ldots\tag{14}$$

The index ℓ can be suppressed if one denotes the bosons by their spectroscopic notation with the additional convention that isovector (isoscalar) bosons are denoted by roman (greek) letters. The three labels 'ℓst' can then be replaced by a single symbol as follows:

$$\begin{aligned}&b^\dagger_{001}\mapsto s^\dagger,\quad b^\dagger_{111}\mapsto p^\dagger,\quad b^\dagger_{201}\mapsto d^\dagger,\ldots\\ &b^\dagger_{010}\mapsto \sigma^\dagger,\quad b^\dagger_{100}\mapsto \pi^\dagger,\quad b^\dagger_{210}\mapsto \delta^\dagger,\ldots\end{aligned}\tag{15}$$

A shell-model Hamiltonian usually is invariant in (JT) but seldom in (LST). Since this symmetry property is transferred to the boson Hamiltonian, it is advantageous to introduce the boson operators

$$b^\dagger_{\ell sjm_jtm_t} = \sum_{m_\ell m_s} (\ell m_\ell\, sm_s|jm_j) b^\dagger_{\ell m_\ell sm_stm_t}, \tag{16}$$

which carry good total angular momentum j. Only an additional index j is needed to uniquely characterise a boson as $b^\dagger_j$ since the (ℓst) values are specified by the spectroscopic notation, as explained above.

4. Boson Images of Fermion Operators

How to construct a mapping of a fermion operator in terms of the bosons defined in the previous section? In this section the procedure is explained for a few simple operators in a single-l scenario and in the next section it is generalised to an arbitrary Hamiltonian and to a space consisting of several orbitals.

The bosonic realisations (9) and (11) of the SU(4) generators have the same commutation relations as those for the fermionic generators (3). Therefore, the same expression (4) is valid for the Casimir operator of $\mathrm{SU}_{ST}(4)$. The SU(4) symmetry, if it is conserved by the Hamiltonian in fermion space, is preserved by the mapping and carried over onto the boson space.

Matters are more complicated for the spin–orbit Hamiltonian $\hat{H}_{\mathrm{so}}$. Its matrix elements between nucleon-pair states are

$$\langle l^2LSJT|\hat{H}_{\mathrm{so}}|l^2L'S'JT\rangle = -g(-)^{L'+S+J+T}[L][S][L'][S'][l]\sqrt{6l(l+1)} \times \begin{Bmatrix} L & L' & 1 \\ S' & S & J \end{Bmatrix} \begin{Bmatrix} L & L' & 1 \\ l & l & l \end{Bmatrix} \begin{Bmatrix} S & S' & 1 \\ 1/2 & 1/2 & 1/2 \end{Bmatrix}, \tag{17}$$

where $[x] \equiv \sqrt{2x+1}$ and the symbol in curly brackets is a Racah coefficient.[18] Since the states $|l^2LSJT\rangle$ correspond to the bosons $b^\dagger_{\ell sjt}$ introduced above, this shows that the spin–orbit Hamiltonian mixes spatially favoured with unfavoured bosons. If one confines oneself to pairs with $L \leq 2$ (bosons with $\ell \leq 2$), the boson version of the spin–orbit Hamiltonian $\hat{H}_{\mathrm{so}}$ takes on the following form:

$$\begin{aligned} \frac{1}{4}g\hat{n}_{\mathrm{b}}\ \frac{1}{2}g\Big[& -s^\dagger_0\cdot\tilde{s}_0 + p^\dagger_0\cdot\tilde{p}_0 - 2p^\dagger_2\cdot\tilde{p}_2 - d^\dagger_2\cdot\tilde{d}_2 - \sigma^\dagger_1\cdot\tilde{\sigma}_1 - \pi^\dagger_1\cdot\tilde{\pi}_1 + 2\delta^\dagger_1\cdot\tilde{\delta}_1 - 3\delta^\dagger_3\cdot\tilde{\delta}_3 \\ & + \sqrt{4l(l+1)}(s^\dagger_0\cdot\tilde{p}_0 + p^\dagger_0\cdot\tilde{s}_0) - \sqrt{\frac{2(2l-1)(2l+3)}{5}}(p^\dagger_2\cdot\tilde{d}_2 + d^\dagger_2\cdot\tilde{p}_2) \\ & - \sqrt{\frac{4l(l+1)}{3}}(\sigma^\dagger_1\cdot\tilde{\pi}_1 + \pi^\dagger_1\cdot\tilde{\sigma}_1) + \sqrt{\frac{2(2l-1)(2l+3)}{3}}(\pi^\dagger_1\cdot\tilde{\delta}_1 + \delta^\dagger_1\cdot\tilde{\pi}_1)\Big], \end{aligned} \tag{18}$$

where $\cdot$ denotes a scalar product in either angular momentum or isospin or both, and $\hat{n}_{\mathrm{b}}$ is the boson-number operator. Not surprisingly, spatially unfavoured (p and π) bosons are required in the mapped spin–orbit Hamiltonian. If the latter bosons are not considered, $\hat{H}_{\rm so}$ is mapped onto a Hamiltonian of at least *two-body* character in the bosons.[19]

The mapping of the Gamow–Teller (GT) operator

$$\hat{M}_{\mu\nu} = \sum_{i=1}^{A} \hat{\sigma}_{\mu,i}\hat{t}_{\nu,i} = \sqrt{\frac{1}{2}}\sum_{i=1}^{A} \hat{\sigma}_{\mu,i}\hat{\tau}_{\nu,i}, \tag{19}$$

can be dealt with similarly. The matrix elements, reduced in angular momentum and isospin, of the GT operator between nucleon-pair states are

$$\begin{aligned}\langle l^2LSJT|||\hat{M}|||l^2LS'J'T'\rangle &= (-)^{S'+J}12[S][S'][J][J'][T][T'] \\ &\quad\times \begin{Bmatrix} S & J & L \\ J' & S' & 1 \end{Bmatrix}\begin{Bmatrix} S & S' & 1 \\ 1/2 & 1/2 & 1/2 \end{Bmatrix}\begin{Bmatrix} T & T' & 1 \\ 1/2 & 1/2 & 1/2 \end{Bmatrix},\end{aligned} \tag{20}$$

leading to its mapped boson equivalent,

$$\begin{aligned}&-\sqrt{18}(\sigma_1^\dagger\tilde{s}_0 + s_0^\dagger\tilde{\sigma}_1)^{(11)}_{\mu\nu} + \sqrt{6}(\pi_1^\dagger\tilde{p}_0 + p_0^\dagger\tilde{\pi}_1)^{(11)}_{\mu\nu} - \sqrt{18}(\pi_1^\dagger\tilde{p}_1 - p_1^\dagger\tilde{\pi}_1)^{(11)}_{\mu\nu} \\ &+ \sqrt{30}(\pi_1^\dagger\tilde{p}_2 - p_2^\dagger\tilde{\pi}_1)^{(11)}_{\mu\nu} + \sqrt{24}(p_0^\dagger\tilde{p}_1 - p_1^\dagger\tilde{p}_0)^{(11)}_{\mu\nu} + \sqrt{18}(p_1^\dagger\tilde{p}_1)^{(11)}_{\mu\nu} \\ &+ \sqrt{30}(p_1^\dagger\tilde{p}_2 - p_2^\dagger\tilde{p}_1)^{(11)}_{\mu\nu} + \sqrt{90}(p_2^\dagger\tilde{p}_2)^{(11)}_{\mu\nu} - \sqrt{18}(\delta_1^\dagger\tilde{d}_2 + d_2^\dagger\tilde{\delta}_1)^{(11)}_{\mu\nu} \\ &- \sqrt{30}(\delta_2^\dagger\tilde{d}_2 - d_2^\dagger\tilde{\delta}_2)^{(11)}_{\mu\nu} - \sqrt{42}(\delta_3^\dagger\tilde{d}_2 + d_2^\dagger\tilde{\delta}_3)^{(11)}_{\mu\nu}.\end{aligned} \tag{21}$$

If the $|l^2LSJT\rangle$ basis is complete [*e.g.*, for $(JT) = (01)$, (11) and (10) if $L \le 2$], the mapping is exact for the two-nucleon system.

5. Boson Image of Fermion Operators for Several Orbitals

The Hamiltonian composed of the operators (4) and (5) in a single-l orbital is too schematic to give an adequate description of nuclei. Realistic shell-model calculations nowadays take place in a larger model space (which shall be referred to below as the 'full' space), typically containing all l orbitals of an entire oscillator shell, if not more. It is therefore of interest to extend the findings reported in the preceding sections to the case of several l orbitals. It will be assumed that the same set of (ℓst) values (with $\ell + s + t$ odd) can be taken for the bosons. This is a reasonable assumption if neutrons and protons occupy the same orbitals and the l^2 components are dominant in the collective pairs. With this assumption the boson algebraic structure remains the same. The underlying fermion structure, however, is different and is important for the definition of the mapped boson Hamiltonian.

A difficulty in the present mapping from fermions to bosons is that the pairs made to correspond to bosons are defined in (LST) coupling whereas the fermion Hamiltonian has good (JT) but not (LST). The separate quantum numbers L and S are *not* conserved by a realistic shell-model Hamiltonian. As a result, the mapping procedure is rather involved but can be summarised with the following steps:

(1) The shell-model Hamiltonian $\hat{H}^{\rm f}$ is decomposed into its central, spin–orbit and tensor parts $\hat{H}^{\rm f}(k)$, $k=L=S=0,1,2$.
(2) Reduced matrix elements of $\hat{H}^{\rm f}(k)$ are calculated between one- and two-pair states. This is possible since $\hat{H}^{\rm f}(k)$ has definite tensor character $(LST)=(kk0)$.
(3) The expressions for the matrix elements in the fermion space are used to derive the boson Hamiltonian $\hat{H}^{\rm b}$.

The remainder of this section gives a detailed account of each of the above steps. The reader not interested in this formal part may wish to move straight to the application section.

(1) An operator $\hat{T}$, scalar in (JT) but not in (LST), can in general be written as

$$\hat{T}=\sum_k \hat{T}_l^{(k)}\cdot\hat{T}_s^{(k)}=\sum_k\sum_{\kappa=-k}^{+k}(-)^\kappa\hat{T}_{l,\kappa}^{(k)}\hat{T}_{s,-\kappa}^{(k)}\equiv\sum_k\hat{T}(k), \tag{22}$$

where the last equation defines a notation. We have the general result[20]

$$\langle\alpha LSJT|\hat{T}(k)|\alpha'L'S'JT\rangle=\frac{(-)^{L'+S+J}}{\sqrt{2T+1}}\left\{\begin{array}{ccc}L&S&J\\S'&L'&k\end{array}\right\}\langle\alpha LST||||\hat{T}(k)||||\alpha'L'S'T\rangle, \tag{23}$$

where the matrix element on the right-hand side is reduced in L, S and T. The inverse result reads

$$\begin{aligned}\langle\alpha LST||||\hat{T}(k)||||\alpha'L'S'T\rangle&=\sqrt{2T+1}(2k+1)\\&\times\sum_J(-)^{L'+S+J}(2J+1)\left\{\begin{array}{ccc}L&S&J\\S'&L'&k\end{array}\right\}\langle\alpha LSJT|\hat{T}(k)|\alpha'L'S'JT\rangle.\end{aligned} \tag{24}$$

Once an operator $\hat{T}$ is decomposed in its parts $\hat{T}(k)$ it is therefore possible to connect its matrix elements in the basis $|LST\rangle$ to those in $|LSJT\rangle$ or *vice versa*. For a one-body (two-body) operator the values of k can be $k=0,1$ $(k=0,1,2)$. A one-plus-two-body shell-model Hamiltonian $\hat{H}^{\rm f}=\hat{H}^{\rm f}_1+\hat{H}^{\rm f}_2$ can therefore be decomposed as

$$\hat{H}^{\rm f}=\hat{H}^{\rm f}_1(0)+\hat{H}^{\rm f}_1(1)+\hat{H}^{\rm f}_2(0)+\hat{H}^{\rm f}_2(1)+\hat{H}^{\rm f}_2(2). \tag{25}$$

The different parts of the Hamiltonian are determined from their one- or two-body matrix elements. The matrix elements of $\hat{H}^{\rm f}_1(k)$ are between the one-nucleon states $|nl{}^1\!/\!_2 j{}^1\!/\!_2\rangle$, with radial quantum number n, orbital angular momentum l, spin $^1\!/\!_2$, total angular momentum j and isospin $^1\!/\!_2$,

$$\begin{aligned}&\langle nl{}^1\!/\!_2 j{}^1\!/\!_2|\hat{H}^{\rm f}_1(k)|nl{}^1\!/\!_2 j{}^1\!/\!_2\rangle\\&=(2k+1)\left\{\begin{array}{ccc}l&1/2&j\\1/2&l&k\end{array}\right\}\sum_{j'}(-)^{j-j'}(2j'+1)\left\{\begin{array}{ccc}l&1/2&j'\\1/2&l&k\end{array}\right\}\epsilon_{nlj'},\end{aligned} \tag{26}$$

where ϵ_{nlj} are the single-particle energies in jj coupling,

$$\epsilon_{nlj}\equiv\langle nl{}^1\!/\!_2 j{}^1\!/\!_2|\hat{H}^{\rm f}_1|nl{}^1\!/\!_2 j{}^1\!/\!_2\rangle. \tag{27}$$

The part $\hat{H}_2^{\rm f}(k)$ is determined by its matrix elements between two-nucleon states

$$\langle abLSJT|\hat{H}_2^{\rm f}(k)|a'b'L'S'JT\rangle = (2k+1)\begin{Bmatrix} L & S & J \\ S' & L' & k \end{Bmatrix}$$
$$\times \sum_{J'}(-)^{J+J'}(2J'+1)\begin{Bmatrix} L & S & J' \\ S' & L' & k \end{Bmatrix}\langle abLSJ'T|\hat{H}_2^{\rm f}|a'b'L'S'J'T\rangle. \qquad (28)$$

where $|abLSJT\rangle \equiv |n_a l_a {}^1\!/\!_2 {}^1\!/\!_2, n_b l_b {}^1\!/\!_2 {}^1\!/\!_2; LSJT\rangle$. The reduced matrix elements of $\hat{H}_i^{\rm f}(k)$ in the basis $|LST\rangle$, needed in (2), can be obtained with the help of Eq. (24).

(2) Next, one calculates the matrix elements of the central, spin–orbit and tensor parts $\hat{H}^{\rm f}(k)$ between one and two pairs defined in an (LST) basis.

(2a) The reduced matrix elements of $\hat{H}_1^{\rm f}(k)$ between *one-pair states* are

$$\langle abLST||||\hat{H}_1^{\rm f}(k)||||a'b'L'S'T\rangle = -\frac{1}{\sqrt{2}}[L][S][L'][S'][T](-)^{l_a+l_b}\delta_{aa'}\delta_{bb'}$$
$$\times\left[(-)^{L'+S'}\begin{Bmatrix} l_a & L & l_b \\ L' & l_a & k \end{Bmatrix}\begin{Bmatrix} 1/2 & S & 1/2 \\ S' & 1/2 & k \end{Bmatrix}\langle a||||\hat{H}_1^{\rm f}(k)||||a\rangle\right.$$
$$\left.+(-)^{L+S}\begin{Bmatrix} l_b & L & l_a \\ L' & l_b & k \end{Bmatrix}\begin{Bmatrix} 1/2 & S & 1/2 \\ S' & 1/2 & k \end{Bmatrix}\langle b||||\hat{H}_1^{\rm f}(k)||||b\rangle\right], \qquad (29)$$

where it is assumed that a certain order in the basis is imposed, $a \le b$ and $a' \le b'$, and where the reduced matrix elements between single-particle states are related as follows to the single-particle energies:

$$\langle nl{}^1\!/\!_2{}^1\!/\!_2||||\hat{H}_1^{\rm f}(k)||||nl{}^1\!/\!_2{}^1\!/\!_2\rangle$$
$$= \sqrt{2}(2k+1)\sum_j(-)^{l+1/2+j}(2j+1)\begin{Bmatrix} l & 1/2 & j \\ 1/2 & l & k \end{Bmatrix}\epsilon_{nlj}, \qquad (30)$$

The reduced matrix elements of $\hat{H}_2^{\rm f}(k)$ are a special case of Eq. (24):

$$\langle abLST||||\hat{H}_2^{\rm f}(k)||||a'b'L'S'T\rangle = \sqrt{2T+1}(2k+1)$$
$$\times\sum_J(-)^{L'+S+J}(2J+1)\begin{Bmatrix} L & S & J \\ S' & L' & k \end{Bmatrix}\langle abLSJT|\hat{H}_2^{\rm f}(k)|a'b'L'S'JT\rangle. \qquad (31)$$

(2b) *Two-pair states* are defined as follows:

$$|abcd[\Gamma_1\Gamma_2]\Gamma\rangle = \sum_{qrst}\sum_{\bar\Gamma_1\bar\Gamma_2}[qr\bar\Gamma_1, st\bar\Gamma_2|\}abcd[\Gamma_1\Gamma_2]\Gamma]\,|qr\bar\Gamma_1, st\bar\Gamma_2;\Gamma\rangle, \qquad (32)$$

where Γ stands for LST, Γ_i for $L_iS_iT_i$ and so on. The symbol $[\cdots|\}\cdots]$ is a $2{\times}2\to 4$ coefficient of fractional parentage,[18] which is non-zero only if $qrst$ is a permutation of $abcd$. The state $|qr\bar\Gamma_1\rangle$ is anti-symmetric in the nucleons 1 and 2 while $|st\bar\Gamma_2\rangle$ is anti-symmetric in the nucleons 3 and 4, and Γ results from the coupling of $\bar{\Gamma}_1$ and $\bar{\Gamma}_2$. The reduced matrix elements of $\hat{H}^{\rm f}_i(k)$ ($i=1,2$) between two-pair states are obtained from the relation

$$\langle abcd[\Gamma_1\Gamma_2]\Gamma||||\hat{H}^{\rm f}_i(k)||||a'b'c'd'[\Gamma'_1\Gamma'_2]\Gamma'\rangle = \binom{2}{i}^{-1}\binom{4}{i}\sum_{qr}\sum_{sts't'}\sum_{\bar{\Gamma}_1\bar{\Gamma}_2}\sum_{\bar{\Gamma}'_2}$$
$$\times\, [qr(\bar{\Gamma}_1)st(\bar{\Gamma}_2)\Gamma|\}abcd[\Gamma_1\Gamma_2]\Gamma]\,[qr(\bar{\Gamma}_1)s't'(\bar{\Gamma}'_2)\Gamma'|\}a'b'c'd'[\Gamma'_1\Gamma'_2]\Gamma']$$
$$\times\,(-)^{\bar{\Gamma}_1+\bar{\Gamma}'_2+\Gamma}[\Gamma][\Gamma']\begin{Bmatrix}\bar{\Gamma}_2 & \Gamma & \bar{\Gamma}_1\\ \Gamma' & \bar{\Gamma}'_2 & (kk0)\end{Bmatrix}\langle st\bar{\Gamma}_2||||\hat{H}^{\rm f}_i(k)||||s't'\bar{\Gamma}'_2\rangle, \qquad (33)$$

with $(-)^\Gamma=(-)^{L+S+T}$, $[\Gamma]=[L][S][T]$ and so on, and where the symbol between curly brackets is a product of three Racah coefficients,

$$\begin{Bmatrix}\bar{\Gamma}_2 & \Gamma & \bar{\Gamma}_1\\ \Gamma' & \bar{\Gamma}'_2 & (kk0)\end{Bmatrix} = \begin{Bmatrix}\bar{L}_2 & L & \bar{L}_1\\ L' & \bar{L}'_2 & k\end{Bmatrix}\begin{Bmatrix}\bar{S}_2 & S & \bar{S}_1\\ S' & \bar{S}'_2 & k\end{Bmatrix}\begin{Bmatrix}\bar{T}_2 & T & \bar{T}_1\\ T' & \bar{T}'_2 & 0\end{Bmatrix}. \qquad (34)$$

The reduced matrix element on the right-hand side of Eq. (33) is between one-pair states, known from the expressions given in (2a).

(3) The collective two-nucleon subspace is defined by the combinations

$$|B_{LM_LSM_STM_T}\rangle = \sum_{a\le b}\beta^{LST}_{ab}|abLM_LSM_STM_T\rangle, \qquad (35)$$

where the β^{LST}_{ab} are obtained from the diagonalisation of $\hat{H}^{\rm f}$ in the space $|abLM_LSM_STM_T\rangle$ with fixed (LST). The notation $|B_{LM_LSM_STM_T}\rangle$ indicates that this part of the two-nucleon subspace is mapped onto bosons, where it is assumed that the labels (LST) uniquely specify the collective pair, that is, that no more than one collective pair is considered with the same (LST). To make contact with the bosons $b^\dagger_{\ell sjt}$, one introduces the combinations

$$|B_{LSJM_JTM_T}\rangle = \sum_{a\le b}\beta^{LST}_{ab}|abLSJM_JTM_T\rangle. \qquad (36)$$

The projections M_J and M_T can be omitted since all results in the following are independent of these labels.

One associates a pair state $|B_{LSJT}\rangle$ with a boson state $|b_{LSJT}\rangle$. The one-body matrix elements between these bosons are deduced from a given shell-model Hamiltonian $\hat{H}^{\rm f}$ with use of the relation

$$\langle b_{LSJT}|\hat{H}^{\rm b}_1|b_{L'S'JT}\rangle \doteq \langle B_{LSJT}|\hat{H}^{\rm f}_1+\hat{H}^{\rm f}_2|B_{L'S'JT}\rangle, \qquad (37)$$

where the notation $\doteq$ indicates that the equality holds by virtue of the mapping. The matrix elements on the right-hand side of Eq. (37) are computed from

$$\langle B_{LSJT}|\hat{H}^{\rm f}|B_{L'S'JT}\rangle = \sum_{a\le b}\sum_{a'\le b'}\beta^{LST}_{ab}\beta^{L'S'T}_{a'b'}\langle abLSJT|\hat{H}^{\rm f}|a'b'L'S'JT\rangle. \qquad (38)$$

Note that the one-body boson operator has non-zero off-diagonal matrix elements with $LS \neq L'S'$, which is a crucial ingredient of the mapping. Reduced matrix elements in an (LST) basis of the separate the central, spin–orbit and tensor parts of the shell-model Hamiltonian can be obtained similarly,

$$\langle B_\Gamma |||| \hat{H}^{\mathrm{f}}(k) |||| B_{\Gamma'} \rangle = \sum_{a \leq b} \sum_{a' \leq b'} \beta^{\Gamma}_{ab} \beta^{\Gamma'}_{a'b'} \langle ab\Gamma |||| \hat{H}^{\mathrm{f}}(k) |||| a'b'\Gamma' \rangle. \tag{39}$$

The determination of the two-body matrix elements between the bosons is more complicated because the collective two-pair states $|B_{\Gamma_1} B_{\Gamma_2} LSJT\rangle$ do not form an orthogonal basis. The equation defining the one-plus-two-body boson Hamiltonian is

$$\langle b_{\Gamma_1} b_{\Gamma_2} LSJT | \hat{H}^{\mathrm{b}} | b_{\Gamma'_1} b_{\Gamma'_2} L'S'JT \rangle \doteq \langle\!\langle B_{\Gamma_1} B_{\Gamma_2} LSJT | \hat{H}^{\mathrm{f}} | B_{\Gamma'_1} B_{\Gamma'_2} L'S'JT \rangle\!\rangle. \tag{40}$$

The bases on both sides of this equation are orthogonal. This property is trivially valid for the bosons (on the left-hand side) and is so by construction for the fermion pairs (on the right-hand side). The procedure to transform from a non-orthogonal basis $|B_{\Gamma_1} B_{\Gamma_2} LSJT\rangle$ to an orthogonal one $|B_{\Gamma_1} B_{\Gamma_2} LSJT\rangle\!\rangle$ relies on the diagonalisation of the overlap matrix, as described in Ref. 21.

Note that Eq. (40) defines the one-plus-two-body boson Hamiltonian while its one-body part is already determined from Eq. (37). The two-body part of the boson Hamiltonian is obtained by substracting the one-body contribution

$$\langle b_{\Gamma_1} b_{\Gamma_2} LSJT | \hat{H}^{\mathrm{b}}_1(0) + \hat{H}^{\mathrm{b}}_1(1) + \hat{H}^{\mathrm{b}}_1(2) | b_{\Gamma'_1} b_{\Gamma'_2} L'S'JT \rangle, \tag{41}$$

where $\hat{H}^{\mathrm{b}}_1(k)$ results from the mapping of $\hat{H}^{\mathrm{f}}_1(k) + \hat{H}^{\mathrm{f}}_2(k)$. The matrix elements of $\hat{H}^{\mathrm{b}}_1(k)$ can be obtained from

$$\begin{aligned} &\langle b_{\Gamma_1} b_{\Gamma_2} LSJT | \hat{H}^{\mathrm{b}}_1(k) | b_{\Gamma'_1} b_{\Gamma'_2} L'S'JT \rangle \\ &\quad = \frac{(-)^{L'+S+J}}{\sqrt{2T+1}} \begin{Bmatrix} L & S & J \\ S' & L' & k \end{Bmatrix} \langle b_{\Gamma_1} b_{\Gamma_2} \Gamma |||| \hat{H}^{\mathrm{b}}_1(k) |||| b_{\Gamma'_1} b_{\Gamma'_2} \Gamma' \rangle, \end{aligned} \tag{42}$$

where the reduced matrix element on the right-hand side equals

$$\begin{aligned} \langle b_{\Gamma_1} b_{\Gamma_2} \Gamma |||| \hat{H}^{\mathrm{b}}_1(k) |||| b_{\Gamma'_1} b_{\Gamma'_2} \Gamma' \rangle &= \left[\frac{(2\Gamma+1)(2\Gamma'+1)}{(1+\delta_{\Gamma_1\Gamma_2})(1+\delta_{\Gamma'_1\Gamma'_2})} \right]^{1/2} \\ \times \Bigg[&(-)^{\Gamma_1+\Gamma_2+\Gamma'} \langle B_{\Gamma_1} |||| \hat{H}^{\mathrm{f}}(k) |||| B_{\Gamma'_1} \rangle \begin{Bmatrix} \Gamma_1 & \Gamma & \Gamma_2 \\ \Gamma' & \Gamma'_1 & (kk0) \end{Bmatrix} \delta_{\Gamma_2\Gamma'_2} \\ &+ (-)^{\Gamma+\Gamma'} \langle B_{\Gamma_2} |||| \hat{H}^{\mathrm{f}}(k) |||| B_{\Gamma'_1} \rangle \begin{Bmatrix} \Gamma_2 & \Gamma & \Gamma_1 \\ \Gamma' & \Gamma'_1 & (kk0) \end{Bmatrix} \delta_{\Gamma_1\Gamma'_2} \\ &+ (-)^{\Gamma_1+\Gamma_2+\Gamma'_1+\Gamma'_2} \langle B_{\Gamma_1} |||| \hat{H}^{\mathrm{f}}(k) |||| B_{\Gamma'_2} \rangle \begin{Bmatrix} \Gamma_1 & \Gamma & \Gamma_2 \\ \Gamma' & \Gamma'_2 & (kk0) \end{Bmatrix} \delta_{\Gamma_2\Gamma'_1} \\ &+ (-)^{\Gamma'_1+\Gamma'_2+\Gamma} \langle B_{\Gamma_2} |||| \hat{H}^{\mathrm{f}}(k) |||| B_{\Gamma'_2} \rangle \begin{Bmatrix} \Gamma_2 & \Gamma & \Gamma_1 \\ \Gamma' & \Gamma'_2 & (kk0) \end{Bmatrix} \delta_{\Gamma_1\Gamma'_1} \Bigg]. \end{aligned} \tag{43}$$

A similar procedure can be followed for any operator, leading, for example, to a mapped boson GT operator $\hat{O}^{\rm b}_{\rm GT}$. Higher-order terms between the bosons are needed in the operators to correct for violations of the Pauli principle. The results presented in the next section are obtained with $\hat{H}^{\rm b}$ with up to two-body interactions. On the other hand, a one-body $\hat{O}^{\rm b}_{\rm GT}$ provides a faithful image of the fermion GT operator although higher-body terms (not considered here) might occur in the boson image of the effective GT operator, discussed below.

6. An Application: Gamow–Teller Strength in $N \sim Z$ Nuclei

The importance of spatially unfavoured bosons can be illustrated with an example, namely their role in the description of GT strength in charge-exchange reactions towards $N = Z$ nuclei. This strength was determined experimentally in nuclei in the $0f_{7/2}$ region by Fujita and co-workers.[22,23] In particular, the ^{42}Ca(^{3}He,t)^{42}Sc reaction defines the 'elementary' GT modes in this region of the nuclear chart. The schematic shell-model analysis of Ref. 24 of the $A = 42$ case in terms of the two spin–orbit partner orbitals $0f_{7/2}$ and $0f_{5/2}$ shows that favoured ($L = 0, 2$) as well as unfavoured ($L = 1$) pairs enter into the discussion. If the nuclear interaction were spin–isospin independent, unfavoured pairs would carry no component in yrast states. However, spin-dependent components mix favoured and unfavoured pairs.

The diagonalisation of $\hat{H}^{\rm f}$ in the full shell-model space leads to eigenvalues and eigenvectors that are different from those obtained with the same $\hat{H}^{\rm f}$ in the collective subspace. With the OLS procedure of Okubo, Lee and Suzuki[25,26] an effective Hamiltonian $\hat{H}^{\rm f}_{\rm eff}$ can be defined in the collective subspace, which exactly reproduces part of the spectrum in the full space. This effective fermion Hamiltonian (or indeed any effective fermion operator) can then also be mapped onto its boson counterpart, leading to the operators $\hat{H}^{\rm b}_{\rm eff}$ and $\hat{O}^{\rm b}_{\rm GTeff}$. For a given shell-model operator one derives therefore two boson images, $\hat{O}^{\rm b}$ and $\hat{O}^{\rm b}_{\rm eff}$, the former mapped from the bare $\hat{O}^{\rm f}$ and the latter from its effective version $\hat{O}^{\rm f}_{\rm eff}$. Given the uncertainties related to the OLS procedure and the extrapolation of effective operators to higher nucleon numbers n, an interpolated boson Hamiltonian is used,

$$\hat{H}^{\rm b} = x\hat{H}^{\rm b} + (1-x)\hat{H}^{\rm b}_{\rm eff}, \tag{44}$$

where x is an interpolation parameter between 0 and 1, as explained in Ref. 21.

The advantage of a boson approximation is that a variety of bosons can be taken to probe what are the important collective degrees of freedom in a given physical process. The method is illustrated here by mapping the modified Kuo-Brown Hamiltonian KB3G,[27] appropriate for pf-shell nuclei,[28] onto a system of s, sp, sd or spd bosons, which have the appropriate intrinsic spin–isospin structure, that is, $(st) = (01)$ and (10) for the s and d bosons, and $(st) = (00)$ and (11) for the p boson. The boson image of the bare and effective KB3G Hamiltonian is calculated and an interpolation parameter x is chosen so as to optimise agreement with the experimental energy spectra. For the calculation of the GT strength, the

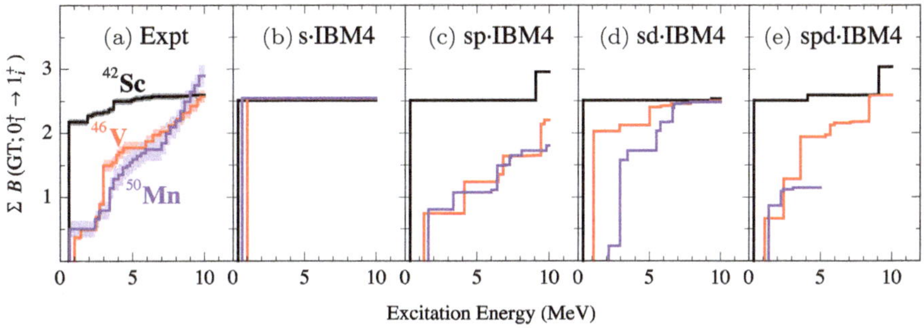

Fig. 1. Summed Gamow–Teller strength for the charge-exchange reactions towards ^{42}Sc (black), ^{46}V (red) and ^{50}Mn (purple). Panel (a) shows the observed strength[22,23] and panels (b) through (e) show results of the various IBM-4 calculations discussed in the text.

starting point is the operator $\sum_i \hat{\sigma}_{\mu,i}\hat{t}_{\pm,i}$, quenched by a factor 0.74 appropriate for the pf shell;[22] subsequently, the effective GT operator in the collective subspace is calculated and mapped onto its boson image.

The results of the various IBM-4 calculations are shown in Fig. 1.[a] Note that the s-IBM-4 calculation includes $\ell = 0$ bosons with $(st) = (01)$ *and* with $(st) = (10)$. Both the standard isovector $J = 0$ pair and the much debated isoscalar $J = 1$ neutron-proton pair (see, *e.g.*, the review 29) are thus represented in this approximation. It is obvious that they fail to reproduce the observed GT strength. In contrast, the unfavoured p bosons are instrumental in suppressing the GT strength at low energy. Insufficient strength is seen in sp-IBM-4 at higher energies — strength that seems to originate from the d bosons.

7. Concluding Remark

In summary, a boson description of the spin–orbit interaction requires the introduction of bosons that correspond to pairs of fermions in spatially unfavoured configurations. Their consideration leads to the most general SU(4)-invariant formulation of the interacting boson model. An application to charge-exchange reactions towards $N = Z$ nuclei reveals that such spatially unfavoured bosons are crucial in the mechanism of the suppression of Gamow–Teller strength at low energy.

Dedication and Acknowledgements

As should be clear from the content of this contribution the interacting boson model has played an important role in my research. I am deeply indebted to Professor

[a]Fijuta *et al.*[22] also have data for ^{54}Co. This nucleus is challenging numerically, even in a boson approximation (*e.g.*, dimension 34 802 372 in *spd*-IBM-4). Further, issues concerning the choice of the parameter x make the extrapolation to $n = 7$ less reliable. Therefore, data concerning ^{54}Co are not included here.

Akito Arima, who has had a major influence on my career as a physicist, and I dedicate this paper to his memory.

I thank Bill Gelletly, who drew my attention to the problem of Gamow–Teller strength in $N \sim Z$ nuclei, and Augusto Macchiavelli, Yoshitaka Fujita and Berta Rubio for stimulating discussions.

References

1. D. C. Tsui, H. L. Störmer, and A. C. Gossard, *Phys. Rev. Lett.* **48**, 1559 (1982).
2. J. K. Jain, *Phys. Rev. Lett.* **63**, 199 (1989).
3. S. T. Belyaev and V. G. Zelevinsky, *Nucl. Phys.* **39**, 582 (1962).
4. T. Marumori, M. Yamamura, and A. Tokunaga, *Progr. Theor. Phys.* **31**, 1009 (1964).
5. T. Marumori, M. Yamamura, A. Tokunaga, and K. Takada, *Progr. Theor. Phys.* **32**, 726 (1964).
6. A. Klein and E. R. Marshalek, *Rev. Mod. Phys.* **63**, 375 (1991).
7. A. Bohr and B. R. Mottelson, *Nuclear Structure. II Nuclear Deformations* (Benjamin, New York, 1975).
8. F. Iachello and A. Arima, *The Interacting Boson Model* (Cambridge University Press, Cambridge, 1987).
9. A. Arima and F. Iachello, *Phys. Rev. Lett.* **35**, 1069 (1975).
10. A. Arima, T. Otsuka, F. Iachello, and I. Talmi, *Phys. Lett. B* **66**, 205 (1977).
11. T. Otsuka, A. Arima, and F. Iachello, *Nucl. Phys. A* **309**, 1 (1978).
12. T. Otsuka, A. Arima, F. Iachello, and I. Talmi, *Phys. Lett. B* **76**, 139 (1978).
13. J. P. Elliott and A. P. White, *Phys. Lett. B* **97**, 169 (1980).
14. J. P. Elliott and J. A. Evans, *Phys. Lett. B* **101**, 216 (1981).
15. E. Wigner, *Phys. Rev.* **51**, 106 (1937).
16. J. P. Elliott, *Proc. Roy. Soc. (London) A* **245**, 128 & 562 (1958).
17. P. Van Isacker, J. Dukelsky, S. Pittel, and O. Juillet, *J. Phys. G* **24**, 1261 (1998).
18. I. Talmi, *Simple Models of Complex Nuclei. The Shell Model and the Interacting Boson Model* (Harwood, Chur, 1993).
19. O. Juillet and S. Josse, *Eur. Phys. J. A* **8**, 291 (2000).
20. J. P. Elliott, A. D. Jackson, H. A. Mavromatis, E. A. Sanderson, and B. Singh, *Nucl. Phys. A* **121**, 241 (1968).
21. P. Van Isacker, J. Engel, and K. Nomura, *Phys. Rev. C* **96**, 064305 (2017).
22. Y. Fujita *et al.*, *Phys. Rev. Lett.* **112**, 112502 (2014).
23. Y. Fujita *et al.*, *Phys. Rev. C* **91**, 064316 (2015).
24. P. Van Isacker and A. O. Macchiavelli, *Phys. Lett. B* **780**, 414 (2018).
25. S. Okubo, *Prog. Theor. Phys.* **12**, 603 (1954).
26. K. Suzuki and S. Y. Lee, *Prog. Theor. Phys.* **64**, 2091 (1980).
27. A. Poves, J. Sánchez-Solano, E. Caurier, and F. Nowacki, *Nucl. Phys. A* **694**, 157 (2001).
28. E. Caurier, G. Martínez-Pinedo, F. Nowacki, A. Poves, and A. P. Zuker, *Rev. Mod. Phys.* **77**, 427 (2005).
29. S. Frauendorf and A. O. Macchiavelli, *Progr. Part. Nucl. Phys.* **78**, 24 (2014).

https://doi.org/10.1142/9789819813353_0009

Alpha-clustering Structure of the Light Nuclei and its Impact on Astrophysics and Cosmology

Toshitaka Kajino[*,†,‡,‖], Xingqun Yao[*], Zhenyu He[*], Motohiko Kusakabe[*], Yuta Yamazaki[†,‡], Yudong Luo[§], and Kanji Mori[¶]

[*] *Beihang University, Beijing 100083, P. R. China*
[†] *The University of Tokyo, Tokyo 113-0033, Japan*
[‡] *National Astronomical Observatory of Japan, Tokyo 181-8588, Japan*
[§] *Peking University, Beijing 100871, Japan*
[¶] *Fukuoka University, Fukuoka 814-0180, Japan*
[‖] *kajino@buaa.edu.cn, kajino.cosmo.only1@gmail.com*

Alpha-clustering structure of the light nuclei ^{7}Be and ^{7}Li and their radiative α-capture reactions play the significant roles in nucleosynthesis of various astrophysical phenomena. We discuss their roles in the missing solar neutrinos, Big-Bang nucleosynthesis, explosive r-process in supernovae, collapsars and binary neutron star mergers, Galactic chemical evolution, neutrino flavor oscillations, and extremely enhanced lithium abundance in red clump stars. We summarize the impact on various astrophysical and cosmological problems.

1. Introduction

Nuclear astrophysics is an interdisciplinary field involving nuclear physics, particle physics, astronomy and astrophysics.[1] Until 1998 when Takaaki Kajita and his collaborators in Super-Kamiokande found a clear signal for the neutrino flavor oscillation,[2] the missing solar neutrino problem had been one of the biggest questions that involve not only astrophysics but also nuclear physics and particle physics. Theoretical calculations of the neutrino flux in the Standard Solar Model predict factor of $\sim$2 larger flux than the observed value. There are many uncertainties involved in the theoretical prediction of the solar neutrino flux. They are the uncertainties which arise from modeling the solar interior, plasma effect at high temperatures, nuclear reactions under extreme stellar conditions, or unknown nature of elementary particles of neutrinos.

Since early 1980s I was very interested in how to eliminate uncertainties from the theoretical estimates of the nuclear reaction cross sections for $^3\mathrm{He}(\alpha,\gamma)^7\mathrm{Be}$ and $^3\mathrm{H}(\alpha,\gamma)^7\mathrm{Li}$ which affect strongly the missing solar neutrino problem and the Big-Bang nucleosynthesis. I chose this subject as my PhD thesis,[3] being supervised by Prof. Akito Arima. This study triggered further extensive studies until today associated with Big-Bang cosmology, Galactic chemical evolution, supernova nucleosynthesis, neutrino astrophysics, extremely enhanced lithium abundance in red clump stars, etc. In this article we discuss how our research works started in

1980s have developed in multiple directions to resolve many unanswered questions in nuclear astrophysics and cosmology.

2. Alpha-clustering Structure of ^{7}Be and ^{7}Li

Intensive efforts of measuring the ^{3}He$(\alpha,\gamma)^7$Be reaction cross sections have ever been made for its significance in the missing solar neutrino problem,[4] while there are only a few available data for its mirror conjugate reaction ^{3}H$(\alpha,\gamma)^7$Li. This is mainly because of the difficulty of treating poisonous tritium as a beam and also the gas target of ^{4}He. Theoretically, however, since they are mirror conjugate to each other, we can calculate both reaction cross sections in equal theoretical method simultaneously.[5,6]

^{7}Li is a stable nucleus and there are a large number of accumulated data of the electromagnetic structure such as longitudinal Coulomb and transverse magnetic form factors of high-energy electron scatterings, B(E2) transition probability between the ground and first excited states, charge and magnetic radii, and electric quadrupole, magnetic dipole and octupole moments of the ground state. These electromagnetic properties of ^{7}Li are used to constrain the radiative α-capture reaction cross sections for both ^{3}H$(\alpha,\gamma)^7$Li and ^{3}He$(\alpha,\gamma)^7$Be by assuming the charge conjugation symmetry of the nuclear force.[7,8]

At energies as low as 1–10 keV, which correspond to the temperature at the center of the Sun, and 100–500 keV, which correspond to the Gamow windows for nuclear reactions in explosive nucleosynthesis in the Big-Bang and supernovae, it is practically impossible to measure the fusion reaction cross sections directly in a laboratory system because they become extremely small at such sub-Coulomb energies. Theoretical calculations based on quantum mechanics are required. Resonating group method (RGM)[9] provides a good description of nuclear clustering correlations of the light nuclei such as ^{7}Be and ^{7}Li. The biggest advantage of this method is that both scattering and bound states are well described by taking account of the Pauli principle between nucleons that consist of composite nuclei ^{3}H, ^{3}He, ^{4}He, ^{7}Li, and ^{7}Be.

Figure 1(a) shows calculated inclusive form factors[7] which consist of inelastic Coulomb quadrupole (C2′) and magnetic dipole (M1′) contributions for the transition $3/2^- \to 1/2^-$ as well as the elastic ones (CO, C2, Ml and M3) for the transition $3/2^- \to 3/2^-$. Since the high-energy electron scattering experiment has high precision, the measured form factors are the useful observables to choose a suitable effective interaction and oscillator strength used in the RGM calculations.[8] Green dashed lines indicate the calculated results by using the other sets of effective interaction and oscillator strength which deviate from the measured form factors by $\pm 2\,\sigma$. Figure 1(b) shows astrophysical S-factor for ^{3}H$(\alpha,\gamma)^7$Li. Our extrapolated S-factor[6] at zero energy is $S_{34}^{Li}(0) = 0.098 \pm 0.006$ keV·b.

Once we fix the effective interaction and oscillator strength which best fit all measured quantities for ^{7}Li, we can calculate the mirror conjugate reaction cross sections. Figure 2 shows the calculated astrophysical S-factor for ^{3}He$(\alpha,\gamma)^7$Be.

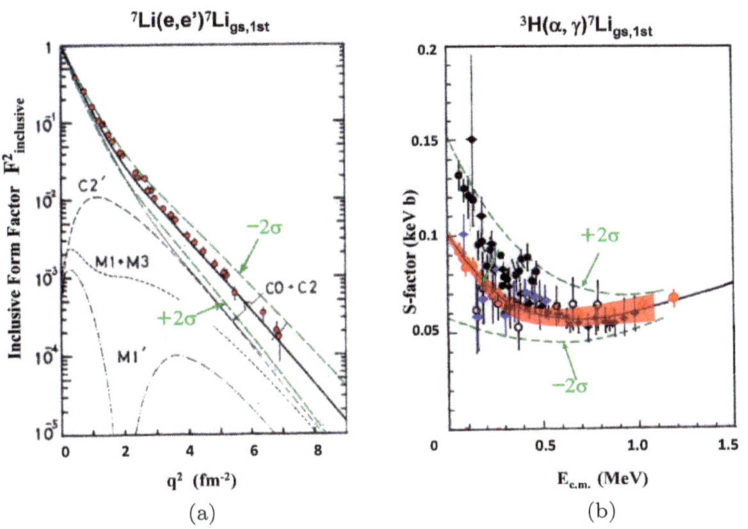

Fig. 1. (a) The square of the inclusive form factors of high-energy elastic ($3/2^-_{gr} \rightarrow 3/2^-_{gr}$) and inelastic ($3/2^-_{gr} \rightarrow 1/2^-_{1st}$) electron scattering on ^{7}Li as a function of squared transferred momentum.[7,8] (b) The astrophysical S-factor for $^3\text{H}(\alpha,\gamma)^7\text{Li}_{gr+1st}$ as a function of incident energy.[5,6] Red band indicates the theoretical result which best fits all measured data of the electromagnetic properties including the form factors. Green dashed lines indicate the calculated results by artificially using effective nuclear force and oscillator strength which deviate from the measured inclusive form factors by $\pm 2\sigma$.

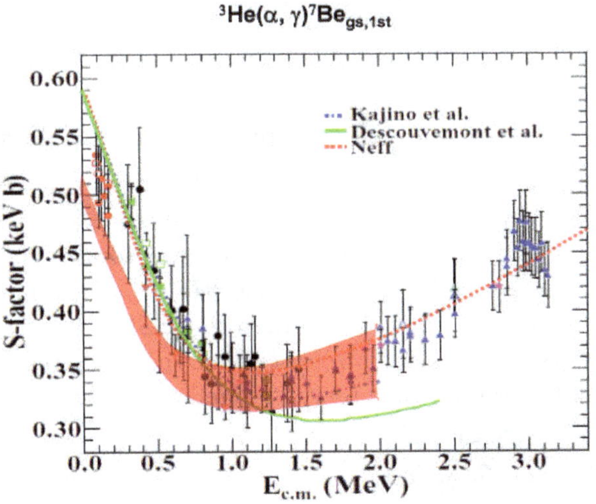

Fig. 2. The astrophysical S-factor for $^3\text{He}(\alpha,\gamma)^7\text{Be}_{gr+1st}$ as a function of incident energy.[5,6] Red band indicates the theoretical result corresponding to the red band in Fig. 1(b) for the mirror conjugate reaction $^3\text{H}(\alpha,\gamma)^7\text{Li}_{gr+1st}$ which best fits all measured data of the electromagnetic properties of ^{7}Li.

Although the scatter of experimental data is very large, our theoretical calculation predicts the astrophysical S-factor[6] at zero energy $S_{34}^{Be}(0) = 0.50 \pm 0.03$ keV·b for ^{3}He$(\alpha,\gamma)^7$Be.

Our recommended astrophysical S-factor[5,6] for ^{3}He$(\alpha,\gamma)^7$Be, $S_{34}^{Be}(0) = 0.50 \pm 0.03$ keV·b, is to be compared with $S_{34}(0) = 0.56 \pm 0.02$(expt) $\pm\, 0.02$(theor) keV·b which is the evaluated value by taking account of all available experimental data.[10] The latter value ignores mirror conjugate symmetry of the A = 7 nuclear systems. Our theoretical estimate is 12% smaller. We will apply our recommended S-factors to various astrophysical and cosmological problems related to the production of ^{7}Li and ^{7}Be.

3. Missing Solar Neutrinos

Solar neutrino flux is estimated to be proportional to $\phi_\nu \propto {S_{34}}^{0.8}$ in the Standard Solar Model.[4] Therefore, there is still a room that the accurate determination of the ^{3}He$(\alpha,\gamma)^7$Be reaction cross section could reduce the solar neutrino flux further by 10% with the use of our recommended S-factor, $S_{34}^{Be}(0) = 0.50 \pm 0.03$ keV·b. However, from these rates without the use of any artificial parameters and a survey of other nuclear reactions that influence the production of solar neutrinos, we conclude that remaining uncertainties in the nuclear reaction rates are not a likely explanation of factor of ~2 lower observed detection rate in various solar neutrino detectors.

We also calculated thermonuclear reaction rates for the two (α,γ) reactions as a function of temperature for various astrophysical applications.[11] This would shift the allowed region[12] of neutrino oscillation parameters θ_{12} and Δm^2_{12}.

4. Big-Bang Cosmology and ^{7}Li Overproduction Problem

The success in the Big-Bang nucleosynthesis (BBN) of the light elements ^{2}H, ^{3}He, and ^{4}He is a piece of evidence to support the standard cosmology.[13] Another success is the well known observational fact of the cosmic microwave background (CMB) anisotropies and fluctuations.[14] Both reactions ^{3}He$(\alpha,\gamma)^7$Be and ^{3}H$(\alpha,\gamma)^7$Li play the significant roles in the production of ^{7}Li in the standard homogeneous BBN,[11] too. There is, however, a serious discrepancy that the predicted ^{7}Li abundance is almost a factor ~3 larger than the observed abundance in metal-poor halo stars.[15]

4.1. *Nuclear Physics Solution*

^{7}Li is produced by ^{3}H$(\alpha,\gamma)^7$Li and ^{3}He$(\alpha,\gamma)^7$Be in the universe models with lower and higher baryon-to-photon number ratios $\eta_{10} \le 2.0$ and $2.0 \le \eta_{10}$, respectively, where $\eta_{10} = n_B/n_\gamma \times 10^{10}$. Produced ^{7}Be decays to ^{7}Li in 53 days by electron capture ^{7}Be$(e^-,\nu_e)^7$Li after the nuclear reactions freeze out. Applying our theoretical analysis of the two (α,γ) reactions to the standard BBN model, ^{7}Li abundance is reduced by ~12% from currently accepted value because of the dominance of

$^3He(\alpha,\gamma)^7Be$ at the observed $\eta_{10} = 6.16 \pm 0.02$ corresponding to the baryon density $\Omega_b h^2 = 0.0224 \pm 0.0001$ in the standard ΛCDM model determined from Planck analysis.[14]

Hayakawa *et al.*[16] have recently measured the destruction reaction cross sections for $^7Be(n, p)^7Li_{gr+1st}$ and found, for the first time, that the transition to the first excited state of ^7Li destroys ^7Be by ≤15%. Thus, the net reduction of ^7Li abundance due to $^3He(\alpha,\gamma)^7Be$ and $^7Be(n, p)^7Li_{gr+1st}$ amounts to ∼25–30%. These precise experimental data and our theoretical studies prove that these nuclear reaction uncertainties are not responsible for the factor of ∼3 overproduction of ^7Li abundance.

4.2. *Cosmological Solution*

The standard BBN model[17] assumes homogeneous matter and photon energy density distributions. Inhomogeneous BBN (IBBN) model[18] was proposed in 1987–1990 by taking account of spatial baryon-number density fluctuations which could arise from various cosmic phase transitions preceding BBN epoch. Although IBBN predicts remarkably enhanced abundances of Be, B and intermediate mass nuclei,[19–22] overproduction of primordial ^7Li still remains.

CMB temperature fluctuations and anisotropies were discovered a few years later in 1992.[24] Inspired by this discovery, we have recently proposed an extensive IBBN model[23] by taking account of the effect of stochastic fluctuations of primordial magnetic field (PMF).[25–27] In this new IBBN model, the domains of different temperatures exist in the BBN epoch due to variations in the local PMF although total energy density is uniform. Local photon temperature in each domain follows a Gaussian distribution of the field strength. Therefore, the effective distribution function of particle velocities averaged over domains of different temperatures deviates from the Maxwell–Boltzmann distribution, as shown in Fig. 3(a) for $^3He(\alpha,\gamma)^7Be$, so that the charged particle reaction rates are quenched. This mimics non-extensive Tsallis distribution.[28,29]

We perform BBN network calculations by taking account of the PMF fluctuations and deduce the element abundances as a function of η_{10}, scale invariant strength of the PMF energy density ρ_{B_C}, and fluctuation variance parameter σ_B. We then find[23] that the fluctuations of the PMF reduce the ^7Be production and enhance D production, leading to significantly reduced ^7Li abundance by $50 \sim 60\%$, as displayed in Fig. 3(b). Although the overproduction problem of ^7Li abundance is not completely solved only by this effect, our proposed new mechanism which is consistent with the observed constraints on CMB anisotropies drastically relaxes the tension of the Big-Bang ^7Li problem.

4.3. *New Physics Solution*

There are many other proposals by introducing the so-called new physics. As an example, we made a hybrid dark matter model[30] with both axions and relic massive

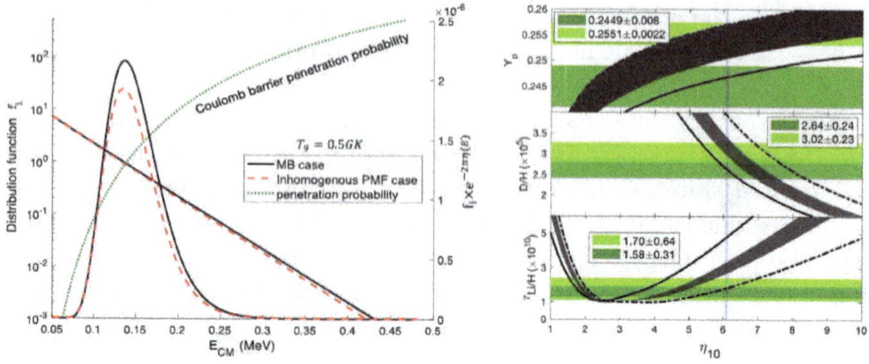

(a) (b)

Fig. 3. (a) Deviation of the Gamow window for $^3He(\alpha,\gamma)^7Be$ in our PMF model[23] (red dashed curve) from Maxwell–Boltzmann distribution (black solid curve) at $T_9 = 0.5$. Dotted green line indicates the Coulomb barrier penetration probability. Although the deviation is small, the Gamow peak in the PMF model is suppressed compared with Maxwell–Boltzmann distribution leading to remarkable suppression of the ^7Be production. (b) Calculated abundances[23] of ^4He (mass fraction Y_p), D/H, and $^7Li/H$ (number abundance ratios) as a function of baryon-to-photon ratio η_{10} in BBN. Green bands show the adopted observational constraints. This figure shows that larger fluctuation variance parameter σ_B of PMF can suppress the production of ^7Li but increase the value of D/H. The vertical blue band shows the Planck constraint on η_{10}.

neutral supersymmetric (SUSY) particles. In this model axion cooling of photons between the BBN and CMB-photon recombination epochs would lead to the different η_{10} values in the early universe. We demonstrate that the ^7Li abundance can be consistent with observations in metal-poor halo stars without destroying the important concordance with ^4He and ^2H.

Another kind of models is to introduce relic long-lived charged particles X^- like stau in SUSY which existed in BBN epoch.[31,32] They bind to ^7Be to form ^7Be_X atom which is destroyed by proton capture reaction through both atomic and nuclear excited state of ^8B_X. Although this scenario provides a solution to the ^7Li overproduction, model parameters, i.e. the lifetime and abundance of the relic particles, must be fine-tuned. New particle search is highly desirable.

5. R-Process in Supernovae, Collapsars and Neutron Star Mergers in Galactic Chemical Evolution

A new era of astronomy has been opened by the historic multi-messenger observations of the supernova SN1987A and the neutron star merger GW170817. These events have a large impact on nuclear astrophysics, in particular on the study of r-process. Binary neutron star mergers (NSMs)[33,34] as well as core collapse supernovae (CCSNe)[35] and collapsars[36,37] are the most viable candidates for the origin of r-process elements among many other possible astrophysical sites.[38] CCSNe include both neutrino-driven wind (ν-wind)[39,40] and magneto-hydrodynamic jet (MHDJ)[41]

SNe, which leave neutron star as a remnant, and collapsar[42,43] is a very massive single star collapsing into a black hole.

The ν-wind SN produces r-process elements of $A \leq 120$ including the first peak elements around $A \sim 90$ but not heavier than this.[38] MHDJ SN and collapsar r-process produces the heavy r-process elements including the second and third peak elements around $A \sim 130$ and 195.[35,37] Nuclear properties of extremely neutron-rich unstable nuclei such as β-decay half lives[45,46] and fission fragment distributions[35] take the important keys to the r-process. In addition, light mass nuclear reactions affect the production of these elements, in particular in the ν-wind SN nucleosynthesis, because the r-process is a primary process which begins from the initial chemical compositions of abundant neutrons with small amounts of protons and α particles without intermediate-to-heavy mass nuclei.[39] Figure 4 illustrates the nuclear reaction network[40] which we constructed for our r-process nucleosynthesis calculations. Inset shows the most relevant light nuclear reaction flows[38,44,47]: ^{3}H$(\alpha,\gamma)^7$Li(n, $\gamma)^8$Li$(\alpha$, n$)^{11}$B(n, $\gamma)^{12}$B; ^{4}He$(\alpha$n, $\gamma)^9$Be(n, $\gamma)^{10}$Be$(\alpha,\gamma)^{11}$B(n, $\gamma)^{12}$B.

The uncertainties of the reaction cross sections for ^{3}H$(\alpha,\gamma)^7$Li have already been discussed in the previous section. ^{8}Li$(\alpha$, n$)^{11}$B reaction rates are subject to large uncertainties[44] because ^{8}Li ($\tau_{1/2} = 838.75$ ms) is an unstable nucleus, which makes direct experimental measurement very difficult, and because the compound nucleus ^{12}B ($\tau_{1/2} = 20.20$ ms) has a complicated nuclear structure near the threshold energy.[47] Inclusive cross sections determined by the Trojan Horse Method are smaller than those of the exclusive cross sections determined in different experimental method.[48] Since these light nuclear reactions affect the final abundances of r-process elements,[40,44,49] further high precision measurement is desirable.

As for the NSM r-process, the basic difference from the CCSN is a different emergent event rate and a delay timescale[50,51] from star formation of their progenitors

Fig. 4. Low-mass region of the extended nuclear reaction network[40] used for the our r-process nucleosynthesis calculations. Inset shows the most relevant light nuclear reaction flows[38,40,44] to affect the r-process nucleosynthesis of heavy nuclei.

until the time when r-process occurs. SNe and collapsars are dying single massive stars that culminate their evolution in explosions in a few My. Therefore, they can enrich r-process elements in the interstellar medium (ISM) from the early Galaxy. On the other hand, the majority of NSMs may not contribute to the early Galaxy because they take cosmologically long timescale due to extremely slow GW radiation to lose kinetic energy of the orbital motion until they merge.[37] The coalescence timescale estimated from observed double pulsars ranges from a few hundred My to the cosmic age or beyond.[51] Their emergent rate is as low as 0.1–1% of the SN rate.[35,38]

Figure 5 shows our calculated Galactic chemical evolution of the r-process elements.[37] It is unlikely for NSMs to emerge in metal-deficient early Galaxy. Many halo stars with extremely low metallicity, which were born in the early Galaxy, are found to show a universal elemental abundance pattern similar to the solar-system abundance pattern.[52,53] Universality holds in the range of atomic number 52 (Te) $\leq$ Z $\leq$ 82 (Pb) corresponding to the range of mass number 128 $\leq$ A $\leq$ 208. Only the CCSNe and collapsars naturally explain the observed universality of the r-process

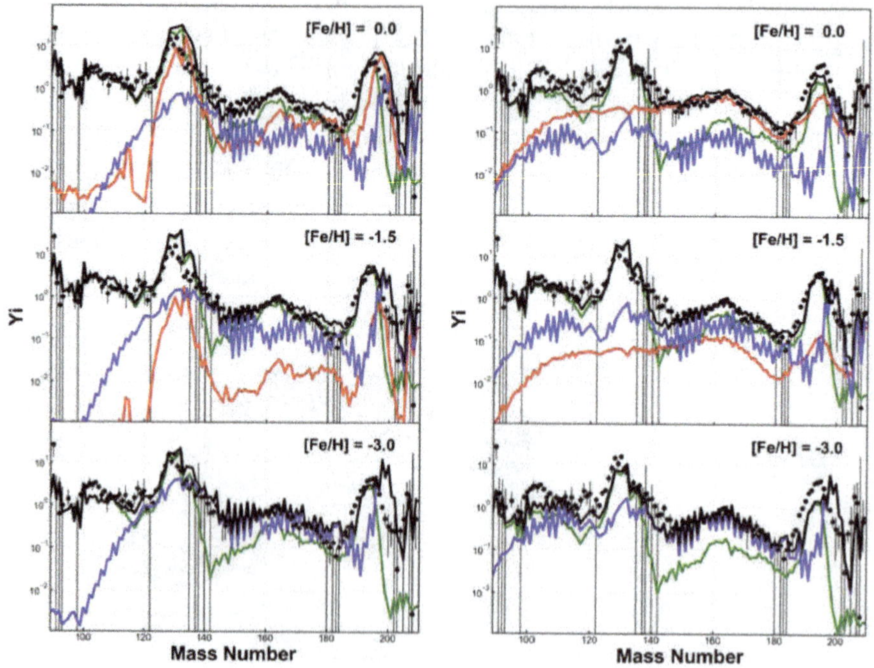

Fig. 5. The time-varying isotopic abundance patterns[37] of heavy r-process nuclei at metallicities [Fe/H] = 0.0, −1.5, and −3.0 from top to bottom, where $[X/H] = \log(n_{Fe}/n_H) - \log(n_{Fe}/n_H)_\odot$. The left panels are for the case in which a symmetric fission fragment distribution is adopted, while the right panels are for an asymmetric fission fragment distribution.[35,57] The black lines show the total calculated r-process abundances. The red and green lines represent the contributions from NSMs and CCSNe (including the ν-wind and MHDJ SNe), respectively. The blue lines show the contribution from collapsars. The minimum coalescence time 10 Myr is used in these calculations.

elements[37] although the variance also is observed in extremely metal-deficient halo stars.[54,55]

Nevertheless, NSMs are very interesting phenomena which exhibit strong gravitational wave radiation to constrain the equation-of-state of neutron stars or the source of production of long-lived actinoids like ^{254}Cf ($\tau_{1/2} = 60.5$ d), ^{232}Th ($\tau_{1/2} = 14.0$ Gy), ^{238}U ($\tau_{1/2} = 4.468$ Gy), etc.[53,56] These could be the observational tracers of NSM events as well as CCSNe and collapsars.

Although the r-process in each of three sites, i.e. CCSN, collapsar or neutron star merger, leads to a similar universal elemental distribution as a function of atomic number Z, the detailed isotopic abundances as a function of mass number A differ from one site to another.[35,37] These differences may allow for the identification of which sources contributed to the evolution of r-process material in the Galaxy.[37] There arises also a growing evidence for the existence of an intermediate neutron-capture process (i-process) in recent spectroscopic observations of carbon-rich metal-poor stars. Namely, several isotopes between the second and third r-process peak elements show an abundance pattern between the r- and s-processes. We have recently found that the collapsar r-process is associated with the both i- and s-processes by fission neutrons and that the strong even-odd effect in the mass abundance pattern as shown by blue lines in Fig. 5 could be a piece of observational evidence.[58] All these conclusions and predictions may be testable if the next generation of observatories like James Web Space Telescope and ground-based Thirty Meter Telescope can deduce isotopic abundances in old metal-deficient halo stars.

6. Supernova Nucleosynthesis and Neutrino Flavor Oscillations

The ν-process nucleosynthesis in CCSNe is a very unique and sensitive probe to still unknown neutrino mass hierarchy through the Mikheyev–Smirnov–Wolfenstein (MSW) matter effect.[59,60] A huge number of neutrinos ejected from the proto-neutron star propagate through the stellar interior, and when the electron number density satisfies the specific MSW high-density resonance condition, the flavor conversion occurs between $\bar{\nu}_e$ and $\bar{\nu}_{\mu,\tau}$ for the inverted hierarchy and ν_e and $\nu_{\mu,\tau}$ for the normal hierarchy, respectively.[12] These neutrinos interact with abundant nuclei in each layer, and various isotopes such as ^{7}Li, ^{11}B, ^{19}F, ^{92}Nb, ^{98}Tc, ^{138}La, ^{180}Ta, etc. are synthesized.[61–73] The neutrino oscillation effect considerably changes the final yields of these nuclei via the charged current interactions.

Figure 6 shows our calculated ^{7}Li+^{7}Be and ^{11}B+^{11}C abundances which depend strongly on neutrino mass hierarchies.[74,75] The A = 11 abundances in normal hierarchy (blue curves) are higher than those calculated in inverted hierarchy (red curves). Since the original neutrino temperatures at the neutrino-spheres are different from one another $T_{\nu_e} < T_{\bar{\nu}_e} < T_{\nu_x}$, where ν_x stands for ν_μ, ν_τ and their anti-neutrinos, the MSW matter effect makes more efficient swapping of the neutrino energy spectra between ν_e and ν_x than the swapping between $\bar{\nu}_e$ and ν_x. As a result, ν_e has

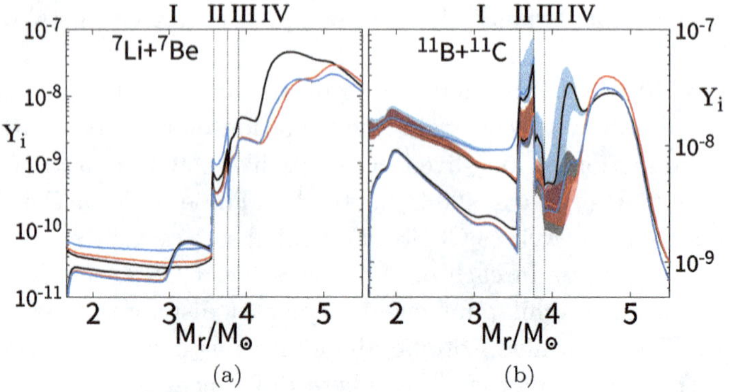

Fig. 6. The calculated abundances[74] of (a) ^{7}Li+^{7}Be and (b) ^{11}B+^{11}C as a function of Lagrangian mass coordinate in units of $M_\odot$. Red and blue lines correspond to inverted and normal hierarchies, respectively. Colored bands reflect the uncertainty of the nuclear reaction rates.[74,75] Solid and dashed lines are the results using Suzuki *et al.*[76] and Hoffman & Woosley[77] rates for ^{16}O+ν, respectively.

the largest enhancement of the high energy component after high-density MSW resonance occurs. This leads to stronger charged current reactions via ν_e than $\bar{\nu}_e$ to enhance the production of ^{7}Be and ^{11}C through the ^{4}He(ν_e, e^-p)^{3}He(α,γ)^{7}Be and ^{12}C(ν_e, e^-p)^{11}C, etc., in the normal hierarchy (blue curves).

Stellar interior is divided into four layers, i.e. O-Ne-Mg layer (region I), O-C layer (II), C-He layer (III) and He layer (IV) in Fig. 6. Most of ^{7}Li, ^{7}Be, ^{11}B and ^{11}C are respectively produced in the regions of Lagrangian masses of 4.5–6.0 $M_\odot$, 3.5–5.5 $M_\odot$, 1.8–5.5 $M_\odot$, and 1.8–5.7 $M_\odot$. Because the lifetimes of ^{11}C and ^{7}Be are shorter than the age of SN remnant, they decay to ^{11}B and ^{7}Li and the calculated abundance ratio (^{7}Li+^{7}Be)/(^{11}B+^{11}C) provides the key information to distinguish neutrino mass hierarchy. SiC-X presolar grains, which condensate from SN ejecta, were discovered to exhibit ^{7}Li/^{11}B < 0.56 at the 2-σ C.L.[78,79] Mixing process of ejecta would be complicated.[75] Assuming here as an example that the materials of SN ejecta are fully mixed, neutrino mass hierarchy is constrained through the comparison between our theoretical result and measured value. Inverted hierarchy is statistically more favored at the 2-σ C.L. Details are discussed in Refs. 74, 75.

7. Enhanced Lithium Abundance in Red Clump Stars

As the temperature becomes high enough in the stellar evolution of low-mass stars, ^{7}Be is produced by ^{3}He(α, γ)^{7}Be. If the mixing between the stellar interior and the surface is slow, produced ^{7}Be is destroyed by ^{7}Be(p, γ)^{8}B(e$^+\nu_e$)^{8}Be. ^{7}Li produced from electron capture ^{7}Be(e$^-$, ν_e)^{7}Li is immediately burned via ^{7}Li(p, α)^{4}He. However, if the mixing occurs sufficiently fast, ^{7}Be is transported to cooler regions near the surface before the electron capture takes place. In this way, the resultant ^{7}Li survives and enrich the stellar envelope (Cameron-Fowler mechanism).[80,81]

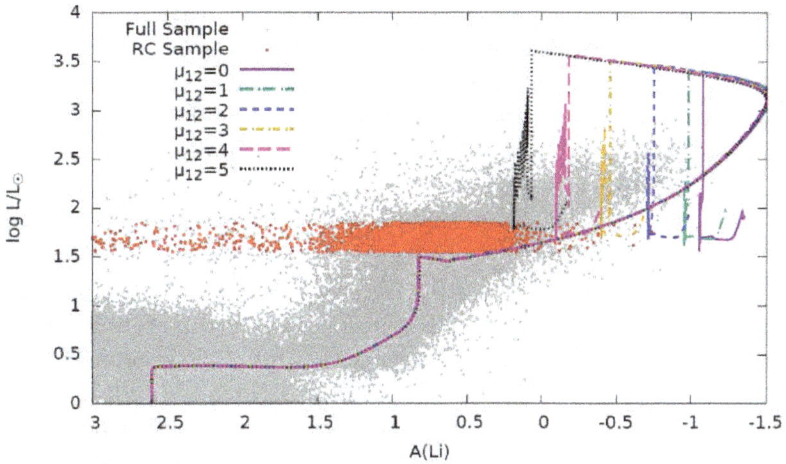

Fig. 7. The lines show the evolution of our models[84] with various neutrino magnetic moments $\mu_\nu = (0 - 5) \times 10^{-12}\mu_B$ in the Luminosity$-$A(Li) plane, where A(Li) $= \log(\mathrm{n(Li)/n(H)}) + 12$. We adopt thermohaline mixing coefficient $\alpha_{thm} = 100$.[84] The grey dots are GALAH DR2 samples[82] with reliable ^{7}Li abundances and the red dots are RC samples selected by Kumar *et al.*[83]

In the atmosphere of most red giants, ^{7}Li is easily destroyed and the surface lithium abundance decreases as stars evolve. However, recent astronomical observations[82,83] show that almost all of red clump stars keep high ^{7}Li abundances $-0.9 \leq$ A(Li), where A(Li) is defined by A(Li) $= \log(\mathrm{n(Li)/n(H)}) + 12$. This is not predicted at all by the standard theory of the low-mass stellar evolution.

In order to reconcile the discrepancy between the observations and the theoretical prediction, we consider additional energy loss channels that may come from physics beyond the Standard Model, i.e. the neutrino magnetic moment.[84] The existence of neutrino masses was established by the detection of the neutrino oscillations. In the Standard Model of elementary particle physics, finite mass neutrinos have magnetic moments although detection in the laboratory is too hard at present and near-future experiments. However, its effect is enlarged in stellar evolution. Figure 7 shows our calculated results assuming the magnetic moment $\mu_\nu = (0 - 5) \times 10^{-12}\mu_B$, where μ_B is the Bohr magneton. We find that the production of ^{7}Be becomes more active if the neutrino magnetic moment is as large as $\geq 1.0 \times 10^{-12}\mu_B$ because thermohaline mixing becomes more efficient and a heavier He core is formed. This is consistent with the constraint[85] from globular clusters $\mu_\nu \leq (1.5 - 2.2) \times 10^{-12}\mu_B$. The observed high ^{7}Li abundance denoted by red dots in Fig. 7 could be explained by these μ_ν values.

8. Summary

In this article we discussed mainly the roles of $^3\mathrm{He}(\alpha,\gamma)^7\mathrm{Be}$ and $^3\mathrm{H}(\alpha,\gamma)^7\mathrm{Li}$ reactions in various astrophysical and cosmological problems.

We first discussed that the microscopic cluster model wave functions in the resonating group method describe very well the alpha-clustering structure of ^{7}Be and ^{7}Li. Assuming charge conjugation symmetry of the nuclear force, we theoretically calculated the reaction cross sections for ^{3}He$(\alpha,\gamma)^7$Be and ^{3}H$(\alpha,\gamma)^7$Li simultaneously and consistently at extremely low energies of astrophysical interest, i.e. $1-10$ keV and $100-500$ keV corresponding to the Gamow window energies at temperatures near the center of the Sun and supernova explosions, respectively. Our recommended astrophysical S-factors at zero energy turn out to be $S_{34}^{Be}(0) = 0.50 \pm 0.03$ keV·b for ^{3}He$(\alpha,\gamma)^7$Be and $S_{34}^{Li}(0) = 0.098 \pm 0.006$ keV·b for ^{3}H$(\alpha,\gamma)^7$Li.

We then applied these S-factors to the nucleosynthetic processes in various astrophysical sites. The first application is the refined estimate the solar neutrino flux. We conclude that there is a room that the accurate determination of the ^{3}He$(\alpha,\gamma)^7$Be reaction cross section could reduce the solar neutrino flux further by 10% with the use of our recommended S-factor, which would shift the allowed region of neutrino oscillation parameters θ_{12} and Δm_{12}^2.

We secondly discussed the Big-Bang nucleosynthesis. Combining our S-factors for ^{3}He$(\alpha,\gamma)^7$Be with the recent new measurement of the ^{7}Be(n, p)^{7}Li$_{gr+1st}$ reaction cross section, we find that the net reduction of ^{7}Li abundance could amount to ~25–30%. These improved nuclear reaction rates are not fully responsible for the factor of ~3 overproduction of primordial ^{7}Li abundance. We also discussed the other proposed mechanisms such as the cosmological effect of stochastic fluctuations of primordial magnetic field, and the effect of so-called new physics on dark matter particles. More efforts are desirable both experimentally and theoretically to find direct evidence for our proposed mechanisms to work in reality.

We thirdly discussed the r-process nucleosynthesis and Galactic chemical evolution. Among several candidate sites, core collapse supernovae and collapsars turn out to be the major contributors over the entire history of cosmic evolution. Neutron star mergers have arrived later and make a minimal contribution to the solar-system r-process elements. Since r-process in both supernovae and collapsars is a primary process which starts from the plasma condition with huge neutron number-density with small number of protons and alpha particles, final yields of heavy elements depend on the light-mass nuclear reactions; ^{3}H$(\alpha,\gamma)^7$Li(n, $\gamma)^8$Li$(\alpha$, n$)^{11}$B(n, $\gamma)^{12}$B; ^{4}He$(\alpha$n, $\gamma)^9$Be(n, $\gamma)^{10}$Be$(\alpha,\gamma)^{11}$B(n, $\gamma)^{12}$B. Precise determination of these reaction cross sections are required.

We fourthly discussed neutrino-process nucleosynthesis in supernovae and proposed a method to constrain still unknown neutrino mass hierarchy by the use of MSW effect. Relatively less abundant nuclei like ^{7}Li, ^{11}B, ^{19}F, ^{92}Nb, ^{98}Tc, ^{138}La, ^{180}Ta, etc. are known to be synthesized in supernova neutrino-process. Neutrino flavor oscillations considerably change the final yields of these nuclei via the charged current interactions: ^{7}Be and ^{11}C are produced by ^{4}He$(\nu_e, e^-$p$)^3$He$(\alpha,\gamma)^7$Be and ^{12}C$(\nu_e, e^-$p$)^{11}$C, respectively. ^{7}Li and ^{11}B are produced by ^{4}He$(\bar{\nu}_e, e^+$n$)^3$H$(\alpha,\gamma)^7$Li and ^{12}C$(\bar{\nu}_e, e^+$n$)^{11}$B, respectively. SiC-X presolar grains, which condensate from SN ejecta, were discovered to exhibit ^{7}Li/^{11}B ratios. Comparing our theoretical prediction of the abundance ratio with observed value, we tried to constrain the neutrino mass hierarchy.

Finally, we discussed that extremely enhanced lithium abundance in red clump stars could be explained by Cameron-Fowler mechanism coupled with new physics of neutrino magnetic moment which triggers efficient extra-mixing of stellar materials due to additional cooling. We point out that the precise determination of the ^{3}He$(\alpha, \gamma)^7$Be reaction cross section takes the key because this reaction competes with ^{7}Be(p, $\gamma)^8$B$(e^+\nu_e)^8$Be and ^{7}Be$(e^-, \nu_e)^7$Li(p, $\alpha)^4$He.

Acknowledgments

I would like to express my sincere thanks to Prof. Akito Arima who continuously encouraged me with his warm message "Open up new frontier field of your own! Never follow the fashion. Seek for truth, work hard, and help each other with friends all over the world." My borderless and endless voyage in nuclear astrophysics and cosmology started with his encouragement when I defended my PhD thesis on missing solar neutrinos and stellar nuclear reactions. Our research works in 1980s have developed in multiple directions with new paradigms in nuclear astrophysics, providing the firm basis in nuclear physics to resolve many unanswered questions in astrophysics and cosmology even today.

References

1. C. A. Bertulani and T. Kajino, *Prog. Part. Nucl. Phys.* **89**, 56–100 (2016).
2. Y. Fukuda and others (The Super-Kamiokande Collaboration), *Phys. Rev. Lett.* **81**, 1562–1567 (1998).
3. T. Kajino, *Solar Neutrino Problem and Nuclear Reactions.* PhD thesis, University of Tokyo (1984).
4. J. N. Bahcall, *Neutrino Astrophysics.* (Cambridge University Press), Cambridge (1989).
5. T. Kajino and A. Arima, *Phys. Rev. Lett.* **52**, 739–742 (1984).
6. T. Kajino, *Nucl. Phys. A* **460**, 559–580 (1986).
7. T. Kajino, T. Matsuse, and A. Arima, *Nucl. Phys. A* **413**, 323–352 (1984).
8. T. Kajino, T. Matsuse, and A. Arima, *Nucl. Phys. A* **414**, 185–205 (1984).
9. K. Ikeda *et al.*, *Prog. Theor. Phys. Suppl.* **68**, 1–385 (1980).
10. E. G. Adelberger *et al.*, *Reviews of Modern Physics* **83**, 195–246 (2011).
11. T. Kajino, H. Toki, and S. M. Austin, *Astrophys. J.* **319**, 531–540 (1987).
12. M. Fukugita and T. Yanagida, *Physics of Neutrinos.* Springer, Heiderberg (2003).
13. E. W. Kolb and M. S. Turner, *The Early Universe.* Addison-Wesley Publishing Company, New York (1990).
14. P. A. R. Ade and others (Planck Collaboration), *Astron. Astrophys.* **594**, A13 (2016).
15. S. G. Ryan, T. Kajino, T. C. Beers, T. K. Suzuki, D. Romano, F. Matteucci, and K. Rosolankova, *Astrophys. J.* **549**, 55–71 (2001).
16. S. Hayakawa *et al.*, *Astrophys. J.* **915**, L13 (2021).
17. D. N. Schramm and R. V. Wagoner, *Ann. Rev. Nucl. Part. Sci.* **27**, 37–74 (1977).
18. C. Alcock, G. M. Fuller, and G. J. Mathews, *Astrophys. J.* **320**, 439 (1987).
19. R. N. Boyd and T. Kajino, *Astrophys. J.* **336**, L55 (1989).

20. T. Kajino and R. N. Boyd, *Astrophys. J.* **359**, 267 (1990).
21. T. Kajino, G. J. Mathews, and G. M. Fuller, *Astrophys. J.* **364**, 7 (1990).
22. M. Orito, T. Kajino, R. N. Boyd, and G. J. Mathews, *Astrophys. J.* **488**, 515–523 (1997).
23. Y. Luo, T. Kajino, M. Kusakabe, and G. J. Mathews, *Astrophys. J.* **872**, 172 (2019).
24. G. F. Smoot and others (COBE Collaboration), *Astrophys. J.* **396**, L1 (1992).
25. D. G. Yamazaki, K. Ichiki, T. Kajino, and G. J. Mathews, *Phys. Rev. D* **77**, 043005 (2008).
26. D. G. Yamazaki, K. Ichiki, T. Kajino, and G. J. Mathews, *Phys. Rev. D* **81**, 023008 (2010).
27. D. G. Yamazaki, T. Kajino, G. J. Mathews, and K. Ichiki, *Phys. Rep.* **517**, 141–167 (2012).
28. C. Tsallis, *J. Statis. Phys.* **52**, 479–487 (1988).
29. S. Q. Hou, J. J. He, A. Parikh, D. Kahl, C. A. Bertulani, T. Kajino, G. J. Mathews, and G. Zhao, *Astrophys. J.* **834**, 165 (2017).
30. M. Kusakabe, A. B. Balantekin, T. Kajino, and Y. Pehlivan, *Phys. Lett. B* **718**, 704–708 (2013).
31. M. Kusakabe, T. Kajino, R. N. Boyd, T. Yoshida, and G. J. Mathews, *Phys. Rev. D* **76**, 121302 (2007).
32. M. Kusakabe, T. Kajino, R. N. Boyd, T. Yoshida, and G. J. Mathews, *Astrophys. J.* **680**, 846–857 (2008).
33. S. Wanajo, Y. Sekiguchi, N. Nishimura, K. Kiuchi, K. Kyutoku, and M. Shibata, *Astrophys. J.* **789**, L39 (2014).
34. Y. Hirai, Y. Ishimaru, T. R. Saitoh, M. S. Fujii, J. Hidaka, and T. Kajino, *Astrophys. J.* **814**, 41 (2015).
35. S. Shibagaki, T. Kajino, G. J. Mathews, S. Chiba, S. Nishimura, and G. Lorusso, *Astrophys. J.* **816**, 79 (2016).
36. D. M. Siegel, J. Barnes, and B. D. Metzger, *Nature* **569**, 241–244 (2019).
37. Y. Yamazaki, Z. He, T. Kajino, G. J. Mathews, M. A. Famiano, X. Tang, and J. Shi, *Astrophys. J.* **933**, 112 (2022).
38. T. Kajino, W. Aoki, A. B. Balantekin, R. Diehl, M. A. Famiano, and G. J. Mathews, *Prog. Part. Nucl. Phys.* **107**, 109–166 (2019).
39. K. Otsuki, H. Tagoshi, T. Kajino, and S.-y. Wanajo, *Astrophys. J.* **533**, 424–439 (2000).
40. M. Terasawa, K. Sumiyoshi, T. Kajino, G. J. Mathews, and I. Tanihata, *Astrophys. J.* **562**, 470–479 (2001).
41. N. Nishimura, T. Kajino, G. J. Mathews, S. Nishimura, and T. Suzuki, *Phys. Rev. C* **85**, 048801 (2012).
42. K. Nakamura, T. Kajino, G. J. Mathews, S. Sato, and S. Harikae, *Int. J. Mod. Phys. E* **22**, 1330022 (2013).
43. M. Famiano, A. B. Balantekin, T. Kajino, M. Kusakabe, K. Mori, and Y. Luo, *Astrophys. J.* **898**, 163 (2020).
44. T. Sasaqui, T. Kajino, G. J. Mathews, K. Otsuki, and T. Nakamura, *Astrophys. J.* **634**, 1173–1189 (2005).
45. T. Suzuki, T. Yoshida, T. Kajino, and T. Otsuka, *Phys. Rev. C* **85**, 015802 (2012).
46. G. Lorusso, S. Nishimura, Z. Y. Xu, A. Jungclaus, Y. Shimizu, G. S. Simpson, P. A. Söderström, H. Watanabe, F. Browne, P. Doornenbal, G. Gey, H. S. Jung, B. Meyer, T. Sumikama, J. Taprogge, Z. Vajta, J. Wu, H. Baba, G. Benzoni, K. Y. Chae, F. C. L. Crespi, N. Fukuda, R. Gernhäuser, N. Inabe, T. Isobe, T. Kajino, D. Kameda, G. D. Kim, Y. K. Kim, I. Kojouharov, F. G. Kondev, T. Kubo, N. Kurz, Y. K. Kwon,

G. J. Lane, Z. Li, A. Montaner-Pizá, K. Moschner, F. Naqvi, M. Niikura, H. Nishibata, A. Odahara, R. Orlandi, Z. Patel, Z. Podolyák, H. Sakurai, H. Schaffner, P. Schury, S. Shibagaki, K. Steiger, H. Suzuki, H. Takeda, A. Wendt, A. Yagi, and K. Yoshinaga, *Phys. Rev. Lett.* **114**, 192501 (2015).

47. K. Sumiyoshi, H. Utsunomiya, S. Goko, and T. Kajino, *Nucl. Phys. A* **709**, 467–486 (2002).
48. S. K. Das, T. Fukuda, Y. Mizoi, H. Ishiyama, H. Miyatake, Y. X. Watanabe, Y. Hirayama, S. C. Jeong, H. Ikezoe, M. Matsuda, K. Nishio, and T. Hashimoto, *Phys. Rev. C* **95**, 055805 (2017).
49. K. Kim, Y. Kim, X. Yao, Z. He, M. Kusakabe, T. Kajino, *et al.* (2022). To be published.
50. P. Beniamini and T. Piran, *Mon. Not. Roy. Astron. Soc.* **487**, 4847–4854 (2019).
51. D. R. Lorimer, *Living Rev. Rel.* **11**, 8 (2008).
52. C. Sneden, J. J. Cowan, and R. Gallino, *Ann. Rev. Astron. Astrophys.* **46**, 241–288 (2008).
53. S. Honda, W. Aoki, T. Kajino, H. Ando, T. C. Beers, H. Izumiura, K. Sadakane, and M. Takada-Hidai, *Astrophys. J.* **607**, 474–498 (2004).
54. A. Frebel, W. Aoki, N. Christlieb, H. Ando, M. Asplund, P. S. Barklem, T. C. Beers, K. Eriksson, C. Fechner, M. Y. Fujimoto, S. Honda, T. Kajino, T. Minezaki, K. Nomoto, J. E. Norris, S. G. Ryan, M. Takada-Hidai, S. Tsangarides, and Y. Yoshii, *Nature.* **434**, 871–873 (2005).
55. W. Aoki, A. Frebel, N. Christlieb, J. E. Norris, T. C. Beers, T. Minezaki, P. S. Barklem, S. Honda, M. Takada-Hidai, M. Asplund, S. G. Ryan, S. Tsangarides, K. Eriksson, A. Steinhauer, C. P. Deliyannis, K. Nomoto, M. Y. Fujimoto, H. Ando, Y. Yoshii, and T. Kajino, *Astrophys. J.* **639**, 897–917 (2006).
56. A. A. Aziz, N. Sofiah Ahmad, S. Ahn, W. Aoki, M. Bhuyan, K.-J. Chen, G. Guo, K. I. Hahn, T. Kajino, H. Abu Kassim, D. Kim, S. Kubono, M. Kusakabe, A. Li, H. Li, Z. H. Li, W. P. Liu, Z. W. Liu, T. Motobayashi, K.-C. Pan, T. S. Park, J.-R. Shi, X. Tang, W. Wang, L. Wen, M.-R. Wu, H.-L. Yan, and N. Yusof, *AAPPS Bulletin.* **31**, 18 (2021).
57. T. Suzuki, S. Shibagaki, T. Yoshida, T. Kajino, and T. Otsuka, *Astrophys. J.* **859**, 133 (2018).
58. Z. He, M. Kusakabe, T. Kajino, S. Zhou, H. Koura, and S. Chiba (2022). To be published.
59. L. Wolfenstein, *Phys. Rev. D* **17**, 2369–2374 (1978).
60. S. P. Mikheyev and A. Y. Smirnov, *Yadernaya Fizika.* **42**, 1441–1448 (1985).
61. T. Yoshida, M. Terasawa, T. Kajino, and K. Sumiyoshi, *Astrophys. J.* **600**, 204–213 (2004).
62. T. Yoshida, T. Kajino, and D. H. Hartmann, *Phys. Rev. Lett.* **94**, 231101 (2005).
63. T. Yoshida, T. Kajino, H. Yokomakura, K. Kimura, A. Takamura, and D. H. Hartmann, *Astrophys. J.* **649**, 319–331 (2006).
64. T. Yoshida, T. Kajino, H. Yokomakura, K. Kimura, A. Takamura, and D. H. Hartmann, *Phys. Rev. Lett.* **96**, 091101 (2006).
65. T. Yoshida, T. Suzuki, S. Chiba, T. Kajino, H. Yokomakura, K. Kimura, A. Takamura, and D. H. Hartmann, *Astrophys. J.* **686**, 448–466 (2008).
66. T. Hayakawa, T. Kajino, S. Chiba, and G. J. Mathews, *Phys. Rev. C* **81**, 052801 (2010).
67. T. Hayakawa, P. Mohr, T. Kajino, S. Chiba, and G. J. Mathews, *Phys. Rev. C* **82**, 058801 (2010).
68. T. Hayakawa, K. Nakamura, T. Kajino, S. Chiba, N. Iwamoto, M. K. Cheoun, and G. J. Mathews, *Astrophys. J.* **779**, L9 (2013).

69. T. Kajino, G. J. Mathews, and T. Hayakawa, *J. Phys. G. Nucl. Phys.* **41**, 044007 (2014).
70. T. Hayakawa, H. Ko, M.-K. Cheoun, M. Kusakabe, T. Kajino, M. D. Usang, S. Chiba, K. Nakamura, A. Tolstov, K. Nomoto, M.-a. Hashimoto, M. Ono, T. Kawano, and G. J. Mathews, *Phys. Rev. Lett.* **121**, 102701 (2018).
71. M. Kusakabe, M.-K. Cheoun, K. S. Kim, M.-a. Hashimoto, M. Ono, K. Nomoto, T. Suzuki, T. Kajino, and G. J. Mathews, *Astrophys. J.* **872**, 164 (2019).
72. H. Ko, M.-K. Cheoun, E. Ha, M. Kusakabe, T. Hayakawa, H. Sasaki, T. Kajino, M.-a. Hashimoto, M. Ono, M. D. Usang, S. Chiba, K. Nakamura, A. Tolstov, K. Nomoto, T. Kawano, and G. J. Mathews, *Astrophys. J.* **891**, L24 (2020).
73. H. Ko, D. Jang, M.-K. Cheoun, M. Kusakabe, H. Sasaki, X. Yao, T. Kajino, T. Hayakawa, M. Ono, T. Kawano, and G. J. Mathews, *Astrophys. J., in press. (arXiv:2203.13365)* (2022).
74. X. Yao, M. Kusakabe, T. Kajino, S. Cherubini, S. Hayakawa, and H. Yamaguchi. In *European Physical Journal Web of Conferences*, vol. 260, p. 01007 (2022).
75. X. Yao, M. Kusakabe, T. Kajino, *et al.* (2022). To be published.
76. T. Suzuki, S. Chiba, T. Yoshida, K. Takahashi, and H. Umeda, *Phys. Rev. C* **98**, 034613 (2018).
77. R. D. Hoffman and S. E. Woosley. Stellar nucleosynthesis data (1992). https://dbserv.pnpi.spb.ru/elbib/tablisot/toi98/www/astro/hw92-1.htm.
78. W. Fujiya, P. Hoppe, and U. Ott, *Astrophys. J.* **730**, L7 (2011).
79. G. J. Mathews, T. Kajino, W. Aoki, W. Fujiya, and J. B. Pitts, *Phys. Rev. D* **85**, 105023 (2012).
80. A. G. W. Cameron, *Astrophys. J.* **121**, 144 (1955).
81. A. G. W. Cameron and W. A. Fowler, *Astrophys. J.* **164**, 111 (1971).
82. S. Buder and others (Galah Collaboration), *MNRAS* **478**, 4513–4552 (2018).
83. Y. B. Kumar, B. E. Reddy, S. W. Campbell, S. Maben, G. Zhao, and Y.-S. Ting, *Nature Astron.* **4**, 1059–1063 (2020).
84. K. Mori, M. Kusakabe, A. B. Balantekin, T. Kajino, and M. A. Famiano, *Mon. Not. Roy. Astron. Soc.* **503**, 2746–2753 (2021).
85. S. Arceo-Díaz, K. P. Schröder, K. Zuber, and D. Jack, *Astroparticle Phys.* **70**, 1–11 (2015).

https://doi.org/10.1142/9789819813353_0010

Arima Sensei's Contribution to Japan's Science and Technology Policies

Minoru Kuniya

Project Researcher, Comprehensive Research Organization for Science and Society (Former Director of the National Institute of Science and Technology Policy, Ministry of Education, Culture, Sports, Science and Technology)

1. Arima Sensei's Evaluation Studies and Science and Technology Policies

I first met Dr. Arima in 1980, when he was still a professor at the School of Science, the University of Tokyo and shortly before he was appointed as Director of the Computer Centre. We met not at a physics meeting, his field of academic specialization, but at a literary meeting (he was also a renowned poet of haiku). He was surprised to learn that I had been working in science and technology and we stayed in contact for close to 40 years after that meeting. During that time, I had the opportunity to observe Arima sensei's activities from a science and technology policy perspective. I would like to use the materials at my disposal to describe Arima sensei's accomplishments (regarding science and technology policy).

(1) Citation Index survey (evaluation)

Arima sensei served as Director of the Computer Centre, the University of Tokyo [Note 1]. He started researching evaluations of research achievements when he received funding from the University of Tokyo Centennial Anniversary Project during the tenure of President Hirano. This was a groundbreaking study that was unprecedented in Japan.

At the same time, in the government, the third report (July 30, 1982) of the second Provisional Commission for Administrative Reform (the so-called Doko Commission) raised the importance of research evaluation involving science and technology. Upon receiving this report, "Implementation of the evaluation of large-scale research and development projects within the Science and Technology Council" pointed out by the Administrative Reform Council (the Doko Reform Council) on July 22, 1985. This was during the period when Japan entered its era of research evaluation.

It was in this period that Arima sensei actively published papers and articles on research evaluation using the Citation Index [Note 2]. His was the first scientific evaluation method proposed in Japan.

(1) "Report on Joint Research within the University of Tokyo for the Academic Years 1981 and 1982 using University of Tokyo Centennial Academic Research Incentive Funds" A. Arima and Y. Kaneda, March 1984.
(2) "Measuring Researcher Activity: An Attempt to Evaluate Performance using the Number of Published Papers and Citations" Arima and Kaneda, June 1984, "Kagaku" published by Iwanami Shoten, Publishers.
(3) "Evaluation and Methodology -13- An Attempt to Evaluate Research Achievements in Scientific Fields at Universities, etc. — An Examination of Evaluation Methods and Test Calculations by Computer" January 1985, Japanese Scientific Monthly
(4) "Evaluation of Research Papers Across a Variety of Science Fields" (FY1985-1987) https://kaken.nii.ac.jp/report/KAKENHI-PROJECT-61234002/612340021986jisseki/
(5) "Academic Papers are Valuable Only when Cited" Arima and Kaneda, August 1987, Questioning Japanese Ingenuity <Special feature> "Kagaku Asahi"

Since the publication of these research papers and articles, Citation Index research from the viewpoint of science and technology administration was conducted also in the government. The influence of Arima's papers has been steadily permeating science and technology administration.

The Science and Technology Agency is particularly interested in the Citation Index. This index has received attention from the National Institute of Resources, and the National Institute of Science and Technology Policy, the institutes responsible for science and technology administration, as one indicator of science and technology, and has also been adopted for use in the annual White Paper on Science and Technology (prepared by the Science and Technology Agency and later the Ministry of Education, Culture, Sports, Science and Technology).

Particularly during the 2000s, papers on evaluations using the Citation Index began to appear frequently at academic conferences.

As interest in the Citation Index has grown at academic conferences and research institutes, awareness of the importance of citation frequency in science and technology policy has also increased. This leads to the establishment of an index for citation frequency in the Science and Technology Basic Plan (of the government).

"Through these efforts, we aim to strengthen both the qualitative and quantitative foundation of knowledge. However, evaluations on the quality of academic papers are difficult, so we will focus on papers with high citation frequencies, a method that is widely used as an alternative evaluation indicator. The aim will be to have 10% of the total number of papers in Japan reach the top 10% most cited papers during the Fifth Basic Plan." (The Fifth Science and Technology Basic Plan, approved by the Cabinet on January 22, 2016)

[Note 1] Arima sensei was called to work in a factory in his childhood (during the WWII period), and was proud of high skill at working with his hands. He was proficient with computers and his first paper after entering university was a paper on old-fashioned computers ("Small Relay Computers" S. Kobayashi and A. Arima, Journal of the Physical Society of Japan, 1959-06-05CiNii).

Japan eventually entered the era of big computers. Based on the Science Council of Japan's recommendations on the introduction of big computers, Arima sensei proposed suitable models of the computers to be introduced at the University of Tokyo (primarily concerning their introduction and the development of domestic computers), and proffered ideas on the appropriate shape of a computer center, which led to a debate with Dr. Shu Ono.

As the budget was granted in 1963, the University of Tokyo created a committee for the preparation of the installation of a major computer, and finally selected a HITAC computer. Subsequently, a subcommittee was convened to move forward with establishing its computer center, with Dr. Shu Ono representing the perspective of the Science Council and Arima sensei representing that of users.

Arima sensei thus played an important role during the introduction of computers at the University of Tokyo.

[Note 2] For Arima's papers, especially those related to citation rate, SCI (Science Citation Index) of the Institute for Scientific Information (ISI) is indispensable. It is said that ISI's history began with a paper by Eugene Garfield in the journal, Science. Garfield successfully commercialized the ISI. However, since 1992, ISI has successively transferred its business to Thomson, Thomson Reuters, and Clarivate Analytics.

1955 Garfield mentions the idea of the Impact Factor in Science.

1958 Garfield founds the Institute for Scientific Information (ISI).

In 1961, ISI receives an NIH grant to develop a prototype of the SCI (Science Citation Index), a citation index of natural science papers, which was published the following year. It was later released as Web of Science.

Arima sensei was first exposed to the Citation Index when he joined the State University of New York at Stony Brook as a professor in 1971 and came across it in their physics laboratory. He said, "Every time I returned to Stony Brook, I checked who were citing which of my papers. Over time, I began to see who was working in what fields." It has been utilized in a variety of ways since that time. Incidentally, Thomson has acknowledged Arima sensei is one of the greatest contributors to the spread of the Citation Index in Japan.

Arima sensei said, "I became interested in the Citation Index when I was about 40 years old. I looked to see what other people were working on and used it to guide my work. I could also see which fields were popular and what were not. I looked at the Citation Index in my 40s out of personal interest, but after I turned 50 I became director of the Computer Centre, president of the Physical Society of Japan, and a member of the Science Council, so I had to consider the future of Japan. I used the Citation Index to see if I could transfer what I was doing directly to policy." (*Science and Technology Evaluations and Reforms: A Critical Biography of Akito Arima and Ideas on Science and Technology Policy* Part 3, Chapter 2, "An Interview with Akito Arima on Policies and Reform" 2019)

(2) Development of evaluation studies

As already mentioned, Arima sensei conducted investigations of the Citation Index upon assuming the role of Director of Computer Centre. He then found that the improvement of productivity and quality in research papers required, as a matter of policy, (1) improving research environments at university faculties and (2) training young researchers. Consequently, Arima sensei expanded his agenda to include university evaluations and self-inspections, university enrichment, and fostering young researchers.

In doing this, Arima sensei himself served as Director of the Computer Centre (April 1981), Dean of the School of Science at the University of Tokyo (April 1985), Special Advisor to the President of the University of Tokyo (April 1987), President of the University of Tokyo (April 1989), and President of RIKEN (October 1993), each of which led to policy proposals at various stages of his career.

These are all intricately interrelated. I would like to review some of Arima sensei's papers from these periods and examine how they connect with management issues at the university and research institute.

○ "Creating Better Educational and Research Environments" ("Kohou" (PR magazine of the University of Tokyo), Vol. 18, No. 1, June 1986)

This paper was written shortly after he moved from Director of the Computer Centre to Dean of the School of Science. The report largely expresses the need to improve research environments of universities such that they are equivalent to those of research institutes, concerns over turning the university observatory into an outside research center, and opposition to the separation between teaching and research. Based on survey analysis of research papers, the paper advocates for the improvement of university research environments and states that universities need to use scientific data to appeal to society at large.

○ "Preface: For a Promising Future for the School of Science ("Kohou", Vol. 18, supplementary issue, January 1987)

This article was published in a supplementary issue of the School of Science's Public Relations Magazine. During Arima's tenure as Dean of the School of Science, in June 1986 and January 1987 faculty and others were asked to write their thoughts, suggestions, and opinions to promote discussion on the future of the School of Science. "Creating a Better Research Environment" appeared in the June 1986 issue and this "Preface: For a Promising Future for the School of Science" appeared in the January 1987 issue. These would ultimately lead to the University of Tokyo White Paper.

○ "Researcher Self-Inspections and Establishing an Evaluation Organization" December 1991, Keidanren Monthly Report

This article is an interview with Arima sensei after he had assumed the position of President and was published as a "<Special Feature> Issues Involving Science and Technology in Japan". Keidanren had made a proposal for the establishment of a 21st-century R&D system (October 1991). Although some complained that such a move should have been issued a little earlier, he had consistently insisted that self-inspection and evaluation are the responsibility of the university. This was followed by the publication of the University of Tokyo White Paper (December 1992) and by the first external evaluation conducted among all national, public, and private universities in Japan, the "Report from the Evaluation Committee of the Department of Physics, School of Science, University of Tokyo" (March 1993).

○ **Publication Address "The University of Tokyo: Current State and Challenges" December 1992**

This is the first report (the so-called "University of Tokyo White Paper") compiled by Arima sensei as President of the University of Tokyo as a self-inspection and evaluation of the University of Tokyo (it was published four times before the University was converted into a corporation). Arima sensei wrote the "Publication Address" and the detailed "Future Prospects" for these reports. "Future Prospects" describes management policies as the President of the University of Tokyo and "Publication Address" discusses the reasons for self-inspection and evaluation.

○ **"The National University Renaissance/University of Tokyo: Concept of the University with Emphasis on Graduate Schools" "The National University Renaissance: Centers of a Revitalized 'Knowledge' 1" A. Arima [extra] edition, Dobunshoin, March 1993**

"The National University Renaissance" analyzes issues facing national universities following the third postwar crisis and attempts to find ways to resolve them. Presidents at national universities were asked to speak about the history, current state, and future aspirations of their universities. For the University of Tokyo, President Arima wrote about "The Concept of the University with Emphasis on Graduate School" and "Current State and Challenges (University of Tokyo White Paper)". Here, he presented not only the concept of such a university in university reforms, but also the need for approval from the larger society was mentioned.

○ **"A Discussion on 'Research Evaluation"', February 1996, Japanese Scientific Monthly**

Appearing in the "Special Issue: Academic Research and Evaluation" of Japanese Scientific Monthly, this paper describes attempts for evaluation at the Computer Centre, the University of Tokyo and lists the diversity of evaluation, evaluation indicators, the necessity of evaluations, the importance of evaluations by outsiders, and the various problems with evaluation systems, as well as future issues. As an expert on evaluation, he comprehensively addressed issues in the evaluation and included a section with his description of the importance of outside evaluations.

(Unlike the papers mentioned in section (1) written by Arima alone or by himself and Kanada, some of the papers and articles listed just above were the product of organizations headed by Arima sensei (Dean of the School of Science and President of the University). They are the following articles (see the list below): (a) "Special Feature: Prospects for the Future" and other articles were edited by the Public Relations Office of the School of Science, the University of Tokyo (Yukimi Sasaki and others were editors and Arima sensei authored two articles), (b) "The University of Tokyo: Current Status and Issues" (the so-called "University of Tokyo White Paper") was compiled from each part in the University of Tokyo (Hiroshi Shiono and others were editors. Arima sensei wrote the "Publication Address" and "Future Prospects"), and (c) "Evaluation of the Department of Physics" was conducted in the School of Science, the University of Tokyo. It is imagined that Arima sensei's contributions were significant, as can be inferred from the various earlier papers mentioned above. It is believed that the statements of those involved in article (c) came about under the leadership of Arima sensei as the President.

(a) 1986–87 Solicited internal and external opinions about the School of Science in the "Public Relations Magazine from the School of Science, University of Tokyo" (TOUKYOU DAIGAKU RIGAKUBU KOUHOU)
(b) December 1992 "The University of Tokyo: Current Status and Issues" (The first University of Tokyo White Paper (TOUDAI HAKUSHO). There were four subsequent editions.)
(c) March 1993 "Report from the Evaluation Committee of the Department of Physics, School of Science, University of Tokyo"

(3) Expanding evaluations

There have been a growing number of papers on fundamental research/science and society, in Arima sensei's later activities covered by this article. A few representative examples are selected here because these papers deal with topics related to administrative reform (particularly the division between fundamental and developmental reforms, which became the standard for ministerial reorganizations), which will be discussed later.

○ **"Between Science and Technology" March 1993, Bungeishunju weekly magazine**

In this article, Arima sensei considers research at universities from a broad perspective. The aforementioned March 1993 "The National University Renaissance/University of Tokyo: Concept of the University with Emphasis on Graduate Schools 1" article in "The National University Renaissance" traced the history of the University of Tokyo. In this article, he examines the academic world from a broader perspective, first describing the creation of schools and departments at German, British, Italian, American, Japanese and Chinese universities in detail.

○ **Scientific Fundamentals and Technological Developments (Lecture) March 1995 Science & Technology Journal/The Japan Foundation of Public Communication on Science and Technology**

He continues the discussion in "Scientific Fundamentals and Technological Developments". The purpose of this paper is the same as the Bungeishunju paper, but after stating, among other things, that it takes time for fundamentals to be linked to technology even while scientific fundamentals advance alongside technology, that there are national characteristics in the way scientific fundamentals and technology progress. He further states that the West currently reflects giving excessive priority to fundamentals and giving low priority to applications. He concludes with the following thesis.

As mentioned above, he noted in the conclusion of the Citation Index study that increasing paper productivity requires (1) increasing university budgets and (2) securing young researchers (of these two issues, he advocated for the former with particular emphasis from the beginning and wrote many papers on the subject.)

What makes Dr. Arima unique is that he achieved the policies described in this paper through focused action and political influence based on his positions as President of the University of Tokyo and President of RIKEN and it seems fair to describe this as an unprecedented achievement among scholars. This paper will now discuss two factors that significantly impacted politics: budget expansions and increasing the number of young researchers. He wrote numerous papers on these subjects, so for the sake of brevity this paper will only focus on the most representative ones.

(1) Expanding budgets

○ "A Crisis Facing National Universities" IDE, October 1989 (later included in "Tales of University Poverty")

This well-known article, which was later included in "Tales of University Poverty," was published in the IDE (a magazine published by the Institute for Development of Higher Education), shortly after Arima sensei was appointed president of the University of Tokyo. It shocked many intellectuals upon publication. In journalism, this article led to a special feature on "University Bankruptcy" in Nikkei Business (December 17, 1990), a special feature on "The Deterioration of National Universities" in AERA (May 28, 1991), a special feature "Is There Any Way to Save Universities from Bankruptcy?" in Nikkei Business (July 22, 1991), resulted in The Sankei Shinbun beginning a series of articles titled "Questioning Universities" (June 24–December 26, 1991), and caused the dilapidation of universities to become widely known throughout society.

A distinctive feature of Arima sensei's activities is that his articles not only offer recommendations, but also paved the way for actual budget increases by encouraging action from universities, politicians, the government, journalism, the business community, and other groups.

One of the largest responses to Arima sensei's activities and papers was a lecture held at the Science Council of Japan on June 6, 1991, titled "Can Japan's Academic Research Environment Cope with the 21st Century?" The series "Questioning Universities" in Sankei Shinbun (see above) details the June 1991 Arima Lecture, describing that "The shocking report from Arima... points out the current state of affairs that is eating away at our heritage".

In addition, the Japan Business Federation (KEIDANREN) issued statements and proposals such as the "Desire to Establish Research and Development Systems for the 21st Century" (October 8, 1991) and the "Proposals for Strengthening University Science and Engineering Functions" (April 28, 1992).

The government also decided to double public investment in the early stages as part of its General Guidelines for Science and Technology Policies (Cabinet decision) for a permanent plan (in this progression, government investment in research and development (a total of 17 trillion yen between 1996 and 2000) was later specified in the Science and Technology Basic Plan based on the Science and Technology Basic Law, but the second phase of the basic plan showed a target aiming towards doubling that amount (a total of 24 trillion yen between 2001 and 2005)).

(2) Increasing the number of young researchers

○ "Characteristics of Japanese Education" A. Arima (public statement), Toyama Prefectural Citizens' College for Life-long Learning, March 1995

Although published in 1995, this paper is a July 22, 1994 lecture on education given at the Toyama Prefectural Citizens' College for Life-long Learning. In this context, it discusses increasing the budget for higher education and the number of PhD students. One part of the publication is titled "The Plan to Double Doctorates". While predicting that "the number of people in Japan with PhDs will soon double", Arima sensei points out that "the problem is getting industries and the government to take in those people with PhDs". This paper and lecture were presented at the very early stages of discussions in the literature on increasing the number of PhDs.

Multiple other papers on young researchers made various sectors aware of the need for science and technology policies that would increase the number of PhDs and improve environments for them. In response, the "Plan to Support 10,000 Doctoral Graduates" was conceived and put into effect.

An examination of studies of the Science and Technology Basic Law was conducted between 1994 and 1995. Although the proposal to submit the bill during a June 1955 ordinary session of the Diet was once abandoned, the bill was immediately submitted later on October 27, passed on November 8, and went into effect on November 15. Based on this, the "Science and Technology Basic Plan" was formulated in accordance with the Science and Technology Basic Law.

On June 30, LDP Policy Research Council chairman Koichi Kato led the three ruling parties, the LDP, the Japan Socialist Party, and the New Party Sakigake, in an announcement of "immediate priority measures to be added based on an agreement between the three parties", which included the "Plan to Support 10,000 Doctoral Graduates".

Following this agreement by the ruling parties, each ministry and agency submitted budget requirements and the plan was launched. Subsequently, the Science and Technology Agency, the Ministry of Education, Culture, Sports, Science and Technology, the Ministry of Agriculture, Forestry and Fisheries, the Ministry of Economy, Trade and Industry, and others enacted budgetary measures and the 1996 budget was compiled and put into effect. The number of doctoral graduates reached 10,000 during the 1999 fiscal year.

2. The Administrative Reform Council

Based on the implementation of Citation Index research and evaluation studies, Arima sensei has had a significant impact on science and technology policy. Administrative reform could even be viewed as a summary of his impact because he was a member of the Administrative Reform Council during the Hashimoto administration, which determined the subsequent administrative structure in the field of science and technology. Of course, a political topic of this magnitude could not be decided by Arima sensei alone and many political compromises had to be made. Rather, it should be seen as the result of intense negotiations between Prime Minister Hashimoto and Arima sensei. I would therefore like to trace the tumultuous history of the establishment of the new Ministry of Education, Culture, Sports, Science and Technology (MEXT) and look at the vision for science and technology administration presented by Arima sensei during that time. Future reviews of the administrative structure of science and technology should consider these points.

(1) The Background of Administrative Reform

At the start of the second Hashimoto Cabinet (established November 7, 1996. The reorganized Cabinet was established in 1997 and continued until July 30, 1998), Prime Minister Hashimoto proposed six major reforms: administrative reforms, fiscal structural reforms, economic structural reforms, financial system reforms, social security structural reforms, and education reforms. The Administrative Reform Council was established to carry out these reforms and was established in the Prime Minister's Office from November 21, 1996, to June 30, 1998, as an organization (chaired by Prime Minister Hashimoto) to study the reorganization of central ministries and agencies.

Central ministries and agencies were reorganized based on the final report from this Administrative Reform Council. The Basic Law on the Administrative Reform of the Central Government was enacted from this final report, which led to the consolidation and reorganization of one agency and 22 ministries into one agency and 12 ministries.

Based on the 1997 Administrative Reform Council final report, in 2001 the Science and Technology Agency that had been established in 1956 merged with the Ministry of Education to form the Ministry of Education, Culture, Sports, Science and Technology. While this fact is well known, details of the circumstances surrounding it are less so. In particular, there was a proposal to have the Science and Technology Agency become the Ministry of Science and Technology, but the agency ultimately merged with the Ministry of Education to become the Ministry of Education, Culture, Sports, Science and Technology, while the nuclear administration portion of science and technology administration was transferred to the Cabinet Office (safety regulation administration was transferred to the Cabinet Office and later transferred to the Ministry of the Environment after the 2011 East Japan Earthquake).

As Commissioner Arima in particular played a central part in defining the role of science and technology administration at the 1997 Administrative Reform Council, I decided to explore the impressions and opinions of Commissioner Arima while presenting as many objective facts as possible.

- **Intensive Deliberation Day 1**

After the chairperson's (Commissioner Fujita) explanation of the draft proposal, the first topic of discussion was science and technology administration. Discussion took up all remaining time of the first day of intensive deliberations.

First, the chairperson's draft and arguments from Commissioner Arima and others in the deliberations strongly advocated for the organizational separation of top-down R&D (particularly in industries that had yet to move, like nuclear fusion, space exploration, new energies, etc.) and bottom-up R&D. This was specifically in support of Chairperson's Proposal A (a proposal for the Ministry of Science and Technology and the Ministry of Education being organized in parallel), which generally received majority support at the meeting.

Prime Minister Hashimoto strongly criticized the trend among the meeting members behind the Ministry of Science and Technology proposal. His criticism was based on a recent accident at the Power Reactor and Nuclear Fuel Development Corporation's Monju plant and the corporation's handling of the incident.

Commissioner Arima's response to the Prime Minister's fierce criticism of the "Ministry of Science and Technology" proposal led to the rapid creation of Chairperson's Proposal B (the Yokogushi (lateral joint) organization proposal). As a concrete context, it indicated the specific need for an organization that would handle both science-and-technology research and academic research as a single entity, which received approval from the Prime Minister and others. The Council for Science and Technology was proposed as the name for this new organization.

Specific ideas invoked by the name Council for Science and Technology were discussed and proposals were made regarding the Council's secretariat (Cabinet Office, Cabinet Secretariat), character, and membership. Incidentally, the Council for Science and Technology present at that time was positioned inside the Prime Minister's Office and was managed by a joint secretariat of the Science and Technology Agency and the Ministry of Education.

- **Intensive Deliberation Day 2**

Discussion on the character of the Council for Science and Technology continued and it was pointed out that the organization should be able to view policies that consider the entire government as a whole. It was also pointed out that the organization in charge of science and technology must have the same characteristics as those for academia, education, and culture and it was agreed that a ministry integrating science and technology, academia, education, and culture would be established. The interim report tentatively designated it as the Ministry of Education, Science and Technology.

(2) Documentation by Arima in administrative reforms

Within this context, I will examine documents submitted by Arima sensei. The first document is an opinion within his general administrative reforms that were submitted before the intensive deliberations on the Science and Technology Agency.

• On Creating a Nation of Science and Technology (Opinion of Akito Arima, Commissioner)

1. Administrative reform must reflect the spirit of a Nation of Science and Technology provided by the Science and Technology Basic Law and the Science and Technology Basic Plan.
2. Activity at national research institutes and special corporate research institutes must be increased.
3. The roles and conditions of existing research institutes must first be examined.
4. We should strive to build the most effective organizations to enhance activities that move Japan toward becoming a nation of science and technology, including the integration of research institutes.

Arima sensei submitted the following document when the conclusion was presented for the integration of several ministries and the founding of the Ministry of Education, Science and Technology (later changed to the Ministry of Education, Culture, Sports, Science and Technology during the legislative phase). This extract only contains the section about the Ministry of Education, Science and Technology.

Opinion (memo) on the Administrative Reform Council's intensive deliberations (November 4, 1997, Akito Arima, Commissioner)

1. The Ministry of Education, Science and Technology must act as a central hub that organically connects academic research at universities (including those in the humanities and social sciences) with the promotion of large-scale projects at national research institutes and special corporations. The Ministry must also continuously promote comprehensive research and development.
2. This will not only stimulate research and development in Japan but also foster education and research at universities and graduate schools.
3. If any of these efforts yield results that can be applied and commercialized soon, they should be transferred to the appropriate ministries and agencies for implementation.
4. For example, in the space field, if technological developments based on the H-IIA rocket currently under development that is intended for the global satellite launch market are used to lower future costs, such developments would be included in industrial administration. For nuclear energy, the various experiments and research using the prototype fast breeder reactor Monju should be conducted as part of science and technology administration, but the development of prototype reactors for the future commercialization of fast breeder reactors should be included in industrial policy.

Research and development for space and research and development systems for Monju in the Ministry of Education, Science and Technology should be organized in such a way that R&D personnel can engage in research without relying on other organizations.

(3) Arima's Draft for Administrative Reform

The above sections overview the developing circumstances behind the Administrative Reform Council and comments from Arima sensei, but he created a draft of science and technology reforms in preparation for the Administrative Reform Council. The content was likely written after the Administrative Reform Council held hearings with various ministries and agencies and before the intensive deliberations on the proposal to establish the Ministry of Education, Culture, Sports, Science and Technology and the Council for Science and Technology.

The document is "Science and Technology in Japan and Aspirations in Administrative Reform" by Akito Arima (The Journal of Science Policy and Research Management, 1997, No. 3 and 4). It consists of five chapters, which are organized as follows.

Chapter 1: The Capabilities and Issues of Science and Technology in Japan
Chapter 2: Improving Japanese Science and Technology
Chapter 3: Academic Research and Developmental Research
Chapter 4: Vertically Divided Administrations, positive and negative aspects
Chapter 5: Aspirations for Science and Technology Administration

Chapters 1 and 2 describe the premise for writing the following argument, based on Arima sensei's recognition of characteristics from a viewpoint of science and technology administration indicators. Chapter 3 entails the differences between academic research and developmental research obtained from the evaluations, which is a pre-requisite for the following examinations into reorganizing ministries and agencies. As concrete proposals, Chapter 4 describes the problems of vertically divided administrations and Chapter 5 suggests the ideal form of science and technology administration based on these considerations.

To be concrete, roles in contributions to Japan's basic science and basic technology are to be clarified, and these contributions should be enlarged. For this purpose, the following points are mentioned: (1) the goal of strengthening the Council for Science and Technology and (2) describing that Japan and Russia adopt different principles between basic research in development and application and basic research for academic studies — while evaluating the degree of vertical division in government offices, followed by simulations on the reorganization of several ministries and agencies. Although it does not immediately suggest what shape these ministries should take, the document can be called a road map for concepts on reorganizing ministries and agencies.

Interestingly, this document summarizes all of Arima sensei's papers and recommendations on science and technology administration. These materials allow us to verify the level of consistency in Arima sensei's thoughts on science and technology policy, and, in turn, how scientific he was in creating science and technology policies. A list of Arima sensei's major papers over the past 40 years and materials showing the systematization of his work are attached as references. They indicate that Arima sensei was an unparalleled science and technology policymaker.

[Reference Material] Arima sensei's major papers on the science and technology policy

Policy	Computer Centre Director April 1981–April 1985	Dean, School of Science April 1985–April 1987	Special Advisor to the President of the University of Tokyo April 1987–April 1989	President of the University of Tokyo April 1989–March 1993	President of RIKEN October 1993–May 1998	Notes [Note 1]
1. Evaluations (1) Investigative research into the Citation Index	○ "Measuring Researcher Activity" June 1984, KAGAKU, Iwanami ○ "Evaluation and Methodology -13-" January 1985, Japanese Scientific Monthly		○ "Academic Papers are Valuable Only when Cited" August 1987, KAGAKU ASAHI			Chapter 1
(2) Evaluation and self-inspection		○ "Creating Better Educational and Research Environments" (Public Relations, June 1986)		○ "Researcher Self-Inspections and Establishing an Evaluation Organization" December 1991 Keidanren Monthly Report. ○ The University of Tokyo: Current State and Challenges, December 1992	○ "A Discussion on 'Research Evaluation'", February 1996, Japanese Scientific Monthly	Chapter 1

(*Continued*)

Policy	Computer Centre Director April 1981–April 1985	Dean, School of Science April 1985–April 1987	Special Advisor to the President of the University of Tokyo April 1987–April 1989	President of the University of Tokyo April 1989–March 1993	President of RIKEN October 1993–May 1998	Notes [Note 1]
(3) Science and technology				o The University of Tokyo: Current State and Challenges, December 1992, "Future Prospects" o "Between Science and Technology" March 1993, Bungeishunju	o "Scientific Fundamentals and Technological Developments" (Lecture) March 1995, Science & Technology Journal/The Japan Foundation of Public Communication on Science and Technology	Chapter 3
2. Expanding university budgets	o A. Arima and S. Yonemoto, "University impoverishment and weakening international exchanges" August 1984, KAGAKU ASAHI	o "On Assuming the Position of Dean" (Public Relations, June 1985)	o "Graduate School Reform — Reviving Universities" December 1988, KAGAKU, Iwanami	o "A Crisis Facing National Universities" IDE, October 1989 (later included in "Tales of University Poverty")		Chapter 2

(*Continued*)

(*Continued*)

Policy	Computer Centre Director April 1981–April 1985	Dean, School of Science April 1985–April 1987	Special Advisor to the President of the University of Tokyo April 1987–April 1989	President of the University of Tokyo April 1989–March 1993	President of RIKEN October 1993–May 1998	Notes [Note 1]
3. Fostering young researchers					○ "Characteristics of Japanese Education" A. Arima (public statement), Toyama Prefectural Citizens' College for Life-long Learning, March 1995 ○ "On the development of young scientists and technical personnel" November 1998, Technology and Economy	Chapter 2

(*Continued*)

Policy	Computer Centre Director April 1981–April 1985	Dean, School of Science April 1985–April 1987	Special Advisor to the President of the University of Tokyo April 1987–April 1989	President of the University of Tokyo April 1989–March 1993	President of RIKEN October 1993–May 1998	Notes [Note 1]
4. Internationalization [Note 2] (Before becoming Director of the Computing Center:) o "The Haneda Shock?" Public Relations, October 1973	o "Valuing Small-Scale International Exchange" March 1983, KAGAKU, Iwanami Shoten	o "Creating Better Educational and Research Environments" (RIGAKUBU KOHOU, June 1986)		o "Interdisciplinarity, Globalization, and a Coffee Break" (Lecture) May 1992, Science & Technology Journal/The Japan Foundation of Public Communication on Science and Technology		Chapter 5

Note 1: The remarks come from the corresponding chapters in "Science and Technology in Japan and Aspirations in Administrative Reform" (The Journal of Science Policy and Research Management, 1997, No. 3 and 4).
Note 2: "Internationalization" is an important element of Arima sensei's science and technology policies but could not be fully described in the text. This table only lists the names of papers.

https://doi.org/10.1142/9789819813353_0011

Effective Operators for Valence Space Calculations from the *Ab Initio* No-Core Shell Model

Zhen Li*, N. A. Smirnova*, A. M. Shirokov†, I. J. Shin‡,
B. R. Barrett§, P. Maris¶, and J. P. Vary¶

*LP2IB (CNRS/IN2P3 — Université de Bordeaux), 33175 Gradignan cedex, France
†Skobeltsyn Institute of Nuclear Physics, Lomonosov Moscow State University, Moscow 119991, Russia
‡Rare Isotope Science Project, Institute for Basic Science, Daejeon 34000, Republic of Korea
§Department of Physics, University of Arizona, Tucson, Arizona 85721, USA
¶Department of Physics and Astronomy, Iowa State University, Ames, Iowa 50011, USA

In recent years, a remarkable progress has been achieved in developing novel non-perturbative techniques for constructing valence space shell model Hamiltonians from realistic internucleon interactions. One of these methods is based on the Okubo–Lee–Suzuki (OLS) unitary transformation applied to no-core shell model (NCSM) solutions. In the present work, we implement the corresponding approach to solve for valence space effective electromagnetic operators. To this end, we use the NCSM results for $A = 16-18$, obtained at $N_{\max} = 4$, to derive a charge-dependent version of the effective interaction for the *sd* shell, which allows us to exactly reproduce selected NCSM spectra of ^{18}O, ^{18}F and ^{18}Ne within the two valence nucleon space. We then deduce effective single-particle matrix elements of electric quadrupole ($E2$) and magnetic dipole ($M1$) operators by matching them to the electromagnetic transitions and moments for ^{17}O and ^{17}F from the NCSM at $N_{\max} = 4$. Thus, effective $E2$ and $M1$ operators are obtained as sets of single-particle matrix elements for the valence space (*sd* shell) which allow us to reproduce the NCSM results for $A = 17$ exactly. Systematic comparison of a large set of *sd* shell results on quadrupole and magnetic dipole moments and transitions for $A = 18$ using effective $E2$ and $M1$ operators that we derive from the full NCSM calculations demonstrates a remarkable agreement.

1. Preamble

The authors wish to express their appreciation for the opportunity to participate in this honorary volume for Prof. Akito Arima. Indeed, Prof. Arima led many pioneering works in the microscopic theory of nuclear forces and effective interactions. Several of the authors had the great pleasure to discuss these fundamental physics topics with him and to receive his sage advice. With this contribution, we propose extensions to microscopic theory of valence interactions including electromagnetic effective operators that follow lines of thinking that can trace their origin to the fundamental works of Prof. Akito Arima.

2. Introduction

Outstanding progress in nuclear many-body computations and in construction of highly accurate realistic internucleon interactions during the last couple of decades has made it possible to describe light nuclei and intermediate-mass nuclei near closed shells from first principles.[1–7] Among all these *ab initio* approaches, the no-core shell model[1] (NCSM), being a full configuration-interaction type method, solves the eigenvalue problem by exact diagonalization of the Hamiltonian matrix constructed in a large-dimensional many-body harmonic-oscillator basis. The latest state-of-the-art computations, such as those made possible by the Many-fermion dynamics code for nuclei (MFDn),[8–12] are capable of providing converged or nearly-converged solutions for nuclei up to $A \sim 18$, setting benchmarks for other nuclear structure models.[7,13,14] However, for heavier nuclei the eigenproblem has still to be solved in a more restricted model space, typically for a few valence nucleons occupying one or two harmonic oscillator shells beyond a closed-shell core. This approach with an assumed core of passive nucleons is known as a traditional interacting shell-model (see, e.g., Refs. 15–17). Severe truncation of the model space requires construction of an appropriate *effective Hamiltonian* and *effective electroweak operators*. From the 1960's, for more than fifty years, the many-body perturbation theory (see, e.g., Refs. 18, 19 and references therein) was the dominant approach to deal with the microscopic derivation of effective Hamiltonians. It was, however, noticed that when derived from two-nucleon interaction, an effective valence-space Hamiltonian could not compete in precision with empirically obtained effective Hamiltonians,[20,21] mainly because of its deficient monopole part.[22] One of possible reasons was identified as the absence of three-nucleon forces.[23]

New non-perturbative techniques for deriving effective valence-space interactions have been developed during the last decade, such as similarity-renormalization group (SRG) inspired methods,[24] or application of Okubo–Lee–Suzuki (OLS) unitary transformation[25–27] to either a NCSM solution[28–30] or to a solution found within the coupled cluster theory.[31,32] Recent inclusion of three-nucleon forces via different approaches[19,24,31,33–35] seems to cure the deficiency of the monopole part of the effective interactions and brings encouraging improvement to the theoretical description.

Construction of appropriate effective electroweak operators to be used in a valence space has been another long-standing challenge. Since the earlier work by Arima and collaborators,[36,37] the importance of accounting for the restricted model space has been realized. Many efforts were devoted to the construction of effective electromagnetic and Gamow–Teller operators both microscopically[38–43] or empirically by a fit of the optimum form of a corresponding operator to the data.[20,44–47]

In this work, we use the NCSM solutions for $A = 16$–18 nuclei in order to construct an effective interaction and effective $E2$ and $M1$ operators for the valence *sd*-shell model space. We construct here a charge-dependent version of the effective interaction presented earlier[30] based on the Daejeon16 nucleon-nucleon (*NN*) potential,[48] using ^{18}O, ^{18}F and ^{18}Ne to derive neutron-neutron, neutron-proton and proton-proton two-body matrix elements (TBMEs). Effective $E2$ and $M1$ single-particle matrix elements are obtained by matching them to the corresponding NCSM results for ^{17}O and ^{17}F obtained with basis spaces including all many-body excitation oscillator quanta up to $N_{\max} = 4$. These constructed effective operators are tested on a large number of transitions and moments for $A = 18$ nuclei.

3. Microscopic Effective Interaction from the NCSM

The derivation of valence-space effective interactions from the NCSM is described in detail in Refs. 29, 30 and references therein. We start with a translationally-invariant non-relativistic Hamiltonian for A point-like nucleons, containing relative kinetic energies and NN interactions:

$$H = \sum_{i<j}^{A} \frac{(\vec{p_i} - \vec{p_j})^2}{2mA} + \sum_{i<j}^{A} V_{ij}^{NN}. \tag{1}$$

Here m is the nucleon mass (approximated here as the average of the neutron and proton mass), $\vec{p_i}$ are nucleonic momenta and V_{ij}^{NN} denotes the bare NN interaction. To our chosen Daejeon16 NN interactions we have added the two-body Coulomb interaction between the protons.

The eigenproblem for H is solved by diagonalizing the Hamiltonian matrix, constructed in a many-body spherical harmonic-oscillator basis, using the MFDn code.[8–12] All nucleons are considered as active particles. The model space is defined by two parameters: (i) by a given oscillator quantum, $\hbar\Omega$, and (ii) by the maximum total many-body harmonic-oscillator excitation quanta, $N_{\max}$. This means that we retain only many-body configurations satisfying the condition $\sum_{i=1}^{A}(2n_i + l_i) \leq N_{\min} + N_{\max}$, where n_i is the single-particle radial harmonic-oscillator quantum number, l_i is the single-particle orbital angular momentum quantum number, while $N_{\min}$ is the minimal many-body oscillator quanta satisfying the Pauli principle for the chosen nucleus. The use of the harmonic-oscillator basis allows us to remove states with center-of-mass excitations by using the Lawson method.[49]

In the present study, we select $\hbar\Omega = 14\,\text{MeV}$, which is close to the empirical value[50] in the vicinity of $A = 16$, and which was also adopted in our previous work.[30]

In principle, one may apply the OLS transformation[25–27] to a "cluster" of nucleons to develop an effective interaction for the chosen NCSM space (the "P-space").[1] This is referred to as the cluster approximation. A compact presentation of the OLS method is presented in Ref. 51. The goal is to achieve an effective interaction at fixed cluster size (conventionally two-nucleon or three-nucleon) for which the NCSM calculations converge more rapidly with increasing P-space than using the original (non-transformed) interaction.

In this work we adopt the Daejeon16 *NN* interaction[48] which is based on a chiral N3LO interaction[52,53] and which has been softened already by an SRG transformation[54,55] that improves convergence. Hence, following our procedure[30] we elect to use Daejeon16 directly, without an additional OLS transformation for our NCSM calculations.

For this work, we adopt the NCSM calculations at $N_{\max} = 4$ for the $A = 16, 17$ and ^{18}F. In addition, to obtain a charge-dependent valence effective interaction for the first time, we perform new NCSM calculations at $N_{\max} = 4$ for ^{18}O and ^{18}Ne.

The resulting NCSM eigenvalues and eigenvectors are then employed to define a new effective interaction for the $N_{\max} = 0$ space (the "P'-space") through the application of the OLS method. The NCSM results dominated by *sd*-shell configurations are appropriate for adoption to define the P'-space. That is, from the NCSM results for $A = 16$–18 we apply the OLS method to obtain an effective valence space (*sd*-shell) Hamiltonian consisting of a core, a single-particle and a two-particle interaction.[29,30] This effective interaction is guaranteed to reproduce the NCSM results for the ^{16}O core and the states of $A = 17$ and 18 selected to be included in the definition of the P'-space.

From the $A = 18$ NCSM results we select the number of eigenpairs (eigenvalues and eigenvectors) for each J governed by the number of needed two-body matrix elements. Let us label the total number of eigenpairs (adding up the dimensions for each J) as d_{nn}, d_{pp} and d_{np} for valence space configurations of ^{18}O, ^{18}Ne and ^{18}F respectively. These are the states, selected from among the NCSM candidates, which have the largest valence space ($0\hbar\Omega$ in the present study) probability. It should be noted that these are not necessarily the states with the lowest NCSM eigenvalues. The exact dimensions are $d_{nn} = 14$ for ^{18}O, $d_{pp} = 14$ for ^{18}Ne, and $d_{np} = 28$ for ^{18}F, since it includes both $T = 0$ and $T = 1$ states.

The OLS transformation leads to a two-nucleon effective Hamiltonian defined in the $N_{\max} = 0$ model space by its TBMEs which in the angular momentum J and isospin T coupled form read $\langle ab; JT|H_{\text{eff}}|cd; JT\rangle$ where a, b, c, d denote a full set of single-particle spherical quantum numbers which label an orbital, e.g. $a = (n_a, l_a, j_a, t_{z_a})$, with t_z distinguishing protons and neutrons.

To obtain the one-body neutron and proton contributions (single-particle energies, ϵ), we subtract the ground-state energy of ^{16}O (core energy, E_{core}) from ^{17}O and ^{17}F, respectively. To get the TBMEs of the effective interaction in the valence space, we have to subtract this core energy and the one-body contributions from diagonal TBMEs of the OLS-derived effective valence space Hamiltonian H_{eff},

$$\langle ab; JT|V_{\text{eff}}|ab; JT\rangle = \langle ab; JT|H_{\text{eff}}|ab; JT\rangle - E_{\text{core}} - (\epsilon_a + \epsilon_b), \tag{2}$$

while for non-diagonal TBMEs, we have

$$\langle ab; JT|V_{\text{eff}}|cd; JT\rangle = \langle ab; JT|H_{\text{eff}}|cd; JT\rangle. \tag{3}$$

Table 1. Neutron-neutron, proton-neutron ($T = 1$) and proton-proton TBMEs of the derived effective interaction $\langle ab;\ J, T = 1|V_{\text{eff}}\ |cd;\ J, T = 1\rangle$ in the *sd* shell from Daejeon16. Note that the Coulomb interaction is included in the NCSM calculations. See text for details.

$2j_a$	$2j_b$	$2j_c$	$2j_d$	J	T	V_{nn}	$V_{pn}^{T=1}$	V_{pp}
1	1	1	1	0	1	−2.018	−1.989	−1.536
1	1	3	3	0	1	−0.713	−0.769	−0.797
1	1	5	5	0	1	−1.446	−1.422	−1.344
3	3	3	3	0	1	−1.350	−1.317	−0.847
3	3	5	5	0	1	−2.377	−2.500	−2.582
5	5	5	5	0	1	−2.618	−2.590	−2.046
1	3	1	3	1	1	0.137	0.136	0.482
1	3	3	5	1	1	−0.151	−0.151	−0.152
3	5	3	5	1	1	0.069	0.062	0.441
1	3	1	3	2	1	−0.500	−0.500	−0.116
1	3	1	5	2	1	−1.519	−1.498	−1.438
1	3	3	3	2	1	−0.090	−0.099	−0.089
1	3	3	5	2	1	0.459	0.452	0.428
1	3	5	5	2	1	−1.183	−1.164	−1.120
1	5	1	5	2	1	−1.358	−1.341	−0.908
1	5	3	3	2	1	−0.854	−0.853	−0.828
1	5	3	5	2	1	0.366	0.360	0.336
1	5	5	5	2	1	−0.549	−0.542	−0.509
3	3	3	3	2	1	−0.187	−0.185	0.180
3	3	3	5	2	1	0.861	0.855	0.824
3	3	5	5	2	1	−0.818	−0.837	−0.836
3	5	3	5	2	1	−0.108	−0.116	0.267
3	5	5	5	2	1	0.252	0.259	0.263
5	5	5	5	2	1	−1.140	−1.132	−0.695
1	5	1	5	3	1	0.435	0.433	0.799
1	5	3	5	3	1	0.310	0.302	0.289
3	5	3	5	3	1	0.186	0.169	0.508
3	5	3	5	4	1	−1.628	−1.621	−1.196
3	5	5	5	4	1	1.388	1.371	1.323
5	5	5	5	4	1	−0.212	−0.211	0.184

In this way, from NCSM calculations for ^{18}O, ^{18}Ne and ^{18}F, we derive a set of neutron-neutron, proton-proton and proton-neutron TBMEs, respectively, to be used in the valence space shell-model calculations.

The resulting core energy, proton and neutron single-particles energies, as well as $T = 0, 1$ proton-neutron TBMEs obtained from Daejeon16 have been published in an earlier study (see column "bare" of Table II and column "bare" of Table XI in Ref. 30), while the neutron-neutron and proton-proton TBMEs obtained specifically for this work. For the convenience of the reader, we present the whole set of TBMEs in Tables 1 and 2. By construction, the *sd*-shell spectra of ^{18}O, ^{18}Ne and ^{18}F obtained with the derived effective Hamiltonian exactly reproduce the results of the NCSM for the chosen P'-space eigenvalues. Comparison of the low-energy $T = 1$ states are shown in Fig. 1 in comparison with experiment[56] and with the

Table 2. Proton-neutron ($T = 0$) TBMEs of the derived effective interaction $\langle ab;\ J, T = 0|V_{\text{eff}}|cd;\ J, T = 0\rangle$ in the sd shell from Daejeon16. See text for details.

$2j_a$	$2j_b$	$2j_c$	$2j_d$	J	T	$V_{pn}^T = 0$
1	1	1	1	1	0	−2.909
1	1	1	3	1	0	0.381
1	1	3	3	1	0	0.573
1	1	3	5	1	0	−2.196
1	1	5	5	1	0	−1.224
1	3	1	3	1	0	−3.565
1	3	3	3	1	0	2.176
1	3	3	5	1	0	1.458
1	3	5	5	1	0	−0.761
3	3	3	3	1	0	−1.343
3	3	3	5	1	0	0.305
3	3	5	5	1	0	2.465
3	5	3	5	1	0	−5.702
3	5	5	5	1	0	−3.375
5	5	5	5	1	0	−1.043
1	3	1	3	2	0	−1.839
1	3	1	5	2	0	3.045
1	3	3	5	2	0	−2.290
1	5	1	5	2	0	−0.497
1	5	3	5	2	0	1.887
3	5	3	5	2	0	−3.370
1	5	1	5	3	0	−3.935
1	5	3	3	3	0	−0.004
1	5	3	5	3	0	−1.537
1	5	5	5	3	0	−2.011
3	3	3	3	3	0	−3.144
3	3	3	5	3	0	−1.743
3	3	5	5	3	0	0.977
3	5	3	5	3	0	−0.646
3	5	5	5	3	0	−2.071
5	5	5	5	3	0	−0.690
3	5	3	5	4	0	−4.218
5	5	5	5	5	0	−4.448

results from a phenomenological interaction USDB.[46] The latter produces identical $T = 1$ spectra for all three $A = 18$ isobars, because the USDB Hamiltonian is charge-independent, including the single-particle energies which are the same for protons and neutrons.

Since there are only small differences among the corresponding proton-proton, neutron-neutron and proton-neutron ($T = 1$) TBMEs, as seen from Table 1, (aside from the expected, approximately constant, diagonal Coulomb shift for the proton-proton TBMEs), the excitation spectra demonstrate the approximate validity of isospin symmetry (see Fig. 1).

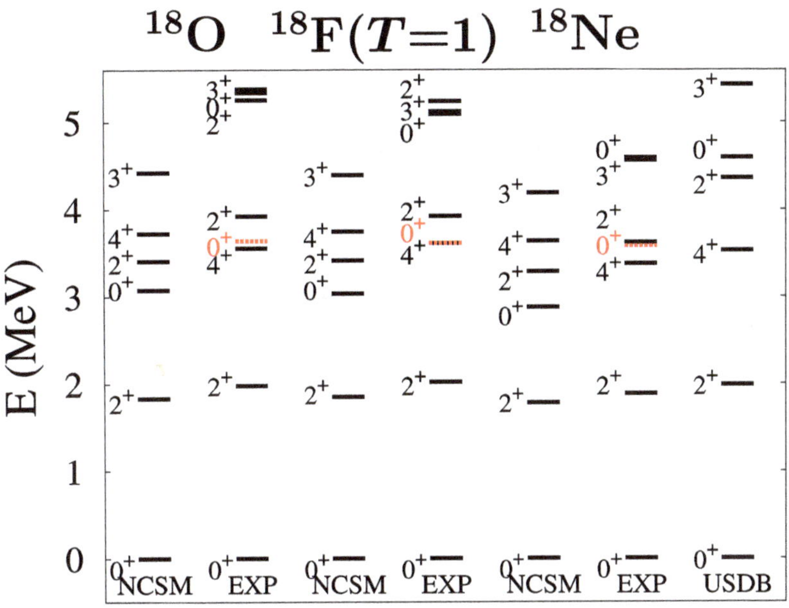

Fig. 1. Low-energy theoretical excitation spectra of ^{18}O, ^{18}Ne and the $T = 1$ part of the ^{18}F spectrum obtained within the NCSM with Daejeon16 at $N_{\max} = 4$ (identical to the *sd*-shell calculation with the proposed effective valence-space interaction). The USDB $T = 1$ spectrum for $A = 18$ and low-energy experimental spectra of positive-parity states[56] are shown for comparison. The 0_2^+ state shown in red in experimental spectrum in ^{18}O is most probably an intruder state as evidenced by the electromagnetic transition pattern.[56] Its isobaric analogue states in ^{18}F and ^{18}Ne are also marked in red, and, therefore, should not be compared with nearby theoretical spectra.

The low-lying $T = 0$ spectrum of ^{18}F, obtained from the NCSM, contains states with the same J^π quantum numbers as the USDB spectrum, however, it is more extended in energy than the USDB spectrum, with different ordering, and the lowest 3^+ state, being almost degenerate with 1^+ state, becomes the ground state (see Fig. 2). Except for the three lowest states, the experimental $T = 0$ spectrum is noticeably more dense and it is challenging to find theoretical counterparts in the *sd* shell calculation. This may hint at a possible intruder character of some of those states. In the NCSM calculation at $N_{\max} = 4$, states dominated by $2\hbar\Omega$ configurations are still located at much higher energy and therefore too far from being converged to shed light on this intruder state issue. Since the present study is focused on the construction of effective operators and to the comparison between the NCSM and valence space results, we do not investigate potential intruder states in the current work. In future applications with larger P-spaces, one may anticipate, however, that as these intruders fall further into the region of the valence-space dominated states, the mixing that is likely to occur will lead to complications in identifying the appropriate eigenpairs for retention in the P' space.

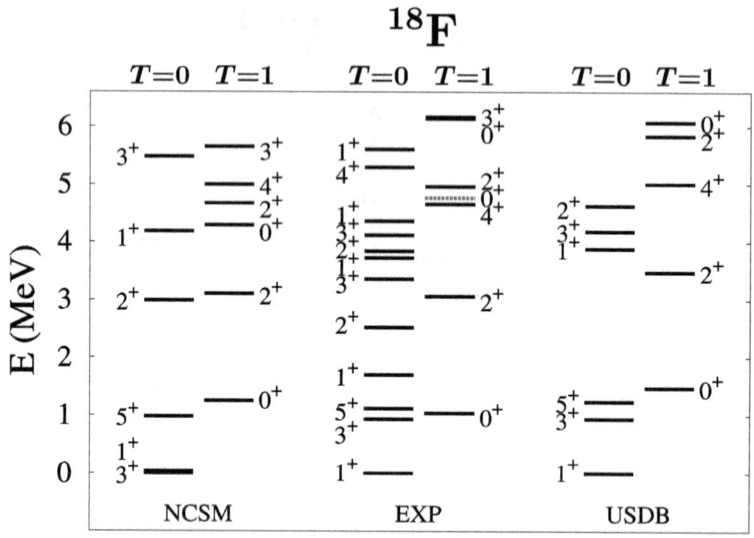

Fig. 2. Low-energy excitation spectra of ^{18}F (up to 6.2 MeV): theoretical spectrum obtained within the NCSM with Daejeon16 at $N_{\max} = 4$ (identical to the *sd*-shell calculation with the effective interaction of Ref. 30), in comparison with the positive-parity experimental spectrum[56] and the one obtained from USDB. The 0_2^+, $T = 1$ state in ^{18}F, analogue to the intruder state in ^{18}O, is shown in red color.

4. Effective Electromagnetic Operators in the *sd* Shell

Considering nucleons as non-relativistic point-like particles and assuming a long-wavelength approximation for emitted or absorbed radiation, one can deduce the following expressions for one-body electric quadrupole and magnetic dipole operators:

$$\hat{O}(E2) = \sum_{k=1}^{A} e_k r_k^2 \hat{Y}_2(\hat{r}_k), \tag{4}$$

and

$$\hat{O}(M1) = \sum_{k=1}^{A} \mu_N [g_k^s \vec{s}_k + g_k^l \vec{l}_k], \tag{5}$$

where $\hat{Y}_2(\hat{r}_k)$ are spherical harmonics of rank 2, $\hat{r}_k = (\theta_k, \phi_k)$ being a spherical angle, $\vec{s}_k$ and $\vec{l}_k$ are spin and orbital angular momenta operators of a nucleon, respectively, e_k denotes the electric charge of the kth nucleon, with the bare values being $e_p = e$ for a proton and $e_n = 0$ for a neutron, while g^s and g^l are spin and orbital gyro-magnetic factors (g-factors) with bare values being $g_p^s = 5.586$, $g_p^l = 1$, $g_n^s = -3.826$, $g_n^l = 0$ for protons and neutrons, respectively. For valence space calculations, e_p and e_n are conventionally understood to be *effective* proton and neutron electric charges, while $g_k^{l,s}$ have to be *effective* g-factors. In principle one could follow the same procedure as we use to develop the valence space effective two-nucleon interaction to develop valence space effective $E2$ and $M1$ operators. The procedures for doing so would follow those demonstrated in Refs. 51, 57. The results, for the P'-space we define here, would be represented as TBMEs for these operators. That is, there are, in theory, both one-body and two-body contributions to such derived effective valence electromagnetic operators. Here, we will obtain approximate theoretical one-body effective valence space operators for $E2$ and $M1$. We then test those approximate operators in a number of ways.

Our starting point is the NCSM calculation of $E2$ and $M1$ transitions and moments in ^{17}O and ^{17}F at $N_{\max} = 4$. The three lowest states are $1/2^+$, $5/2^+$ and $3/2^+$ states which are mapped into single-particle $1s_{1/2}$, $0d_{5/2}$ and $0d_{3/2}$ states in the *sd*-shell calculation. Therefore, we require that the matrix elements of the $E2$ and $M1$ operator with bare electric charges and g-factors between the many-body eigenstates from the NCSM at $N_{\max} = 4$ be exactly reproduced by effective single-particle matrix elements in the valence space.

4.1. *Effective E2 Operator*

To obtain an effective one-body valence-space $E2$ operator, we use the NCSM results for $E2$ transitions and quadrupole moments for ^{17}O and ^{17}F. For ^{17}F, we define a valence-state-dependent proton effective charge such that it exactly reproduces the NCSM values for transitions and moments. Next, we develop a state-dependent effective charge for the valence neutron. We note that the NCSM uses bare electric charges, so only protons contribute to the matrix elements of the $E2$ operator. In contrast, in the valence space ^{17}O has only one valence neutron and therefore we require that it carries an effective charge to be able to reproduce electromagnetic transitions and moments by the valence space calculations. These rules can be summarized as follows:

$$\begin{aligned} e_n(a,b)\langle b\|r^2\hat{Y}_2(\hat{r})\|a\rangle &= \langle J_f\|\hat{O}(E2)\|J_i\rangle \text{ (from } ^{17}\text{O}), \\ e_p(a,b)\langle b\|r^2\hat{Y}_2(\hat{r})\|a\rangle &= \langle J_f\|\hat{O}(E2)\|J_i\rangle \text{ (from } ^{17}\text{F}), \end{aligned} \tag{6}$$

where $e_n(a,b)$ and $e_p(a,b)$ are neutron and proton effective state-dependent charges, J_i and J_f are the initial and final state obtained within the NCSM, which are mapped onto single-particle states a and b, while the $E2$ operator $\hat{O}(E2)$ is defined by Eq. (4). Double bars in this equation and below refer to the matrix element reduced in angular momentum.

Matching the NCSM results for $E2$ transitions and quadrupole moments for ^{17}O and ^{17}F, we have deduced the values for effective electric charges which are summarized in Table 3. These effective charges $e_p(a,b)$ and $e_n(a,b)$ depend on the quantum numbers of the initial and final states, a and b, so the effective $E2$ operator is deduced as a set of individually parameterized single-particle matrix elements $\langle b|\hat{O}(E2)|a\rangle$. However, the state dependence of the effective charges is not strong as is seen from Table 3: the difference between neutron effective charges $e_n(a,b)$ does not exceed $\sim 0.12e$ and the maximal difference between various proton effective

Table 3. Effective charges (in units of e) and effective g-factors in the sd-shell valence space found by the respective NCSM calculations for ^{17}O and ^{17}F. The bare values used in the NCSM calculations are also indicated for reference. Hermiticity requires symmetry of these effective quantities under the interchange of a and b. See text for details.

(a,b)	$e_n(a,b)$	$e_p(a,b)$	$g_n^s(a,b)$	$g_n^l(a,b)$	$g_p^s(a,b)$	$g_p^l(a,b)$
bare	0.0	1.0	−3.826	0.0	5.586	1.0
$(0d_{5/2}, 1s_{1/2})$	0.181	1.171				
$(0d_{5/2}, 0d_{3/2})$	0.281	1.236	−3.608	0.020	5.252	0.916
$(1s_{1/2}, 0d_{3/2})$	0.168	1.297				
$(0d_{5/2}, 0d_{5/2})$	0.179	1.060	−3.751	0.026	5.499	0.976
$(0d_{3/2}, 0d_{3/2})$	0.172	1.248	−3.690	0.033	5.332	0.957
$(1s_{1/2}, 1s_{1/2})$			−3.729		5.468	
	$\bar{e}_n$	$\bar{e}_p$	$\bar{g}_n^s$	$\bar{g}_n^l$	$\bar{g}_p^s$	$\bar{g}_p^l$
average	0.196	1.202	−3.695	0.026	5.388	0.950

charges $e_p(a,b)$ is $\sim 0.25e$. The weak state dependence of effective charges tends to support the use of state-independent effective neutron and proton charges as is conventionally done in the valence-space calculations, see, e.g., Ref. 47. We present in Table 3 average theoretical effective neutron, $\bar{e}_n$, and proton, $\bar{e}_p$, charges which are seen to be smaller than the commonly accepted values[47] for $0\hbar\Omega$ model space calculations, $e_p = 1.36(5)e$ and $e_n = 0.45(5)e$.

To assess the quality of the operator constructed above, we calculate reduced transition matrix elements and quadrupole moments in $A = 18$ sd-shell nuclei using both state-dependent and average effective charges.

Reduced transition matrix elements, $\langle \alpha_f J_f \| \hat{O}(E2) \| \alpha_i J_i \rangle$, are required to express the corresponding $B(E2)$ values of a transition from an initial state $|\alpha_i J_i\rangle$ to the final state $|\alpha_f J_f\rangle$ as

$$B(E2; J_i \to J_f) = \frac{1}{2J_i + 1} |\langle \alpha_f J_f \| \hat{O}(E2) \| \alpha_i J_i \rangle|^2, \tag{7}$$

where J_i (J_f) are the initial (final) angular momenta and α_i (α_f) are other quantum numbers of the initial (final) states. The standard definition of a quadrupole moment in a state $|\alpha J\rangle$ reads

$$Q(J) = \sqrt{\frac{16\pi}{5}} \langle \alpha J, M_J = J | \hat{O}(E2, M_L = 0) | \alpha J, M_J = J \rangle. \tag{8}$$

We have calculated 66 $E2$ transitions in ^{18}O, 66 $E2$ transitions in ^{18}Ne and 269 $E2$ transitions in ^{18}F within the NCSM at $N_{\max} = 4$ using the bare $E2$ operator and in the sd-shell model space using the effective operators. The results are compared in Figs. 3 and 4 where we present absolute values of reduced $E2$ matrix elements

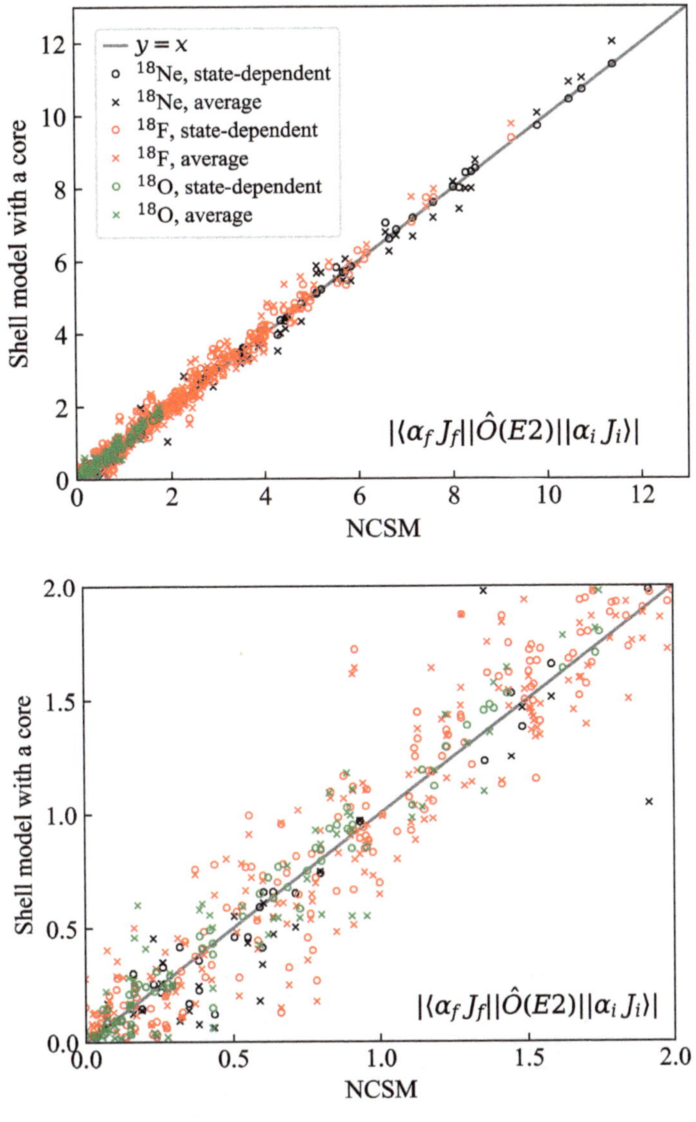

Fig. 3. Comparison of the *sd*-shell results for the absolute values of reduced $E2$ transition matrix elements between various states in ^{18}O, ^{18}F and ^{18}Ne obtained using effective $E2$ operators within shell model with a core with those obtained by the NCSM (top); the same in the vicinity of zero in order to provide details on an enlarged scale for the weaker transitions (bottom).

and signed electric quadrupole moments for different states obtained in the *sd*-shell valence space calculations with the effective operators versus those obtained by the NCSM. We observe an excellent correlation between the NCSM and valence-space values with the root-mean-square (rms) differences between them being very small, especially for the state-dependent parameterization of the operator, as can be seen from Table 4. The $B(E2)$ values, used to calculate the rms difference, have been

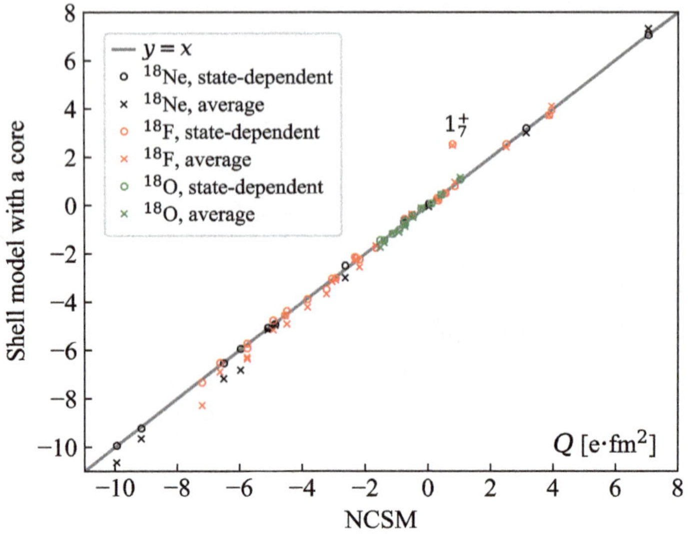

Fig. 4. Comparison of the *sd*-shell results for electric quadrupole moments of various states in ^{18}O, ^{18}F and ^{18}Ne obtained using effective $E2$ with those obtained by the NCSM. An outlier in ^{17}F, which is discussed in the text, is flagged with its spin and sequence in the spectrum indicated.

Table 4. Root-mean-square (rms) differences between the results on $E2$ and $M1$ transitions and moments from valence-space calculations with either the state-dependent (st-dep) or the average parameterizations of the effective $E2$ or effective $M1$ operator and the respective NCSM results. See text for details.

	^{18}O		^{18}F		^{18}Ne	
rms difference	st-dep	average	st-dep	average	st-dep	average
Red $E2$ M.E. $[e \cdot \text{fm}^2]$	0.072	0.168	0.114	0.335	0.221	0.267
$B(E2)[e^2 \cdot \text{fm}^4]$	0.017	0.039	0.189	0.240	0.208	0.583
$Q[e \cdot \text{fm}^2]$	0.063	0.118	0.370	0.478	0.064	0.443
Red $M1$ m.e. $[\mu_N]$	0.070	0.088	0.050	0.071	0.108	0.118
$B(M1)[\mu_N^2]$	0.063	0.081	0.092	0.106	0.060	0.085
$\mu[\mu_N]$	0.023	0.032	0.189	0.182	0.020	0.099

obtained for $J_i \geq J_f$. Table 5 shows $B(E2)$ values for transitions between the lowest states and electric quadrupole moments for the lowest $J \neq 0$ states in ^{18}O, ^{18}Ne and ^{18}F in comparison with the experimental data, when available. We clearly see missing strength in the $E2$ transitions between the lowest $T = 1$ states, which may signify a need for larger N_{max} values, because of the very slow convergence of $E2$ observables.[59] At the same time, the relative order of magnitude of theoretical $B(E2)$ values qualitatively resemble experimental findings. In particular, our calculation suggests that the quadrupole moment of the 5_1^+ state in ^{18}F is negative.

Table 5. $B(E2)$ values for transitions among the lowest states and electric quadrupole moments of the lowest states in ^{18}O, ^{18}F and ^{18}Ne: NCSM versus valence-space results with either state-dependent or average effective charges. The experimental values are shown when available (only the latest result is retained in the table below when a few experimental values are given in Ref. 58).

			sd shell	
	Exp[56,58]	NCSM	st-dep	average
^{18}O				
$B(E2; 2_1^+ \to 0_1^+)[e^2 \cdot \text{fm}^4]$	9.3(3)	0.527	0.535	0.634
$B(E2; 2_2^+ \to 0_1^+)[e^2 \cdot \text{fm}^4]$	3.6(6)	0.001	0.001	0.001
$B(E2; 4_1^+ \to 2_1^+)[e^2 \cdot \text{fm}^4]$	3.3(2)	0.339	0.359	0.434
$Q(2_1^+)[e \cdot \text{fm}^2]$	−3.6(9)	−1.40	−1.46	−1.55
^{18}Ne				
$B(E2; 2_1^+ \to 0_1^+)[e^2 \cdot \text{fm}^4]$	49.6(50)	21.94	21.76	23.79
$B(E2; 2_2^+ \to 0_1^+)[e^2 \cdot \text{fm}^4]$	1.9(9)	0.07	0.07	0.007
$B(E2; 4_1^+ \to 2_1^+)[e^2 \cdot \text{fm}^4]$	24.9(34)	14.43	14.42	16.07
$Q(2_1^+)[e \cdot \text{fm}^2]$		−9.16	−9.23	−9.65
^{18}F				
$B(E2; 3_1^+ \to 1_1^+)[e^2 \cdot \text{fm}^4]$	16.3(6)	8.29	8.51	9.05
$B(E2; 3_2^+ \to 1_1^+)[e^2 \cdot \text{fm}^4]$	1.9(3)	0.014	0.010	0.030
$B(E2; 5_1^+ \to 3_1^+)[e^2 \cdot \text{fm}^4]$	16.2(7)	7.81	7.95	8.64
$Q(1_1^+)[e \cdot \text{fm}^2]$		0.32	0.19	0.28
$Q(3_1^+)[e \cdot \text{fm}^2]$		−5.75	−5.93	−6.36
$Q(5_1^+)[e \cdot \text{fm}^2]$	±7.7(5)	−7.19	−7.34	−8.29

It is worth noting here that the agreement is worse for a couple of cases. In particular, one of them is seen in Fig. 4 where the quadrupole moment of the 1_7^+ state in ^{18}F with $Q = 0.78\ e \cdot \text{fm}^2$ from the NCSM and $Q = 2.54\ e \cdot \text{fm}^2$ from the valence-space calculations. The discrepancy is most likely related to the fact that this state is the highest lying 1^+ state (at around 25 MeV excitation energy) which has only 30% of the valence-space component in the NCSM calculations. Without this point, the rms error for Q in ^{18}F reduces to from 0.370 to 0.114 $e \cdot \text{fm}^2$ for the state-dependent and from 0.478 to 0.339 $e \cdot \text{fm}^2$ for the average parameterization of the operator.

4.2. *Effective $M1$ Operator*

We proceed in a like manner for the effective $M1$ operator. From comparison of the NCSM results on magnetic dipole moments and transitions for $A = 17$ at $N_{\max} = 4$

with single-particle values, we have deduced state-dependent matrix elements of the orbital and spin contribution of the $M1$ operator as well as respective state-dependent effective g-factors $g_k^j(a,b)$, $k = n$, p and $j = l, s$. The rules can be summarized as follows:

$$\begin{aligned} g_n^s(a,b)\langle b\|\mu_N\vec{s}\|a\rangle &= \langle J_f\|\hat{O}(M1)_{\text{spin}}\|J_i\rangle \text{ (from } ^{17}\text{O)}, \\ g_n^l(a,b)\langle b\|\mu_N\vec{l}\|a\rangle &= \langle J_f\|\hat{O}(M1)_{\text{orbit}}\|J_i\rangle \text{ (from } ^{17}\text{O)}, \\ g_p^s(a,b)\langle b\|\mu_N\vec{s}\|a\rangle &= \langle J_f\|\hat{O}(M1)_{\text{spin}}\|J_i\rangle \text{ (from } ^{17}\text{F)}, \\ g_p^l(a,b)\langle b\|\mu_N\vec{l}\|a\rangle &= \langle J_f\|\hat{O}(M1)_{\text{orbit}}\|J_i\rangle \text{(from } ^{17}\text{F)}, \end{aligned} \tag{9}$$

where spin and orbital part are referring to the first and second term of the $M1$ operator, given by Eq. (5), while J_i and J_f are the initial and final states obtained within the NCSM which are mapped onto the single-particle states a and b in the valence space. Note that these equations help to exactly reproduce $M1$ matrix elements between the NCSM states for $A = 17$ within the valence space, except for those which involve the $s_{1/2}$ orbital due to the fact that it is characterized by $l = 0$ orbital quantum number and therefore there is no way to reproduce a small, but non-zero value of $\langle J_f\|\hat{O}(M1)_{\text{orbit}}\|J_i\rangle$. To overcome this difficulty, we can require that in the case when a or b refers to the $s_{1/2}$ single-particle state, the valence space matrix element due to non-zero spin part would exactly reproduce the NCSM matrix element of the full $M1$ operator. Because of the smallness of the orbital contribution to the NCSM results, both definitions lead to almost identical results for magnetic moments and, therefore, their use is practically equivalent. We choose the latter method in the present work. The corresponding effective g-factors are summarized in Table 3. We notice that there is a more modest quenching of the spin operators than the quenching obtained in the previous studies in the sd-shell.[41,42,47,60] We notice also that the modifications to orbital g-factors found (positive for neutrons and negative for protons) are opposite in sign from the corresponding corrections found in the same cited works. For reference, the best fitted empirical values found for the same form of the $M1$ operator using the USDB wave functions in the sd shell are $g_n^s = -3.55(10)$, $g_n^l = -0.09(26)$, $g_p^s = 5.15(9)$, $g_p^l = 1.159(23)$.[47]

As a next step, we have calculated reduced $M1$ transition probabilities and magnetic moments in $A = 18$ sd-shell nuclei. A reduced $M1$ transition matrix element, $\langle\alpha_f J_f\|\hat{O}(M1)\|\alpha_i J_i\rangle$, is required to express the corresponding $B(M1)$ value of a transition from an initial state $|\alpha_i J_i\rangle$ to the final state $|\alpha_f J_f\rangle$ as

$$B(M1; J_i \to J_f) = \frac{1}{2J_i+1}|\langle\alpha_f J_f\|\hat{O}(M1)\|\alpha_i J_i\rangle|^2, \tag{10}$$

while the magnetic moment reads

$$\mu(J) = \sqrt{\frac{4\pi}{3}}\langle\alpha J, M_J = J|\hat{O}(M1, M_L = 0)|\alpha J, M_J = J\rangle. \tag{11}$$

The absolute values of reduced $M1$ matrix elements of 43 $M1$ transitions in ^{18}O, 43 $M1$ transitions in ^{18}Ne and 212 $M1$ transitions in ^{18}F obtained within the NCSM at $N_{\max} = 4$ using the bare $M1$ operator, and in the sd shell model space using the effective operators are compared in Fig. 5. The magnetic moments of

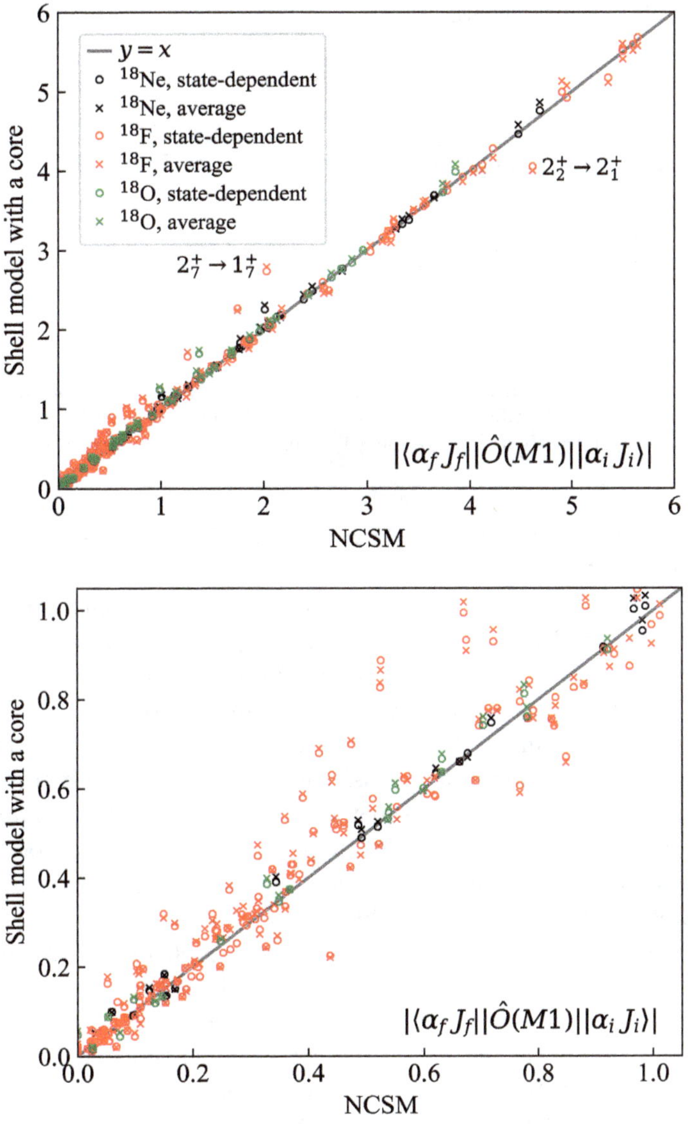

Fig. 5. Comparison of the sd-shell results for the reduced $M1$ matrix elements between various states in ^{18}O, ^{18}F and ^{18}Ne obtained using effective $M1$ operators within shell model with a core with those obtained by the NCSM (top); the same in the vicinity of zero in order to provide details on an enlarged scale for the weaker transitions (bottom). Two outliers in ^{18}F are flagged in the top panel with their state identifications and are discussed in the text.

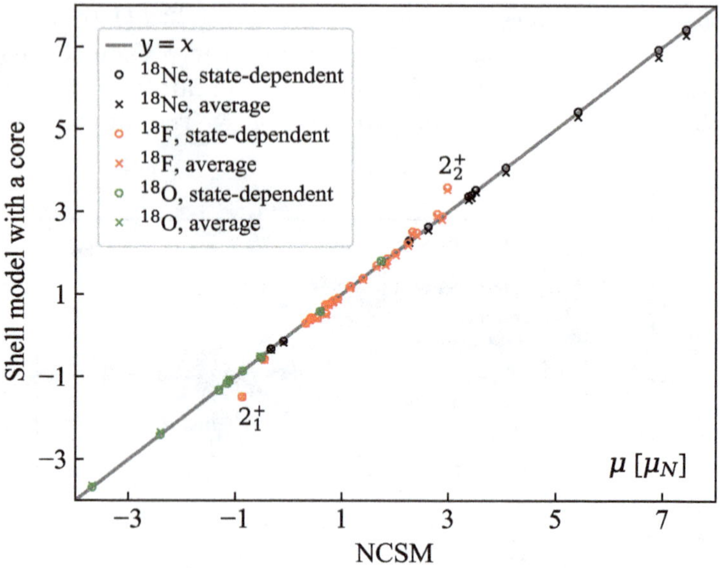

Fig. 6. Comparison of the *sd*-shell results for magnetic dipole moments of various states in ^{18}O, ^{18}F and ^{18}Ne obtained using effective $M1$ operators with those obtained by the NCSM. Two outliers in ^{18}F are flagged with their state identifications and are discussed in the text.

all $J \neq 0$ states (11 states in ^{18}O, 11 states in ^{18}Ne and 25 states in ^{18}F) are compared in Fig. 6. We notice again a remarkable agreement between the two sets of results. The state-dependent parameterization of the $M1$ single-particle matrix elements leads to smaller rms differences as compared to those provided by the average parameterization with an exception of the ^{18}F magnetic moments where both parameterizations provide nearly the same rms differences (see Table 4). The $B(M1)$ values, used to compute the rms difference, have been generated for $J_i \geq J_f$.

Table 6 shows $B(M1)$ values for transitions between the lowest states and magnetic dipole moments of the lowest states in ^{18}O, ^{18}Ne and ^{18}F in comparison with the experimental data, when available. In particular, there is a good agreement with the measured magnetic moments values. We attribute this to the fact that the $M1$-observables converge faster in the NCSM calculations.

Two outlying ^{18}F points are clearly seen in Fig. 6 and these are magnetic moments of the first two 2^+ states: they are only 130 keV separated in energy and appear to be mixed in isospin. The 2_1^+ state contains 88% of $T = 1$ component and 11% $T = 0$ component and vice verse for 2_2^+. The values of the reduced $M1$ transition matrix elements between these two states are also significantly different between the NCSM ($4.258\,\mu_N^2$) and valence space calculations ($3.297\,\mu_N^2$) as seen from Fig. 5.

Table 6. $B(M1)$ values for transitions between the lowest states and magnetic moments of the lowest states in ^{18}O, ^{18}F and ^{18}Ne: NCSM versus valence-space results obtained using either state-dependent or average g-factors. The experimental values are shown when available (only the latest result is retained in the table below when a few experimental values are given in Ref. 58).

			sd shell	
	Exp[56,58]	NCSM	st-dep	average
^{18}O				
$B(M1; 2_2^+ \to 2_1^+)[\mu_N^2]$	0.25(4)	0.072	0.072	0.071
$B(M1; 1_1^+ \to 0_1^+)[\mu_N^2]$		0.101	0.119	0.125
$\mu(2_1^+)[\mu_N]$	−0.57(3)	−0.53	−0.54	−0.52
$\mu(4_1^+)[\mu_N]$	±2.5(4)	−2.41	−2.41	−2.35
^{18}Ne				
$B(M1; 2_2^+ \to 2_1^+)[\mu_N^2]$		0.092	0.092	0.090
$B(M1; 1_1^+ \to 0_1^+)[\mu_N^2]$		0.079	0.089	0.093
$\mu(2_1^+)[\mu_N]$		2.62	2.63	2.55
^{18}F				
$B(M1; 1_1^+ \to 0_1^+)[\mu_N^2]$		5.428	5.366	5.256
$\mu(3_1^+)[\mu_N]$	+1.77(12)	1.857	1.852	1.803
$\mu(5_1^+)[\mu_N]$	+2.86(3)	2.878	2.878	2.798

5. Conclusions and Prospects

In this contribution we present a novel method for deriving effective one-body electromagnetic operators for valence space calculations from the NCSM results. The effective $E2$ and $M1$ one-body operators are obtained as a set of parameterized effective single-particle matrix elements to be used with many-body states in the valence space. Comparison of a large number of reduced $E2$ and $M1$ matrix elements, as well as electric quadrupole and magnetic dipole moments in $A = 18$ *sd*-shell nuclei shows a very good agreement with the NCSM values. Further exploration of these operators in systems with more valence particles is in progress. One also anticipates systematically increasing the space for the NCSM calculations as computational resources allow.

Acknowledgments

Z. Li and N. A. Smirnova acknowledge the financial support of CNRS/IN2P3 via ENFIA and ABI-CONFI Master projects, France. The work of A. M. Shirokov is supported by the Russian Foundation for Basic Research under Grant No. 20-02-00357. The work of I. J. Shin was supported by the Rare Isotope Science Project of Institute for Basic Science funded by Ministry of Science and ICT and National Research Foundation of Korea (2013M7A1A1075764). This work was supported in part by the US Department of Energy (DOE) under Grant Nos. DE-SC0023692 and DE-SC0023495 (SciDAC5/NUCLEI). Computational resources were provided by the National Energy Research Scientific Computing Center (NERSC), which is supported by the US DOE Office of Science under Contract No. DE-AC02-05CH11231, and also provided by the National Supercomputing Center, Republic of Korea, including technical support (KSC-2020-CRE-0027). N. A. Smirnova and A. M. Shirokov thank the Institute for Basic Science, Daejeon, for a hospitality and financial support of their visits.

References

1. B. R. Barrett, P. Navrátil, and J. P. Vary, *Prog. Part. Nucl. Phys.* **69**, 131 (2013).
2. G. Hagen, T. Papenbrock, M. Hjorth-Jensen, and D. J. Dean, *Rept. Prog. Phys.* **77**, 096302 (2014).
3. J. Carlson, S. Gandolfi, F. Pederiva, S. C. Pieper, R. Schiavilla, K. E. Schmidt, and R. B. Wiringa, *Rev. Mod. Phys.* **87**, 1067 (2015).
4. H. Hergert, S. K. Bogner, T. D. Morris, A. Schwenk, and K. Tsukiyama, *Phys. Rept.* **621**, 165 (2016).
5. W. H. Dickhoff and C. Barbieri, *Prog. Part. Nucl. Phys.* **52**, 377 (2004).
6. A. Tichai, R. Roth, and T. Duguet, *Front. Phys.* **8**, 164 (2020).
7. P. Maris, E. Epelbaum, R. J. Furnstahl, J. Golak, K. Hebeler, T. Hüther, H. Kamada, H. Krebs, U.-G. Meißner, J. A. Melendez, A. Nogga, P. Reinert, R. Roth, R. Skibiński, V. Soloviov, K. Topolnicki, J. P. Vary, Y. Volkotrub, H. Witała, and T. Wolfgruber, *Phys. Rev. C* **103**, 054001 (2021).
8. P. Sternberg, E. G. Ng, C. Yang, P. Maris, J. P. Vary, M. Sosonkina, and H. V. Le, *Procedia of the 2008 ACM/IEEE Conference on Supercomputing (SC 2008), (IEEE Press, Piscataway, NJ, 2008)*. **15**, 1 (2008).
9. P. Maris, M. Sosonkina, J. P. Vary, E. G. Ng, and C. Yang, *Procedia Computer Science (May 2010, ICCS 2010) (Elsevier, Amsterdam, 2010)*. **1**, 97 (2010).
10. H. M. Aktulga, C. Yang, E. G. Ng, P. Maris, and J. P. Vary, *Lecture Notes in Computer Science (Springer, Heidelberg, 2012)*. **7484**, 830 (2012).
11. H. M. Aktulga, C. Yang, E. G. Ng, P. Maris, and J. P. Vary, *Concurrency Computat.: Pract. Exper.* **26**, 2631 (2014).
12. M. Shao, H. Aktulga, C. Yang, E. G. Ng, P. Maris, and J. P. Vary, *Comput. Phys. Commun.* **222**, 1 (2018).
13. A. Saxena and P. C. Srivastava, *J. Phys. G* **47**, 055113 (2020).
14. I. J. Shin, N. A. Smirnova, A. M. Shirokov, Z. Yang, B. R. Barrett, Z. Li, Y. Kim, P. Maris, and J. P. Vary, *Phys. Rev. C* **110**, 034306 (2024).
15. B. A. Brown, *Prog. Part. Nucl. Phys.* **47**, 517 (2001).
16. T. Otsuka, M. Honma, T. Mizusaki, N. Shimizu, and Y. Utsuno, *Prog. Part. Nucl. Phys.* **47**, 319 (2001).
17. E. Caurier, G. Martínez-Pinedo, F. Nowacki, A. Poves, and A. P. Zuker, *Rev. Mod. Phys.* **77**, 427 (2005).
18. M. Hjorth-Jensen, T. T. S. Kuo, and E. Osnes, *Phys. Rep.* **261**, 125 (1995).

19. S. R. Stroberg, H. Hergert, S. K. Bogner, and J. D. Holt, *Ann. Rev. Nucl. Part. Sci.* **69**, 307 (2019).
20. B. A. Brown and B. H. Wildenthal, *Ann. Rev. Nucl. Part. Sci.* **38**, 29 (1988).
21. M. Honma, T. Otsuka, B. A. Brown, and T. Mizusaki, *Phys. Rev. C* **69**, 034335 (2004).
22. A. Poves and A. P. Zuker, *Phys. Rep.* **70**, 235 (1981).
23. A. P. Zuker, *Phys. Rev. Lett.* **90**, 042502 (2003).
24. S. R. Stroberg, A. Calci, H. Hergert, J. D. Holt, S. Bogner, R. Roth, and A. Schwenk, *Phys. Rev. Lett.* **118**, 032502 (2017).
25. S. Okubo, *Prog. Theor. Phys.* **12**, 603 (1954).
26. K. Suzuki and S. Y. Lee, *Prog. Theor. Phys.* **64**, 2091 (1980).
27. K. Suzuki, *Prog. Theor. Phys.* **68**, 246 (1982).
28. A. F. Lisetskiy, M. K. G. Kruse, B. R. Barrett, P. Navrátil, I. Stetcu, and J. Vary, *Phys. Rev. C* **80**, 024315 (2009).
29. E. Dikmen, A. F. Lisetskiy, B. R. Barrett, P. Maris, A. M. Shirokov, and J. P. Vary, *Phys. Rev. C* **91**, 064301 (2015).
30. N. A. Smirnova, B. R. Barrett, Y. Kim, I. J. Shin, A. M. Shirokov, E. Dikmen, P. Maris, and J. P. Vary, *Phys. Rev. C* **100**, 054329 (2019).
31. G. R. Jansen, M. Schuster, A. Signoracci, G. Hagen, and P. Navrátil, *Phys. Rev. C* **94**, 011301 (2016).
32. Z. H. Sun, T. D. Morris, G. Hagen, G. R. Jansen, and T. Papenbrock, *Phys. Rev. C* **98**, 054320 (2018).
33. T. Otsuka, T. Suzuki, J. D. Holt, A. Schwenk, and Y. Akaishi, *Phys. Rev. Lett.* **105**, 032501 (2010).
34. J. D. Holt, T. Otsuka, A. Schwenk, and T. Suzuki, *J. Phys. G* **39**, 085111 (2012).
35. T. Fukui, L. D. Angelis, Y. Z. Ma, L. Coraggio, A. Gargano, N. Itaco, and F. R. Xu, *Phys. Rev. C* **98**, 044305 (2018).
36. A. Arima and H. Horie, *Prog. Theor. Phys.* **11**, 509 (1954).
37. A. Arima and H. Horie, *Prog. Theor. Phys.* **12**, 623 (1954).
38. T. T. S. Kuo and E. Osens, *Nucl. Phys. A* **205**, 1 (1973).
39. B. A. Brown, A. Arima, and J. B. McGrory, *Nucl. Phys. A* **77**, 277 (1977).
40. E. Oset and M. Rho, *Phys. Rev. Lett.* **42**, 47 (1979).
41. A. Arima, K. Shimizu, W. Bentz, and H. Hyuga, *Adv. Nucl. Phys.* **18**, 1 (1987).
42. I. Towner, *Phys. Rep.* **155**, 264 (1987).
43. N. M. Parzuchowski, S. R. Stroberg, P. Navrátil, H. Hergert, and S. K. Bogner, *Phys. Rev. C* **96**, 034324 (2017).
44. B. A. Brown, B. H. Wildenthal, W. Chung, S. E. Massen, M. Bernas, A. M. Bernstein, R. Miskimen, V. R. Brown, and V. A. Madsen, *Phys. Rev. C* **26**, 2247–2272 (1982).
45. M. Carchidi, B. H. Wildenthal, and B. A. Brown, *Phys. Rev. C* **34**, 2280 (1986).
46. B. A. Brown and W. A. Richter, *Phys. Rev. C* **74**, 034315 (2006).
47. W. A. Richter, S. Mkhize, and B. A. Brown, *Phys. Rev. C* **78**, 064302 (2008).
48. A. M. Shirokov, I. J. Shin, Y. Kim, M. Sosonkina, P. Maris, and J. P. Vary, *Phys. Lett. B* **761**, 87 (2016).
49. D. H. Gloeckner and D. R. Lawson, *Phys. Lett. B* **53**, 313 (1974).
50. M. W. Kirson, *Nucl. Phys. A* **781**, 350 (2007).
51. J. P. Vary, R. Basili, W. Du, M. Lockner, P. Maris, S. Pal, and S. Sarker, *Phys. Rev. C* **98**, 065502 (2018).
52. D. R. Entem and R. Machleidt, *Phys. Lett. B* **524**, 93 (2002).
53. D. R. Entem and R. Machleidt, *Phys. Rev. C* **68**, 041001 (2003).
54. S. D. Glazek and K. G. Wilson, *Phys. Rev. D* **48**, 5863 (1993).
55. F. Wegner, *Annalen der Physik.* **506**, 77 (1994).

56. NNDC. http://www.nndc.bnl.gov.
57. I. Stetcu, B. R. Barrett, P. Navrátil, and J. P. Vary, *Phys. Rev. C* **71**, 044325 (2005).
58. N. J. Stone, *At. Data Nucl. Data Sci.* **90**, 75 (2005).
59. I. J. Shin, Y. Kim, P. Maris, J. P. Vary, C. Forssén, J. Rotureau, and N. Michel, *J. Phys. G* **44**, 075103 (2017).
60. B. A. Brown and B. H. Wildenthal, *Nucl. Phys. A* **474**, 290–306 (1987).

https://doi.org/10.1142/9789819813353_0012

Akito Arima's Growing Legacy in Numerical Lattice QCD

Shigemi Ohta
Theory Center, Institute of Particle and Nuclear Studies, KEK
Oho 1-1, Tsukuba, 3050801, Japan
shigemi.ohta@kek.jp

Akito Arima impacted supercomputing in general and its application in science in particular, not only in Japan but worldwide. In 1985 as the director of large-scale computing at the University of Tokyo, he introduced a vector supercomputer for use by the nationwide science community in Japan. This and a similar installation that soon followed at the National Laboratory for High Energy Physics (KEK) provided the foundation for large-scale numerical lattice field theory research in Japan. In 1997, as the president of the RIKEN he helped found the RIKEN-BNL Research Center (RBRC) that installed QCDSP and QCDOC purpose-built parallel supercomputers in succession for use in numerical lattice quantum chromodynamics (QCD). These computers allowed the RIKEN-BNL-Columbia (RBC) Collaboration to pioneer the first numerical calculations of lattice QCD with domain wall fermions (DWF) quarks that preserve both flavor and chiral symmetries. Based on these pioneering works RBC and UKQCD Collaborations have been jointly generating dynamical 2+1-flavor DWF numerical lattice-QCD ensembles with strange and degenerate up-and-down quarks at their physical mass values for a decade. Their achievement includes constraining the standard model, explaining $\Delta I = 1/2$ rule and indirect CP violation in kaon decays, and is beginning to help searching for beyond-the-standard-model (BSM) physics. These DWF numerical lattice-QCD ensembles also help in understanding the nucleon. Akito Arima's legacy in supercomputing is still growing through the expanding influence of numerical lattice QCD.

1. Lattice QCD and Supercomputing

Wilson's lattice formulation of quantum chromodynamics (QCD)[1] provides a non-perturbative definition of the theory. It allows lattice-regularized finite numerical models of the theory to be constructed and analyzed on numerical computers at any finite lattice spacing. Lattice regularization is then lifted by sending the lattice spacing to zero through asymptotic scaling of QCD,[2–4] and QCD observables in the physical continuum are calculated. This requires a lot of numerical computations but is feasible.

Computing resources necessary for conducting such numerical lattice-QCD research were not immediately available in Japan before 1985 when Akito Arima, then the director of large-scale computing, installed a vector supercomputer at the University of Tokyo. This and a similar vector supercomputer installation that soon followed at the National Laboratory of High Energy Physics (KEK) in Tsukuba helped to found numerical lattice-QCD research in Japan.

2. Domain-Wall Fermions

However earlier numerical lattice-QCD research was hampered by the fermion species doubling[5] that could naively result in $2^D = 16$-fold increase of the number of light quarks in $D = 4$ space-time dimensions. Thus earlier numerical lattice-QCD studies gave up either chiral or flavor symmetry at their raw numerical level. Though such symmetry breaking can be controlled to disappear in extrapolating to the physical continuum, such control procedures add extra burden such as disentangling unwanted operator mixing that does not happen if continuum-like symmetries were preserved.

Domain-wall fermions (DWF)[6–9] resolve this problem by introducing a fictitious fifth dimension in which the chiral symmetry breaking is exponentially suppressed. However, conventional vector supercomputers were not up to this challenge because of an order of magnitude larger computing demand due to the fifth dimension.

Around the time RIKEN-BNL Research Center (RBRC) was established in 1997,[10] under Akito Arima's presidency at RIKEN. RBRC installed QCDSP (QCD with DSP, digital signal processor), its first purpose-built parallel supercomputer for numerical lattice QCD.[11,12] RIKEN-BNL Columbia Collaboration was formed using this computer to pursue this opportunity to study QCD with DWF. The collaboration quickly demonstrated that the lattice artifacts in the linear $O(a)$ order of the lattice spacing a that violate chiral and flavor symmetries are indeed exponentially suppressed in the fifth dimension.[13]

Following this success RIKEN quickly approved the development of another purpose-built parallel supercomputer, QCDOC (QCD on a chip, with a full computing and communicating ASIC node integrated on a single chip) for lattice QCD,[14,15] though by the time Akito Arima had left RIKEN to become a member of the upper house of Japanese Diet, and then a cabinet minister in charge of education and science. UKQCD Collaboration also decided to join the development and construction of this new parallel supercomputer. Three QCDOC computers were built for each of RBRC, USQCD Collaboration, and UKQCD Collaboration and contributed a lot in advancing QCD research using DWF. Part of UKQCD Collaboration joined this DWF project and the two collaborations, RBC and UKQCD, have been jointly generating dynamical 2+1-flavor DWF lattice-QCD numerical ensembles with physical-mass strange quark and successively light degenerate up-and-down quarks since then with successively finer lattice spacings and larger lattice volumes.[16–22]

3. DWF Achievements

Joint RBC and UKQCD Collaborations have been generating a sequence of dynamical 2+1-flavor DWF numerical lattice ensembles with physical strange- and degenerate up-and-down-quark mass for about a decade now.[22] The various lattice spacings and volumes of these ensembles allow us to control the systematic errors to levels below the Monte-Carlo statistical errors, which are typically a percent or even less.

By looking at mass and electroweak transition matrix elements of pions and kaons with such precisions, we have contributed a lot to constrain the standard model parameters such as quark mass, CKM mixing matrix, and chiral lagrangians.[23–25] By looking more closely with better precision calculations we could help discover beyond-the-standard model (BSM) physics:

Our calculations of indirect CP violation in $K \to \pi\pi$ decay, *i.e.* those of the kaon bag parameter, B_K, using the dynamical 2+1-flavor DWF numerical lattice-QCD ensembles, agree well with the experiment.[22, 26] These ensembles also provide a detailed quantitative understanding of $\Delta I = 1/2$ rule in this decay, that two main terms in the real part of the decay amplitude interfere constructively if the final-state pions isospin is zero and destructively if it is two.[27] Similar precision results are being obtained for the imaginary part of the same amplitude that is relevant for direct CP violation in this decay.[28] A discrepancy between this calculation and the experiment would indicate a BSM contribution to the process.

We are also beginning to contribute to BSM search in muon anomalous magnetic moment, $(g-2)_\mu$, by calculating hadronic contributions of hadronic vacuum polarization and hadronic light-by-light scattering.[29, 30] Here again, a discrepancy between theoretical calculation and experiment would indicate a BSM contribution. Indeed such a discrepancy with about four standard-deviation significance has been reported.[31] To clarify whether some of this is caused by BSM we need better-precision numerical lattice-QCD calculations of the hadronic contributions.

4. Nucleon in DWF Lattice QCD

We have been calculating not only nucleon mass but also its isovector couplings such as vector-, g_V, and axialvector-current, g_A, charges, tensor, g_T, and scalar, g_S, couplings that are important for BSM search, vector- and axialvector-current form factors that provide mean-squared charge radius or anomalous magnetic moment, and low moments of structure functions.

Statistical errors for nucleon in numerical lattice QCD are severer than those for pion or kaon as nucleon is heavier: Signal-to-noise ratio for nucleon should degrade exponentially as $\exp\left((3m_\pi/2 - m_N)T\right)$ as the calculations approach physical mass values, because the signal, $\langle N^\dagger O N\rangle$, should depend as $\sim\exp(-m_N T)$ on time while the noise, $\sqrt{\langle N^\dagger O N N^\dagger O N\rangle}$ can stay longer, $\sim\sqrt{\exp(-3m_\pi T)}$, through different contractions of quark fields, and nucleon mass approaches its experimental value more slowly than pion mass because of Gell'Mann-Oakes-Renner relation,[51] $m_\pi^2 f_\pi^2 \propto m_{\rm ud}\langle \overline{u}u + \overline{d}d\rangle$. Nevertheless, there has been steady and interesting progress.[32, 33]

Figure 1 summarizes nucleon mass as calculated on recent RBC and RBC+UKQCD dynamical 2+1-flavor DWF numerical lattice ensembles. In the earlier two calculations[34] using heavier than physical values for the degenerate up- and down-quarks, pion mass was calculated to be about 0.3294(13) and 0.4164(12) GeV and corresponding nucleon mass values were 1.154(7) and 1.216(7) GeV. These two right-most points would significantly overshoot the physical value in linear

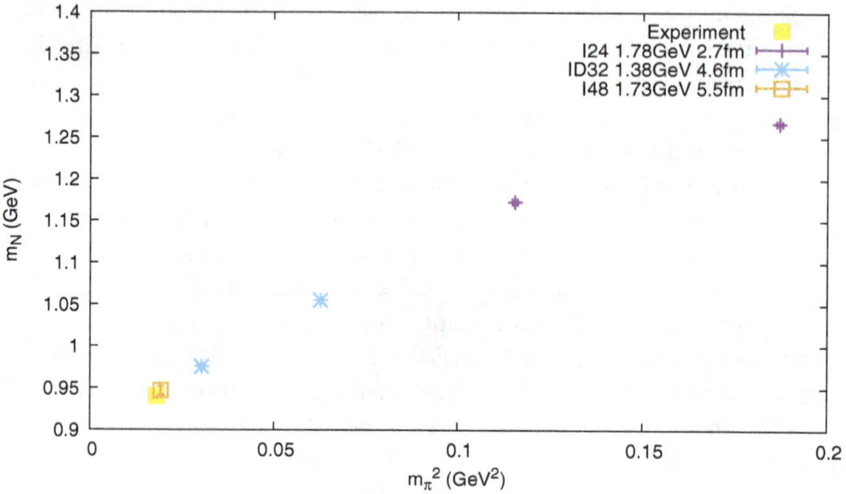

Fig. 1. Nucleon mass as calculated on a few recent RBC and RBC+UKQCD dynamical 2+1-flavor DWF numerical lattice ensembles. The dependence on pion mass squared is convex upward, possibly because of a chiral logarithm, $m_\pi^2 \log m_\pi^2 \sim m_{\rm ud} \log m_{\rm ud}$.

extrapolation in pion mass squared to the physical point. The more recent two intermediate points[32] with calculated pion mass of about 0.2494(3) and 0.1723(3) GeV the corresponding nucleon mass estimates were 1.0550(3) and 0.9752(11) GeV. Linear extrapolation of these two to physical pion mass squared gives 0.950(5) GeV. Though these calculations are at single lattice-spacing values and yet to be extrapolated to the physical continuum, the downward trending of nucleon mass appears to be approaching the physical value as the pion mass approaches its physical value. Indeed the left-most calculation with pion mass of about 0.1392(4) GeV[22] gives an estimate of 0.947(6) GeV for nucleon mass, almost capturing the physical value within the statistical error.[35] This convex-upward nucleon-mass dependence on pion or quark mass may arise from much expected chiral logarithm,[36] $m_\pi^2 \log m_\pi^2 \sim m_{\rm ud} \log m_{\rm ud}$.

Let's turn attention to the isovector axialvector-current charge, g_A, of the nucleon. As is well known, this charge determines neutron life, and through Goldberger-Treiman relation,[37] $m_N g_A \propto m_\pi g_{\pi NN}$, influences much of nuclear physics such as nucleosynthesis and nuclear abundance or (n, p) reaction off nuclei in which Akito Arima took much interest. It is also historically an interesting quantity, because the reported experimental value has been almost monotonically increasing as the observed neutron life has been almost monotonically shortening. Indeed, Maurice Goldhaber joked that a physicist's age can be guessed from the value e remembers for the quantity. The current value, as recommended by Particle Data Group is 1.2754(13)[38] as the latest "ultra-cold" neutron experiments dominate.[39] While the experimental difficulty is easy to understand as counting the initial number of neutrons, neutral objects, is hard, theoretical guesstimates have been more

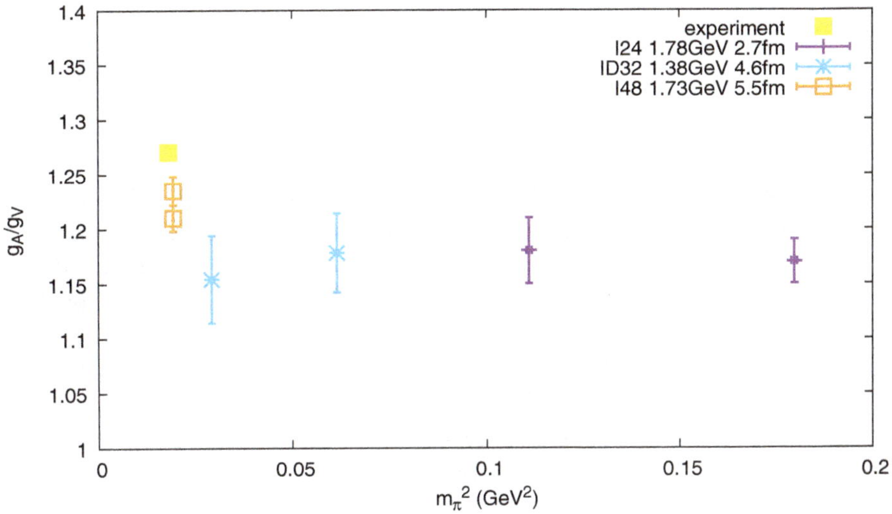

Fig. 2. Ratio g_A/g_V of nucleon isovector axialvector-, g_A to vector-, g_V, current charges, as calculated on a few recent RBC and RBC+UKQCD dynamical 2+1-flavor DWF numerical lattice ensembles. Large, ∼10 % deficit in comparison with the experiment at pion mass heavier than about 0.17 GeV has been confirmed by almost all numerical lattice-QCD calculations with similar lattice spacings. The dependence on pion mass squared or quark mass is quite different from what teh non-relativistic quark model, bag models, or skyrmion model predicted.

embarrassing: good old non-relativistic quark model gave a value of 5/3 from a simple-minded $\langle\sigma\tau\rangle$ estimate. MIT bag model gave a value of about 1.09[40] without pion, and the skyrmion model gave about 0.61.[41]

Figure 2 summarizes our calculations using a few recent RBC and RBC+ UKQCD dynamical 2+1-flavor DWF numerical lattice ensembles.[32,34,35,42] At heavier mass, the calculations show about 10% deficit in comparison with the experiment. Other lattice numerical calculations eventually agreed with this deficit at similar mass and lattice spacings.[43–47] The deficit shrinks in approaching the physical mass but grows in statistical significance. This quark-mass dependence is nothing like what was expected from the non-relativistic quark model, bag models, or the skyrmion model. The cause of this deficit at heavier-than-physical pion mass of down to about 0.17 GeV is not yet understood.

Recent calculations at lighter mass report results closer to the experiment, a few of them capturing the experimental value within their rather large statistical and systematic errors[48–50] as calculations without one or both of chiral and flavor symmetries suffer larger errors from extrapolation to remove the symmetry breaking and thus can end up hiding small effects. Here again, the better-preserved chiral and flavor symmetries in DWF lattice QCD help us resolve a smaller but statistically more significant deficit. We plan to improve our statistics to clarify the situation. Note it is possible the deficit is caused by our not taking into account isospin breaking, both from mass and charge differences between up and down quarks.

Calculated form factors and structure functions are, albeit with larger statistical errors, consistent with experiments.[33,34,52] Signals for tensor and scalar couplings that could eventually help BSM search are also obtained.[32]

5. Future

In the majority of current joint RBC and UKQCD numerical DWF lattice-QCD ensembles, 1) up and down quark mass are set degenerate, 2) quark electric charges do not affect the ensembles, and 3) charm and heavier quark seas do not affect the ensembles. Each of these is estimated to be below one-percent correction for the majority of observables we are reporting and hence is smaller than the respective statistical errors. However for numerical lattice QCD to contribute to the search for BSM physics these systematic errors must be better controlled: thus we are generating more refined numerical lattice-QCD ensembles with DWF quarks, that take into account the isospin breaking by splitting up- and down-quark mass, and by including QED corrections, and also including charm sea. Such refined ensembles would also help understanding nucleon including its isovector axialvector charge as described above. Also important would be to extend the reaches of numerical lattice QCD to larger hadronic systems beyond a single nucleon. Fortunately, research in this direction started a decade ago[53] and is now well underway. Akito Arima's legacy in supercomputing is still growing today with the expanding influence of numerical lattice QCD.

Note added in proof: There has been some progress in nucleon form factor calculations with domain-wall lattice QCD since the time of the original manuscript in 2021. Please refer to the following URLs, https://pos.sissa.it/430/116 and https://pos.sissa.it/453/325, and the references cited there in.

References

1. K. G. Wilson, Confinement of quarks, *Phys. Rev. D* **10**, 2445–2459 (1974). doi: 10.1103/PhysRevD.10.2445.
2. D. J. Gross and F. Wilczek, Ultraviolet behavior of nonabelian Gauge theories, *Phys. Rev. Lett.* **30**, 1343–1346 (1973). doi: 10.1103/PhysRevLett.30.1343.
3. H. D. Politzer, Reliable perturbative results for strong interactions? *Phys. Rev. Lett.* **30**, 1346–1349 (1973). doi: 10.1103/PhysRevLett.30.1346.
4. S. R. Coleman and E. J. Weinberg, Radiative corrections as the origin of spontaneous symmetry breaking, *Phys. Rev. D* **7**, 1888–1910 (1973). doi: 10.1103/PhysRevD.7.1888.
5. H. B. Nielsen and M. Ninomiya, No go theorem for regularizing chiral fermions, *Phys. Lett. B* **105**, 219–223 (1981). doi: 10.1016/0370-2693(81)91026-1.
6. P. H. Ginsparg and K. G. Wilson, A remnant of chiral symmetry on the lattice, *Phys. Rev. D* **25**, 2649 (1982). doi: 10.1103/PhysRevD.25.2649.
7. D. B. Kaplan, A method for simulating chiral fermions on the lattice, *Phys. Lett. B* **288**, 342–347 (1992). doi: 10.1016/0370-2693(92)91112-M.
8. Y. Shamir, Chiral fermions from lattice boundaries, *Nucl. Phys. B* **406**, 90–106 (1993). doi: 10.1016/0550-3213(93)90162-I.

9. V. Furman and Y. Shamir, Axial symmetries in lattice QCD with Kaplan fermions, *Nucl. Phys. B* **439**, 54–78 (1995). doi: 10.1016/0550-3213(95)00031-M.
10. S. Ohta, RIKEN-BNL research center, *Nucl. Phys. A* **670**, 22–31 (2000). doi: 10.1016/S0375-9474(00)00064-6.
11. R. D. Mawhinney, The One teraflops QCDSP computer, *Parallel Comput.* **25**, 1281 (1999). doi: 10.1016/S0167-8191(99)00051-4.
12. T. D. Lee, The Columbia-RIKEN-BNL QCDSP supercomputer project, *Subnucl. Ser.* **37**, 366–377 (2001). doi: 10.1142/9789812811585_0012.
13. T. Blum *et al.*, Quenched lattice QCD with domain wall fermions and the chiral limit, *Phys. Rev. D* **69**, 074502 (2004). doi: 10.1103/PhysRevD.69.074502.
14. D. Chen *et al.*, QCDOC: A 10-teraflops scale computer for lattice QCD, *Nucl. Phys. B Proc. Suppl.* **94**, 825–832 (2001). doi: 10.1016/S0920-5632(01)01014-3.
15. P. A. Boyle *et al.*, Hardware and software status of QCDOC, *Nucl. Phys. B Proc. Suppl.* **129**, 838–843 (2004). doi: 10.1016/S0920-5632(03)02729-4.
16. D. Antonio *et al.*, First results from 2+1 flavor domain wall QCD: Mass spectrum, topology change and chiral symmetry with L(s) = 8, *Phys. Rev. D* **75**, 114501 (2007). doi: 10.1103/PhysRevD.75.114501.
17. C. Allton *et al.*, 2+1 flavor domain wall QCD on a (2 fm)**3 lattice: Light meson spectroscopy with L(s) = 16, *Phys. Rev. D* **76**, 014504 (2007). doi: 10.1103/PhysRevD.76.014504.
18. C. Allton *et al.*, Physical results from 2+1 flavor domain wall QCD and SU(2) chiral perturbation theory, *Phys. Rev. D* **78**, 114509 (2008). doi: 10.1103/PhysRevD.78.114509.
19. Y. Aoki *et al.*, Continuum limit physics from 2+1 flavor domain wall QCD, *Phys. Rev. D* **83**, 074508 (2011). doi: 10.1103/PhysRevD.83.074508.
20. Y. Aoki *et al.*, Continuum limit of B_K from 2+1 flavor domain wall QCD, *Phys. Rev. D* **84**, 014503 (2011). doi: 10.1103/PhysRevD.84.014503.
21. R. Arthur *et al.*, Domain wall QCD with near-physical pions, *Phys. Rev. D* **87**, 094514 (2013). doi: 10.1103/PhysRevD.87.094514.
22. T. Blum *et al.*, Domain wall QCD with physical quark masses, *Phys. Rev. D* **93**(7), 074505 (2016). doi: 10.1103/PhysRevD.93.074505.
23. P. A. Boyle *et al.*, Low energy constants of SU(2) partially quenched chiral perturbation theory from N_f=2+1 domain wall QCD, *Phys. Rev. D* **93**(5), 054502 (2016). doi: 10.1103/PhysRevD.93.054502.
24. S. Aoki *et al.*, FLAG review 2019: Flavour lattice averaging group (FLAG), *Eur. Phys. J. C* **80**(2), 113 (2020). doi: 10.1140/epjc/s10052-019-7354-7.
25. Y. Aoki *et al.*, FLAG review 2021 (11, 2021).
26. D. Antonio *et al.*, Neutral kaon mixing from 2+1 flavor domain wall QCD, *Phys. Rev. Lett.* **100**, 032001 (2008). doi: 10.1103/PhysRevLett.100.032001.
27. T. Blum *et al.*, Lattice determination of the $K \to (\pi\pi)_{I=2}$ decay amplitude A_2, *Phys. Rev. D* **86**, 074513 (2012). doi: 10.1103/PhysRevD.86.074513.
28. R. Abbott *et al.*, Direct CP violation and the $\Delta I = 1/2$ rule in $K \to \pi\pi$ decay from the standard model, *Phys. Rev. D* **102**(5), 054509 (2020). doi: 10.1103/PhysRevD.102.054509.
29. T. Blum, P. A. Boyle, V. Gülpers, T. Izubuchi, L. Jin, C. Jung, A. Jüttner, C. Lehner, A. Portelli, and J. T. Tsang, Calculation of the hadronic vacuum polarization contribution to the muon anomalous magnetic moment, *Phys. Rev. Lett.* **121**(2), 022003 (2018). doi: 10.1103/PhysRevLett.121.022003.
30. T. Blum, N. Christ, M. Hayakawa, T. Izubuchi, L. Jin, C. Jung, and C. Lehner, Hadronic light-by-light scattering contribution to the Muon anomalous magnetic

moment from lattice QCD, *Phys. Rev. Lett.* **124**(13), 132002 (2020). doi: 10.1103/PhysRevLett.124.132002.

31. B. Abi *et al.*, Measurement of the positive Muon anomalous magnetic moment to 0.46 ppm, *Phys. Rev. Lett.* **126**(14), 141801 (2021). doi: 10.1103/PhysRevLett.126.141801.
32. M. Abramczyk, T. Blum, T. Izubuchi, C. Jung, M. Lin, A. Lytle, S. Ohta, and E. Shintani, Nucleon mass and isovector couplings in 2+1-flavor dynamical domain-wall lattice QCD near physical mass, *Phys. Rev. D* **101**(3), 034510 (2020). doi: 10.1103/PhysRevD.101.034510.
33. S. Ohta, Nucleon isovector form factors from physical-mass 2+1-flavor dynamical domain-wall QCD. In *38th International Symposium on Lattice Field Theory* (11, 2021).
34. T. Yamazaki, Y. Aoki, T. Blum, H.-W. Lin, S. Ohta *et al.*, Nucleon form factors with 2+1 flavor dynamical domain-wall fermions, *Phys. Rev. D* **79**, 114505 (2009). doi: 10.1103/PhysRevD.79.114505.
35. S. Ohta, Nucleon isovector axial charge in 2+1-flavor domain-wall QCD with physical mass, *PoS* **LATTICE2018**, 128 (2018). doi: 10.22323/1.334.0128.
36. A. A. Vladimirov and J. Bijnens, Leading chiral logarithms for the nucleon mass, *AIP Conf. Proc.* **1701**(1), 040019 (2016). doi: 10.1063/1.4938636.
37. M. Goldberger and S. Treiman, Decay of the pi meson, *Phys. Rev.* **110**, 1178–1184 (1958). doi: 10.1103/PhysRev.110.1178.
38. Particle Data Group, Review of particle physics, *Progress of Theoretical and Experimental Physics* **2020**(8) (08, 2020). ISSN 2050-3911. doi: 10.1093/ptep/ptaa104. URL https://doi.org/10.1093/ptep/ptaa104. 083C01.
39. A. Czarnecki, W. J. Marciano, and A. Sirlin, Neutron lifetime and axial coupling connection, *Phys. Rev. Lett.* **120**(20), 202002 (2018). doi: 10.1103/PhysRevLett.120.202002.
40. A. Chodos, R. L. Jaffe, K. Johnson, and C. B. Thorn, Baryon structure in the bag theory, *Phys. Rev. D* **10**, 2599 (1974). doi: 10.1103/PhysRevD.10.2599.
41. G. S. Adkins, C. R. Nappi, and E. Witten, Static properties of nucleons in the Skyrme model, *Nucl. Phys. B* **228**, 552 (1983). doi: 10.1016/0550-3213(83)90559-X.
42. T. Yamazaki *et al.*, Nucleon axial charge in 2+1 flavor dynamical lattice QCD with domain wall fermions, *Phys.Rev.Lett.* **100**, 171602 (2008). doi: 10.1103/PhysRevLett.100.171602.
43. J. Liang, Y.-B. Yang, K.-F. Liu, A. Alexandru, T. Draper, and R. S. Sufian, Lattice calculation of nucleon isovector axial charge with improved currents, *Phys. Rev. D* **96**(3), 034519 (2017). doi: 10.1103/PhysRevD.96.034519.
44. J. Dragos, R. Horsley, W. Kamleh, D. B. Leinweber, Y. Nakamura, P. E. L. Rakow, G. Schierholz, R. D. Young, and J. M. Zanotti, Nucleon matrix elements using the variational method in lattice QCD, *Phys. Rev. D* **94**(7), 074505 (2016). doi: 10.1103/PhysRevD.94.074505.
45. T. Bhattacharya, V. Cirigliano, S. Cohen, R. Gupta, H.-W. Lin, and B. Yoon, Axial, scalar and tensor charges of the nucleon from 2+1+1-flavor lattice QCD, *Phys. Rev. D* **94**(5), 054508 (2016). doi: 10.1103/PhysRevD.94.054508.
46. K.-I. Ishikawa, Y. Kuramashi, S. Sasaki, N. Tsukamoto, A. Ukawa, and T. Yamazaki, Nucleon form factors on a large volume lattice near the physical point in 2+1 flavor QCD, *Phys. Rev. D* **98**(7), 074510 (2018). doi: 10.1103/PhysRevD.98.074510.
47. C. C. Chang *et al.*, A per-cent-level determination of the nucleon axial coupling from quantum chromodynamics, *Nature* **558**(7708), 91–94 (2018). doi: 10.1038/s41586-018-0161-8.

48. E. Shintani, K.-I. Ishikawa, Y. Kuramashi, S. Sasaki, and T. Yamazaki, Nucleon form factors and root-mean-square radii on a $(10.8\ \mathrm{fm})^4$ lattice at the physical point, *Phys. Rev. D* **99**(1), 014510 (2019). doi: 10.1103/PhysRevD.99.014510.
49. N. Hasan, J. Green, S. Meinel, M. Engelhardt, S. Krieg, J. Negele, A. Pochinsky, and S. Syritsyn, Nucleon axial, scalar, and tensor charges using lattice QCD at the physical pion mass, *Phys. Rev.* **D99**(11), 114505 (2019). doi: 10.1103/PhysRevD.99.114505.
50. T. Harris, G. von Hippel, P. Junnarkar, H. B. Meyer, K. Ottnad, J. Wilhelm, H. Wittig, and L. Wrang, Nucleon isovector charges and twist-2 matrix elements with $N_f = 2 + 1$ dynamical Wilson quarks, *Phys. Rev. D* **100**(3), 034513 (2019). doi: 10.1103/PhysRevD.100.034513.
51. M. Gell-Mann, R. J. Oakes, and B. Renner, Behavior of current divergences under SU(3) x SU(3), *Phys. Rev.* **175**, 2195–2199 (1968). doi: 10.1103/PhysRev.175.2195.
52. Y. Aoki, T. Blum, H.-W. Lin, S. Ohta, S. Sasaki, *et al.*, Nucleon isovector structure functions in (2+1)-flavor QCD with domain wall fermions, *Phys. Rev. D* **82**, 014501 (2010). doi: 10.1103/PhysRevD.82.014501.
53. T. Yamazaki, K.-i. Ishikawa, Y. Kuramashi, and A. Ukawa, Helium nuclei, deuteron and dineutron in 2+1 flavor lattice QCD, *Phys. Rev. D* **86**, 074514 (2012). doi: 10.1103/PhysRevD.86.074514.

https://doi.org/10.1142/9789819813353_0013

Spin-Flavor $SU(6)$ Symmetry for Baryon-Baryon Interactions

Makoto Oka

Advanced Science Research Center, Japan Atomic Energy Agency,
Tokai 319-1195, Japan
and
Nishina Center for Accelerator-Based Science, RIKEN,
Wako 351-0198, Japan
makoto.oka@riken.jp

Short-range parts of the baryon-baryon (BB) interactions are analyzed from the spin-flavor $SU(6)_{sf} \supset SU(3)_f \times SU(2)_s$ symmetry viewpoint. Due to the Pauli principle of quarks, the symmetry structure of the wave functions is restricted at short distances. Consequently, the BB states with the same spin-flavor quantum numbers may be reduced into one or a few spin-flavor states. Such reduction causes repulsion and/or suppression of transitions at short distances. We show that the observed suppression of the $\Xi N \to \Lambda\Lambda$ conversion can be explained following the above argument. It is also applied to the suppression of the ΣN to ΛN conversion in the spin 1 and isospin 1/2 channel. Furthermore, the effects of the color-magnetic interaction (CMI), which prefers flavor antisymmetric states, to the Pauli-allowed states are discussed.

1. Introduction

The nuclear force is a fundamental and central subject of nuclear physics and has been intensively studied for several decades. Recent developments both in experiment and computational physics allow us to access generalized nuclear forces, that is, interactions involving baryons with strangeness (hyperons, designated in general by Y), such as Λ, Σ, Ξ and Ω baryons. Studies of the generalized nuclear force play important roles in the several aspects as follows. (1) Extension to flavor $SU(3)_f$ will help us to reveal the origin and structure of the various components of the nuclear force.[1,2] (2) YN and YY interactions are the most basic elements of the dynamics of hyper-nuclei.[3,4] (3) Strange baryons may play critical roles in dense nuclear matter that can be realized in compact stars in the universe.[5]

Analyses of the generalized nuclear force are carried out by several different approaches. Most traditional approach is to construct a phenomenological potential based on the meson exchange mechanism.[6,7] There the long-range part of the potential is given by the exchange of the low lying mesons, while the short-range part is mostly phenomenological. The coupling constants and the short-range parameters are either fitted to experimental data or determined from the NN parameters by extrapolation using the $SU(3)_f$ symmetry.

A modern alternative is to formulate the effective field theory,[8] where the systematic expansion in terms of the transferred momentum is introduced. The bound or scattering states of two-baryon systems are calculated order by order according to the chiral perturbation theory. The parameters of the effective Lagrangian are fitted to the experimental data.

The first-principle calculation by the lattice QCD is also available. The HAL-QCD group has established a new method to extract the phase-shift equivalent (non-local) potential for two hadron systems and has applied to general BB interactions.[9] This method has an advantage in calculating BB interactions which are currently not directly accessible by experiment, such as multi-strange systems like $N\Omega$, $\Omega\Omega$ or $\Xi\Xi$.[10]

The results of these approaches can be compared with experimental data of YN scattering experiments,[11] femtoscopy study in heavy ion collisions[12] and spectroscopy of hyper-nuclei.[3,4]

The flavor $SU(3)_f$ symmetry is an approximate symmetry of the quantum chromodynamics (QCD). In the baryon spectrum, the lowest-lying baryons, N, Λ, Σ and Ξ, form the octet representation of flavor $SU(3)_f$. The $SU(3)_f$ is broken by the quark mass term, where the strange quark is heavier than the up and down quarks significantly. The mass spectrum (a lá Gell-Mann and Okubo) and the magnetic moments are given consistently with the perturbative estimate of the $SU(3)_f$ breaking mass term and the wave functions based on the $SU(6)_{sf} \supset SU(3)_f \times SU(2)_s$ symmetry. Because the quark obeys the Fermi-Dirac statistics, the $SU(6)_{sf}$ symmetry is directly connected to the symmetry of the color and orbital motion for systems consisting of quarks. The ground state baryons belong to the **56** representation, that is the totally symmetric representation designated by the Young partition number [3]. Note that the color-singlet state of 3 quarks is totally antisymmetric in $SU(3)_c$.

The $SU(6)_{sf}$ symmetry plays an important role also for two-baryon systems. Now, we consider two-baryon systems consisting of the ground-state baryons belonging to the [3] symmetric, **56**-plet $SU(6)_{sf}$ representation.[13] There the symmetry of 6 quarks will form symmetries given by the Young partition numbers,

$$[3] \otimes [3] = [6]_S \oplus [42]_S \oplus [51]_A \oplus [33]_A, \tag{1}$$

where $[6]_S$ and $[42]_S$ are symmetric under the exchange of two baryons, while $[51]_A$ and $[33]_A$ are antisymmetric. The positive-parity two-baryon states, with $L = 0$ or even partial waves, must belong to the antisymmetric $[33]_A$ or $[51]_A$.

In order to satisfy the Fermi-Dirac statistics, the symmetry combinations are restricted. Table 1 shows possible symmetry combinations for the positive-parity color-singlet states of 6 quarks. One sees that the color-singlet 6-quark state takes the [222] symmetry, and the lowest-energy state with the orbital [6] symmetry must have [33] $SU(6)_{sf}$ symmetry. Such a totally symmetric orbital state will dominate over the mixed symmetric [42] configurations, when two baryons come very close and their wave functions overlap completely (in the relative $L = 0$ channel). Namely, all

Table 1. Possible symmetry combinations of positive-parity and color-singlet 6-quark systems.

Orbital symmetry	Color symmetry	$SU(6)_{sf}$ symmetry
[6]	[222]	[33]
[42]	[222]	[33]
[42]	[222]	[51]

the quarks reside in the lowest orbit within a single potential well, that is the limit of quark shell model.[14] On the other hand, the [51] $SU(6)_{sf}$ symmetry is forbidden for the totally symmetric orbital state by the Pauli principle. This will cause that the [51] state acquires a strong short-range repulsion. In this way, we can study the short-range interactions of two-baryon systems from their $SU(6)_{sf}$ symmetry structure.

When the overlap of two baryons is not complete, the orbital mixed symmetry [42] with $SU(6)_{sf}$ [33] and [51] representations will contribute for the relative $L = 0$ motion. The roles of the orbital [42] and $SU(6)_{sf}$ states have been discussed in the context of NN repulsion obtained in the quark cluster model.[14–16] The relative wave function between the two baryons in the orbital [42] symmetry will have a node at the interaction range so that it is orthogonal to the totally symmetric [6] wave function. The kinetic energy becomes larger, such as $2\hbar\omega$ excitation in the harmonic oscillator potential, and thus an effective repulsion is induced.

In this paper, we discuss the structures of the two-baryon systems in the $SU(6)_{sf} \supset SU(3)_f \times SU(2)_s$ scheme and their significance for the baryon-baryon interactions. In Sect. 2, we classify the two-baryon states in the symmetry basis states. In order to fully realize the $SU(6)_{sf}$ basis states, we need to include all possible states of two baryons consisting with the spin 1/2 octet (N, Λ, Σ and Ξ) baryons as well as the spin 3/2 decuplet (Δ, Σ^*, Ξ^* and Ω) baryons.

In Sect. 3, we discuss the effects of the Pauli forbidden states for several two baryon systems. We discuss how the short-range repulsion is induced by the dominating Pauli-forbidden states. We also present the cases where the transition between different two-baryon configurations at short distances.

In Sect. 4, we discuss the effect of $SU(6)_{sf}$ breaking, which is important because the different $SU(3)_f$ representations within the same $SU(6)_{sf}$ multiplet will split due to the color-spin dependent interaction of quarks. We here study the $SU(6)_{sf}$ breaking effects for the two-baryon systems.

Summary and conclusion is given in Sect. 5.

2. Classification According to $SU(6)_{sf} \supset SU(3)_f \times SU(2)_s$

We consider systems of two $SU(6)_{sf}$ **56**-plet baryons based on the constituent quark model. The quark model hamiltonian is $SU(6)_{sf}$ invariant if we neglect both the quark mass differences and the spin dependences.

The mass differences of u, d and s quarks breaks the $SU(3)_f$ symmetry. In the quark model, their main contribution is a constant shift of the hadron masses, which is supplemented by the kinetic energy term proportional to $1/m_q$. The spin-independent interactions, such as the quark confinement force and the color-electric gluon exchange, do not depend on flavor. Thus the effect of the $SU(3)_f$ breaking to the baryon-baryon interaction, where the number of strange quarks remains constant, plays a minor role. So, in the present analysis, we neglect the effect of the $SU(3)_f$ symmetry breaking.

On the other hand, the $SU(6)_{sf}$ symmetry is broken by the spin dependent interaction among the quarks. It is known that the color-magnetic gluon exchange interaction induces the color-spin dependent interaction, which gives a splitting between the spin-1/2 octet (N, Λ, Σ and Ξ) baryons and the spin-3/2 decuplet (Δ, Σ^*, Ξ^*, and Ω) baryons. We discuss the roles of the color-magnetic interaction in Sect. 4.

In analyzing the baryon-baryon interaction, we first consider the $SU(6)_{sf}$ and $SU(3)_f$ symmetric limit. To apply the $SU(6)_{sf}$ bases to two-baryon configurations, we need to decompose them into $SU(3)_f$ representations. The ground state baryons belong to the $SU(6)_{sf}$ [3]-symmetric, 56-dim. representation, **8** ($S = 1/2$) and **10** (3/2). Then the two-baryon systems are classified into the representations given by

$$\mathbf{8} \otimes \mathbf{8} = \mathbf{1}_S \oplus \mathbf{8}_S \oplus \mathbf{27}_S \oplus \mathbf{8}_A \oplus \mathbf{10}_A \oplus \overline{\mathbf{10}}_A, \tag{2}$$

$$\mathbf{8} \otimes \mathbf{10} = \mathbf{8} \oplus \mathbf{10} \oplus \mathbf{27} \oplus \mathbf{35}, \tag{3}$$

$$\mathbf{10} \otimes \mathbf{10} = \mathbf{27}_S \oplus \mathbf{28}_S \oplus \overline{\mathbf{10}}_A \oplus \mathbf{35}_A, \tag{4}$$

where the subscript S (A) stands for symmetric (antisymmetric) states under the exchange of two baryons. The NN interactions belong to either $\overline{\mathbf{10}}_A$ (for $I = 0$) or $\mathbf{27}_S$ (for $I = 1$) representations. All the other 4 representations, **1**, $\mathbf{8}_S$ and $\mathbf{8}_A$ and **10** in Eq. (2) are realized only when one or two strange baryons are involved. In particular, the flavor singlet **1** representation is formed by the total strangeness -2 channels, $\Lambda\Lambda$, $\Sigma\Sigma(I = 0)$ and $N\Xi(I = 0)$. With the decuplet baryons, we also have new representations, **28** and **35**.

In Tables 2 and 3, we tabulate the possible two baryon states of $SU(6)_{sf}$ [33] and [51] symmetric representations, respectively. There one sees some quantum numbers appear only in the [51] $SU(6)_{sf}$ representation. For such quantum numbers, the totally symmetric [6] orbital state is forbidden and therefore we expect a strong short-range repulsion induced by the Pauli principle. Such channels include $BB'(S, I) = N\Delta(1, 1)$, $N\Delta(2, 2)$, $\Delta\Delta(2, 3)$, $\Delta\Delta(3, 2)$, $(\Delta\Sigma{+}\Delta\Sigma^*)(2, \frac{5}{2})$, $\Delta\Sigma^*(3, \frac{3}{2})$, $\Delta\Sigma^*(3, \frac{5}{2})$, and so on.

On the other hand, many combinations of quantum numbers, particularly with lower spins, appear both in the [33] and [51] $SU(6)_{sf}$ representations. There a part of the two-baryon channels is forbidden at short distances and we need to decompose the two-baryon states into the $SU(6)_{sf} \supset SU(3)_f$ representations. We will see that the spin-dependent interaction that breaks the $SU(6)_{sf}$ symmetry

Table 2. Possible $SU(3)_f$ (designated by the dimension of the representation) and spin S states for the $SU(6)_{sf}$ [33] (490-dim.) representation. See Table 4 for the dimension $\boldsymbol{D}$ of $SU(3)_f$ and $\Delta = \Delta(S, F)$ defined by Eq. (38). The corresponding combinations of the strangeness $\mathcal{S}$ and the isospin I are shown in the last column.

$\boldsymbol{D}$	S	Δ	$(\mathcal{S}, I)$
1	0	-24	$(-2,0)$
27	0	8	$(0,1), (-1,\frac{1}{2}), (-1,\frac{3}{2}), (-2,0), (-2,1), (-2,2), (-3,\frac{1}{2}), (-3,\frac{3}{2}), (-4,1)$
28	0	48	$(0,3), (-1,\frac{5}{2}), (-2,2), (-3,\frac{3}{2}), (-4,1), (-5,\frac{1}{2}), (-6,0)$
8	1	$-28/3$	$(-1,\frac{1}{2}), (-2,0), (-2,1), (-3,\frac{1}{2})$
10	1	$8/3$	$(-1,\frac{3}{2}), (-2,1), (-3,\frac{1}{2}), (-4,0)$
$\overline{\mathbf{10}}$	1	$8/3$	$(0,0), (-1,\frac{1}{2}), (-2,1), (-3,\frac{3}{2})$
35	1	$80/3$	$(0,2), (-1,\frac{3}{2}), (-1,\frac{5}{2}), (-2,1), (-2,2), (-3,\frac{1}{2}), (-3,\frac{3}{2}), (-4,0), (-4,1), (-5,\frac{1}{2})$
8	2	4	$(-1,\frac{1}{2}), (-2,0), (-2,1), (-3,\frac{1}{2})$
27	2	16	$(0,1), (-1,\frac{1}{2}), (-1,\frac{3}{2}), (-2,0), (-2,1), (-2,2), (-3,\frac{1}{2}), (-3,\frac{3}{2}), (-4,1)$
$\overline{\mathbf{10}}$	3	16	$(0,0), (-1,\frac{1}{2}), (-2,1), (-3,\frac{3}{2})$

Table 3. Possible $SU(3)_f$ and spin states for the $SU(6)_{sf}$ [51] (1050-dim.) representation.

$\boldsymbol{D}$	S	Δ	$(\mathcal{S}, I)$
8	0	-12	$(-1,\frac{1}{2}), (-2,0), (-2,1), (-3,\frac{1}{2})$
27	0	8	$(0,1), (-1,\frac{1}{2}), (-1,\frac{3}{2}), (-2,0), (-2,1), (-2,2), (-3,\frac{1}{2}), (-3,\frac{3}{2}), (-4,1)$
8	1	$-28/3$	$(-1,\frac{1}{2}), (-2,0), (-2,1), (-3,\frac{1}{2})$
10	1	$8/3$	$(-1,\frac{3}{2}), (-2,1), (-3,\frac{1}{2}), (-4,0)$
$\overline{\mathbf{10}}$	1	$8/3$	$(0,0), (-1,\frac{1}{2}), (-2,1), (-3,\frac{3}{2})$
27	1	$32/3$	$(0,1), (-1,\frac{1}{2}), (-1,\frac{3}{2}), (-2,0), (-2,1), (-2,2), (-3,\frac{1}{2}), (-3,\frac{3}{2}), (-4,1)$
35	1	$80/3$	$(0,2), (-1,\frac{3}{2}), (-1,\frac{5}{2}), (-2,1), (-2,2), (-3,\frac{1}{2}), (-3,\frac{3}{2}), (-4,0), (-4,1), (-5,\frac{1}{2})$
10	2	8	$(-1,\frac{3}{2}), (-2,1), (-3,\frac{1}{2}), (-4,0)$
27	2	16	$(0,1), (-1,\frac{1}{2}), (-1,\frac{3}{2}), (-2,0), (-2,1), (-2,2), (-3,\frac{1}{2}), (-3,\frac{3}{2}), (-4,1)$
35	2	32	$(0,2), (-1,\frac{3}{2}), (-1,\frac{5}{2}), (-2,1), (-2,2), (-3,\frac{1}{2}), (-3,\frac{3}{2}), (-4,0), (-4,1), (-5,\frac{1}{2})$
28	2	56	$(0,3), (-1,\frac{5}{2}), (-2,2), (-3,\frac{3}{2}), (-4,1), (-5,\frac{1}{2}), (-6,0)$
35	3	40	$(0,2), (-1,\frac{3}{2}), (-1,\frac{5}{2}), (-2,1), (-2,2), (-3,\frac{1}{2}), (-3,\frac{3}{2}), (-4,0), (-4,1), (-5,\frac{1}{2})$

becomes important and then the lower symmetric $SU(3)_f$ representations happen to be dominant. (In fact, the states with the lower values of $\Delta(S, F)$ in Table 2 are favored.) Thus the flavor singlet **1** and octet **8** states will have a good chance to make a compact dibaryon state.

In the following, we take four cases and analyze them from the $SU(6)_{sf} \supset SU(3)_f$ symmetry viewpoint. (1) BB with strangeness $\mathcal{S} = -2$, spin $S = 0$, and isospin $I = 0$, (2) BB with $\mathcal{S} = -1$, $S = 0$, $I = 1/2$, (3) BB with $\mathcal{S} = -1$, $S = 1$, $I = 1/2$ and (4) non-strange BB with $S = 0$, $I = 1$ and $S = 1$, $I = 0$.

2.1. *Strangeness $\mathcal{S} = -2$, spin $S = 0$, isospin $I = 0$ states*

First we consider the states with strangeness $\mathcal{S} = -2$, spin $S = 0$, and isospin $I = 0$. In the particle basis, they consist of $\Lambda\Lambda$, $(N\Xi)_S$, $\Sigma\Sigma$, and $\Sigma^*\Sigma^*$. This system has attracted much attention because of possible existence of the dibaryon, a compact 6-quark state predicted by Jaffe in 1970.[17] The H dibaryon has attracted much attention, which are searched by experiment[18] and also is studied by lattice QCD.[19] The quark cluster model was applied to the H dibaryon system.[20]

For the $SU(3)_f$ basis, they belong to the symmetric BB representations out of $\mathbf{8} \times \mathbf{8}$, and $\mathbf{10} \times \mathbf{10}$, given in terms of the particle bases as

$$|\mathbf{1}\rangle = -\sqrt{\frac{1}{8}}|\Lambda\Lambda\rangle + \sqrt{\frac{1}{2}}|N\Xi\rangle_S + \sqrt{\frac{3}{8}}|\Sigma\Sigma\rangle \tag{5}$$

$$|\mathbf{8}\rangle = -\sqrt{\frac{1}{5}}|\Lambda\Lambda\rangle + \sqrt{\frac{1}{5}}|N\Xi\rangle_S - \sqrt{\frac{3}{5}}|\Sigma\Sigma\rangle \tag{6}$$

$$|\mathbf{27}\rangle_1 = \sqrt{\frac{27}{40}}|\Lambda\Lambda\rangle + \sqrt{\frac{12}{40}}|N\Xi\rangle_S - \sqrt{\frac{1}{40}}|\Sigma\Sigma\rangle \tag{7}$$

$$|\mathbf{27}\rangle_2 = |\Sigma^*\Sigma^*\rangle \tag{8}$$

In the $SU(6)_{sf} \supset SU(3)_f \times SU(2)_s$ basis, they are classified into the four states, ($[33]\,\mathbf{1}$, $[33]\,\mathbf{27}$, $[51]\,\mathbf{8}$, $[51]\,\mathbf{27}$), where the Young partition $[f_1 f_2]$ represents the $SU(6)_{sf}$ representation and the bold number is for the $SU(3)_f$ multiplet. The relation between the $SU(6)_{sf}$ basis and the particle basis is given by

$$\begin{pmatrix} [33]\,\mathbf{1} \\ [33]\,\mathbf{27} \\ [51]\,\mathbf{8} \\ [51]\,\mathbf{27} \end{pmatrix} = \frac{1}{\sqrt{360}} \begin{pmatrix} -\sqrt{45} & \sqrt{180} & \sqrt{135} & 0 \\ \sqrt{135} & \sqrt{60} & -\sqrt{5} & -\sqrt{160} \\ -\sqrt{72} & \sqrt{72} & -\sqrt{216} & 0 \\ -\sqrt{108} & -\sqrt{48} & \sqrt{4} & -\sqrt{200} \end{pmatrix} \begin{pmatrix} \Lambda\Lambda \\ N\Xi_S \\ \Sigma\Sigma \\ \Sigma^*\Sigma^* \end{pmatrix}, \tag{9}$$

or conversely,

$$\begin{pmatrix} \Lambda\Lambda \\ N\Xi_S \\ \Sigma\Sigma \\ \Sigma^*\Sigma^* \end{pmatrix} = \frac{1}{\sqrt{360}} \begin{pmatrix} -\sqrt{45} & \sqrt{135} & -\sqrt{72} & -\sqrt{108} \\ \sqrt{180} & \sqrt{60} & \sqrt{72} & -\sqrt{48} \\ \sqrt{135} & -\sqrt{5} & -\sqrt{216} & \sqrt{4} \\ 0 & -\sqrt{160} & 0 & -\sqrt{200} \end{pmatrix} \begin{pmatrix} [33]\,\mathbf{1} \\ [33]\,\mathbf{27} \\ [51]\,\mathbf{8} \\ [51]\,\mathbf{27} \end{pmatrix}. \tag{10}$$

These classifications can be obtained as the eigenstates of the quark exchange operators. More concretely, the above expression can be obtained by diagonalizing the matrix elements of the P_{36}^{sf} operator, that exchanges a quark from the first baryon and a quark from the second. In Appendix A, the P_{36}^{sf} matrices for the particle basis states are given explicitly.

2.2. $\mathcal{S} = -1$, $S = 0$, $I = 1/2$ states

Next, we consider the strangeness -1 two baryon systems with $S = 0$ and $I = 1/2$. They consist of the symmetric combinations, $((N\Lambda)_S, (N\Sigma)_S, (\Delta\Sigma^*)_S)$, which are rearranged into the $SU(3)_f$ basis states as

$$|\mathbf{8}\rangle = -\sqrt{\frac{1}{10}}|N\Lambda\rangle_S + \sqrt{\frac{9}{10}}|N\Sigma\rangle_S \tag{11}$$

$$|\mathbf{27}\rangle_1 = \sqrt{\frac{9}{10}}|N\Lambda\rangle_S + \sqrt{\frac{1}{10}}|N\Sigma\rangle_S \tag{12}$$

$$|\mathbf{27}\rangle_2 = |\Delta\Sigma^*\rangle_S \tag{13}$$

Note that the $\mathbf{8} \otimes \mathbf{10}$ combinations are not allowed for $S = 0$ states.

Now similarly to the strangeness -2 system, we relate them to the $SU(6)_{sf}$ basis states, giving

$$\begin{pmatrix} [33]\,\mathbf{27} \\ [51]\,\mathbf{27} \\ [51]\,\mathbf{8} \end{pmatrix} = \frac{1}{\sqrt{90}} \begin{pmatrix} \sqrt{45} & \sqrt{5} & -\sqrt{40} \\ \sqrt{36} & \sqrt{4} & \sqrt{50} \\ -\sqrt{9} & \sqrt{81} & 0 \end{pmatrix} \begin{pmatrix} N\Lambda_S \\ N\Sigma_S \\ \Delta\Sigma^*_S \end{pmatrix}, \tag{14}$$

and conversely,

$$\begin{pmatrix} N\Lambda_S \\ N\Sigma_S \\ \Delta\Sigma^*_S \end{pmatrix} = \frac{1}{\sqrt{90}} \begin{pmatrix} \sqrt{45} & \sqrt{36} & -\sqrt{9} \\ \sqrt{5} & \sqrt{4} & \sqrt{81} \\ -\sqrt{40} & \sqrt{50} & 0 \end{pmatrix} \begin{pmatrix} [33]\,\mathbf{27} \\ [51]\,\mathbf{27} \\ [51]\,\mathbf{8} \end{pmatrix}. \tag{15}$$

2.3. $\mathcal{S} = -1$, $S = 1$, $I = 1/2$ states

Now we consider the strangeness -1, $S = 1$, $I = 1/2$ states, $((N\Lambda)_A, (N\Sigma)_A, (N\Sigma^*)_A, (\Delta\Sigma)_A, (\Delta\Sigma^*)_A)$. The $SU(3)_f$ basis states from $\mathbf{8}\times\mathbf{8}$, $\mathbf{8}\times\mathbf{10}$, and $\mathbf{10}\times\mathbf{10}$ combinations are given by

$$|\mathbf{8}\rangle_1 = \frac{1}{\sqrt{2}}|N\Lambda\rangle_A + \frac{1}{\sqrt{2}}|N\Sigma\rangle_A \tag{16}$$

$$|\overline{\mathbf{10}}\rangle_1 = \frac{1}{\sqrt{2}}|N\Lambda\rangle_A - \frac{1}{\sqrt{2}}|N\Sigma\rangle_A \tag{17}$$

$$|\mathbf{8}\rangle_2 = \frac{1}{\sqrt{5}}|N\Sigma^*\rangle_A + \frac{4}{\sqrt{5}}|\Delta\Sigma\rangle_A \tag{18}$$

$$|\mathbf{27}\rangle = \frac{4}{\sqrt{5}}|N\Sigma^*\rangle_A - \frac{1}{\sqrt{5}}|\Delta\Sigma\rangle_A \tag{19}$$

$$|\overline{\mathbf{10}}\rangle_2 = |\Delta\Sigma^*\rangle_A. \tag{20}$$

Again the correspondent $SU(6)_{sf}$ basis states are given by

$$\begin{pmatrix} [33]\,\mathbf{8} \\ [33]\,\overline{\mathbf{10}} \\ [51]\,\mathbf{8} \\ [51]\,\overline{\mathbf{10}} \\ [51]\,\mathbf{27} \end{pmatrix} = \frac{1}{\sqrt{90}} \begin{pmatrix} \sqrt{20} & \sqrt{20} & -\sqrt{10} & -\sqrt{40} & 0 \\ \sqrt{25} & -\sqrt{25} & 0 & 0 & -\sqrt{40} \\ \sqrt{25} & \sqrt{25} & \sqrt{8} & \sqrt{32} & 0 \\ \sqrt{20} & -\sqrt{20} & 0 & 0 & \sqrt{50} \\ 0 & 0 & \sqrt{72} & -\sqrt{18} & 0 \end{pmatrix} \begin{pmatrix} N\Lambda_A \\ N\Sigma_A \\ N\Sigma^*_A \\ \Delta\Sigma_A \\ \Delta\Sigma^*_A \end{pmatrix}, \tag{21}$$

and

$$\begin{pmatrix} N\Lambda_A \\ N\Sigma_A \\ N\Sigma^*_A \\ \Delta\Sigma_A \\ \Delta\Sigma^*_A \end{pmatrix} = \frac{1}{\sqrt{90}} \begin{pmatrix} \sqrt{20} & \sqrt{25} & \sqrt{25} & \sqrt{20} & 0 \\ \sqrt{20} & -\sqrt{25} & \sqrt{25} & -\sqrt{20} & 0 \\ -\sqrt{10} & 0 & \sqrt{8} & 0 & \sqrt{72} \\ -\sqrt{40} & 0 & \sqrt{32} & 0 & -\sqrt{18} \\ 0 & -\sqrt{40} & 0 & \sqrt{50} & 0 \end{pmatrix} \begin{pmatrix} [33]\,\mathbf{8} \\ [33]\,\overline{\mathbf{10}} \\ [51]\,\mathbf{8} \\ [51]\,\overline{\mathbf{10}} \\ [51]\,\mathbf{27} \end{pmatrix}. \tag{22}$$

2.4. *Non-strange* $S = 0$, $I = 1$ *and* $S = 1$, $I = 0$ *states*

Finally, we give the classifications of the NN and $\Delta\Delta$ states with $(S, I) = (0, 1)$ and $(1, 0)$. The $(S, I) = (0, 1)$ $NN - \Delta\Delta$ states belong to the flavor **27** representation and can be classified by the $SU(6)_{sf}$ symmetry as

$$\begin{pmatrix} NN_S \\ \Delta\Delta_S \end{pmatrix} = \frac{1}{3} \begin{pmatrix} \sqrt{5} & \sqrt{4} \\ -\sqrt{4} & \sqrt{5} \end{pmatrix} \begin{pmatrix} [33]\,\mathbf{27} \\ [51]\,\mathbf{27} \end{pmatrix}. \tag{23}$$

The $(S, I) = (1, 0)$ $NN - \Delta\Delta$ states belong to the flavor $\overline{\mathbf{10}}$ representation and are given by

$$\begin{pmatrix} NN_A \\ \Delta\Delta_A \end{pmatrix} = \frac{1}{3} \begin{pmatrix} \sqrt{5} & \sqrt{4} \\ -\sqrt{4} & \sqrt{5} \end{pmatrix} \begin{pmatrix} [33]\,\overline{\mathbf{10}} \\ [51]\,\overline{\mathbf{10}} \end{pmatrix}. \tag{24}$$

3. Short-Range Interactions of Baryons

Now it is straightforward to see the short-range behavior of the baryon-baryon interaction in the $SU(6)_{sf}$ symmetry limit. We remember that the Fermi-Dirac statistics allows only the [33] symmetric states when two baryons sit on top of each other. Thus at the very short distance, the relative wave function of two baryons should be dominated by the [33] component.

As an example, in the case of the $NN - \Delta\Delta(I = 1)$ system (see Sect. 2.4), there is only one [33] state, [33]**27**. The [51] component, [51]**27**, is forbidden by the Pauli principle and thus will disappear at short distances. In other words, the NN and $\Delta\Delta$ components, which are independent in the asymptotic region, are reduced into a single component and are not distinguishable at short distances. There 55% of the NN wave function is [33]**27**, but the remaining 45% is forbidden by the Pauli principle.

It is natural to ask whether this effect of the Pauli principle explains the short-range repulsion between two nucleons that was established experimentally and was also demonstrated by the quark cluster model calculations. It, however, is found that the 45% suppression of the NN short-range wave function by the Pauli principle is not enough to induce the short-range repulsion. To understand the short-range repulsion, we need the $SU(6)_{sf}$ breaking interaction, which splits the NN and $\Delta\Delta$, as is discussed in the next section.

On the other hand, the effect of antisymmetrization becomes prominent in the $\mathcal{S} = -1$, $(S, I) = (0, 1/2)$ channel (Sect. 2.2), where only one [33] state, [33]**27** is allowed at short distances. From Eq. (15), one sees that the probability of the allowed state is 50%, 5.6% and 10% for $N\Lambda$, $N\Sigma$ and $\Delta\Sigma^*$, respectively. Therefore the $N\Sigma$ wave function is strongly suppressed (almost forbidden) at short distances. This is a "quasi"-Pauli-forbidden state, which effectively induces a strong repulsion between the baryons. Thus it is expected that the $N\Sigma$ interaction for the S-wave spin $S = 1/2$ and isospin $I = 0$ channel is strongly repulsive.

A similar quasi-Pauli-forbidden state appears also in the $N\Sigma$ state with $\mathcal{S} = -1$, $(S, I) = (1, 3/2)$ channel. There are five S-wave states, $N\Sigma$, $N\Sigma^*$, $\Delta\Lambda$, $\Delta\Sigma$ and $\Delta\Sigma^*$, and they form [33]**10**, [33]**35**, [51]**10**, [55]**27** and [55]**35** representations. The only state with two octet baryons, $N\Sigma$ contains only 11% of [33] state,

$$|N\Sigma(1, 3/2)\rangle = \sqrt{\frac{1}{9}}|[33]\mathbf{10}\rangle + |[51] \text{ states}\rangle \ldots. \tag{25}$$

Therefore the S-wave interaction in the $(S, I) = (1, 3/2)$ $N\Sigma$ state is also strongly repulsive. In contrast, the $N\Sigma$ $(S, I) = (0, 3/2)$ channel has only two S-wave states, $N\Sigma$ and $\Delta\Sigma^*$, which form [33]**27** and [51]**27** representations. The lowest $N\Sigma$ state is decomposed as

$$|N\Sigma(0, 3/2)\rangle = \sqrt{\frac{5}{9}}|[33]\mathbf{27}\rangle + \sqrt{\frac{4}{9}}|[51]\mathbf{27}\rangle, \tag{26}$$

and the Pauli principle in this case does not induce repulsion.

The short-range repulsions between N and Σ in the $(S, I) = (0, 1/2)$ and $(1, 3/2)$ channels were demonstrated by the quark cluster model calculation.[1] The strengths of the repulsions are much larger than the repulsion between two nucleons, showing that the effect of the Pauli forbidden state is dominant compared with the $SU(6)_{sf}$ breaking effect. The repulsions in these channels were confirmed also in the lattice computation of the potential.[24,25] The E40 experiment at J-PARC was very successful in measuring the $p\Sigma^{\pm}$ scattering cross sections recently.[11,26] The data are analyzed to extract the S-wave phase shifts, which show moderate $N\Sigma$ repulsion at low and medium energies. The repulsion of $\Sigma - N$ interaction is also consistent with the experimental observation of the Σ-nucleus potential.[24,27]

When there are more than one [33] states available, the $SU(6)_{sf}$ classification will give us non-trivial relations. We consider the strangeness $\mathcal{S} = -2$, $(S, I) = (0, 0)$

case (Sect. 2.1). Eq. (10) shows us that the wave functions at short distances are dominated by the [33] states, as

$$|\Lambda\Lambda\rangle = -\sqrt{\frac{1}{8}}|[33]\mathbf{1}\rangle + \sqrt{\frac{3}{8}}|[33]\mathbf{27}\rangle + [51]\ldots \tag{27}$$

$$|N\Xi\rangle_S = \sqrt{\frac{1}{2}}|[33]\mathbf{1}\rangle + \sqrt{\frac{1}{6}}|[33]\mathbf{27}\rangle + [51]\ldots \tag{28}$$

$$|\Sigma\Sigma\rangle = \sqrt{\frac{3}{8}}|[33]\mathbf{1}\rangle - \sqrt{\frac{1}{72}}|[33]\mathbf{27}\rangle + [51]\ldots \tag{29}$$

$$|\Sigma^*\Sigma^*\rangle = -\frac{2}{3}|[33]\mathbf{27}\rangle + [51]\ldots \tag{30}$$

One sees that the probabilities of the [33] states are 50%, 67%, 39% and 44% for $\Lambda\Lambda$, $N\Xi$, $\Sigma\Sigma$ and $\Sigma^*\Sigma^*$, respectively, and no significant suppression is seen for these channels. On the other hand, we see that the overlap between $|\Lambda\Lambda\rangle$ and $|N\Xi_S\rangle$ states is zero. Namely, the [33] components of $|\Lambda\Lambda\rangle$ and $|N\Xi_S\rangle$ are orthogonal to each other. Within the approximation of $SU(6)_{sf}$ invariance of the inter-quark interaction, there is no mixing between $\Lambda\Lambda$ and $N\Xi$ at short distances in this quantum number. We here conclude that the $N\Xi$ to $\Lambda\Lambda$ conversion is strongly suppressed at the distances where the quark wave functions of two baryons overlap significantly. It explains why the transition potential from $N\Xi$ to $\Lambda\Lambda$ is suppressed at short distances. This property of the $N\Xi$-$\Lambda\Lambda$ system has been pointed out by the lattice QCD calculation.[22] It is also consistent with the discoveries of Ξ-hypernuclei in the J-PARC E07 experiment.[4,23]

Another interesting case is $\mathcal{S} = -1$, $(S, I) = (1, 1/2)$ (Sect. 2.3) system. One sees two allowed [33] states, $[33]\mathbf{8}$ and $[33]\overline{\mathbf{10}}$. The two baryon systems contain these states as

$$|N\Lambda\rangle_A = \sqrt{\frac{4}{18}}|[33]\mathbf{8}\rangle + \sqrt{\frac{5}{18}}|[33]\overline{\mathbf{10}}\rangle + [51]\ldots \tag{31}$$

$$|N\Sigma\rangle_A = \sqrt{\frac{4}{18}}|[33]\mathbf{8}\rangle - \sqrt{\frac{5}{18}}|[33]\overline{\mathbf{10}}\rangle + [51]\ldots \tag{32}$$

$$|N\Sigma^*\rangle_A = -\sqrt{\frac{1}{9}}|[33]\mathbf{8}\rangle + [51]\ldots \tag{33}$$

$$|\Delta\Sigma\rangle_A = -\sqrt{\frac{4}{9}}|[33]\mathbf{8}\rangle + [51]\ldots \tag{34}$$

$$|\Delta\Sigma^*\rangle_A = -\sqrt{\frac{4}{9}}|[33]\overline{\mathbf{10}}\rangle + [51]\ldots \tag{35}$$

Now, one sees that the $N\Sigma^*$ state is almost 90% suppressed by the Pauli principle so that a strong repulsion is expected there. We also see that the [33] part of $N\Lambda$ and $N\Sigma$ are nearly orthogonal. Again, the $SU(6)_{sf}$ invariance will suppress the transition between these states at short distances, which again is consistent with the lattice QCD calculation of the transition potential.[25]

4. Color Magnetic Interaction

So far, we have classified the two-baryon systems in terms of the $SU(6)_{sf}$ symmetry. However, this symmetry is broken as the splittings between the octet (N, Λ, ...) and decuplet (Δ, Σ^*, ...) baryons indicate. For a compact state of quarks, the most important dynamics that breaks $SU(6)_{sf}$ invariance is the spin-color dependent force, called color magnetic interaction(CMI). Its origin may be a gluon exchange between constituent quarks,[28] or the instanton induced interaction for light quarks.[29] The CMI is given by

$$V_{\rm CMI} = -\sum_{i<j} (\boldsymbol{\sigma}_i \cdot \boldsymbol{\sigma}_j)(\lambda_i \cdot \lambda_j) v_0(r_{ij}) \tag{36}$$

where r_{ij} is the distance between the i-th and j-th quarks, $\lambda_{i(j)}$ denotes the color Gell-Mann matrix for the $i(j)$-th quark. Here we neglect the mass difference between u, d and s quarks, namely considering the $SU(3)_f$ limit. Assuming for the compact 6-quark state that all the quarks are in the same orbital state, the expectation value is given by

$$\langle V_{\rm CMI} \rangle = \langle v_0 \rangle \Delta(S, F) \tag{37}$$

where $\Delta(S, F)$ depends on the total spin S and the flavor representation, $F = (p, q)$. It is given by the formula[a]

$$\begin{aligned} \Delta(S, F = (p, q)) &= \left\langle -\sum_{i<j} (\boldsymbol{\sigma}_i \cdot \boldsymbol{\sigma}_j)(\lambda_i \cdot \lambda_j) \right\rangle \\ &= n(n-10) + \frac{4}{3} S(S+1) + 4C_2[(p, q)], \end{aligned} \tag{38}$$

where n is the total number of quarks ($n = 6$ for two baryons), C_2 is the $SU(3)$ quadratic Casimir invariant of the flavor representation $F = (p, q)$ given in Table 4. The values of $\Delta(S, F)$ are tabulated for the relevant spin and flavor in Tables 2 and 3.

The strength factor, $\langle v_0 \rangle$ in Eq. (37), is positive and thus the smaller $SU(3)$ Casimir eigenvalue is preferred. As the splitting between the octet and decuplet

[a] A general formula for fully antisymmetric n quarks with spin S, flavor F and color C is given by

$$\begin{aligned} \Delta(S, F, C) &= -\frac{2}{3} \frac{n(n-1)}{2} + \frac{4}{3} \left\langle \sum_{i<j} P^s_{ij} \right\rangle + 2 \left\langle \sum_{i<j} P^c_{ij} \right\rangle + 4 \left\langle \sum_{i<j} P^f_{ij} \right\rangle \\ &= n(n-10) + \frac{4}{3} S(S+1) + 2C_2[C] + 4C_2[F], \end{aligned}$$

where $P^{s,f,c}_{ij}$ are the exchange operators for the spin (s), flavor (f) and color (c) part of the wave function, respectively. $C_2[C]$ vanishes for color-singlet states.

Table 4. Casimir eigenvalues of the $SU(3)$ representations. $[f] = [f_1, f_2, f_3]$ denotes the Young partition numbers for the symmetry of the representation, which specifies $p = f_1 - f_2$ and $q = f_2 - f_3$. Then p and q give the dimension $\boldsymbol{D} = (p+1)(q+1)(p+q+2)/2$ and the Casimir eigenvalue $C_2 = (p^2 + pq + q^2)/3 + p + q$ of the representation.

$\boldsymbol{D}(p,q)$	$[f]$	(p,q)	$C_2(p,q)$
1	[111]	$(0,0)$	0
8	[21]	$(1,1)$	3
10	[3]	$(3,0)$	6
$\overline{\mathbf{10}}$	[33]	$(0,3)$	6
27	[42]	$(2,2)$	8
35	[51]	$(4,1)$	12
28	[6]	$(6,0)$	18

baryons, N and Δ, is given by $16\langle v_0\rangle$, we can make a rough estimate, $\langle v_0\rangle \sim 300/16$ MeV. The flavor symmetric states have larger Casimir eigenvalues than flavor antisymmetric states, the dibaryon states with antisymmetric flavor eigenstates are favored by CMI. Among the [33] symmetric states, the most favored state by CMI is the $\mathbf{1}(S=0)$ state, which appears Eq. (10). According to Table I, $\Delta(S,F)$ is -24. This is the state relevant for the possible H dibaryon, *uuddss* system with strangeness $\mathcal{S} = -2$, $(S,I) = (0,0)$. The other [33] state in Eq. (10) is $\mathbf{27}(S=0)$, for which $\Delta(S,F)$ is $+8$, *i.e.*, ~ 600 MeV above the $\mathbf{1}(S=0)$ state. Thus we conclude that the compact low-energy state is dominated by the $\mathbf{1}(S=0)$ state.

Similar analysis can be done for the $\mathcal{S} = -1$, $(S,I) = (1,1/2)$ states, Eq. (22). The two [33] states, $\mathbf{8}(S=1)$ and $\overline{\mathbf{10}}(S=1)$, have $\Delta(S,F) = -28/3$ and $8/3$, and thus the $\mathbf{8}(S=1)$ is about 225 MeV above the $\overline{\mathbf{10}}(S=1)$. Both the $N\Lambda$ and $N\Sigma$ states contain the $\mathbf{8}(S=1)$ and $\overline{\mathbf{10}}(S=1)$ states by 22 and 28%, respectively, though the relative signs are opposite. So, here one of the combinations of $N\Lambda$ and $N\Sigma$ becomes dominant at short distances, while the overall mixing of the $N\Lambda$ and $N\Sigma$ is suppressed. We can check such behaviors using the lattice QCD calculation.

5. Summary and Conclusions

We have studied the short-range interaction of two-baryon systems with u, d and s quarks from the viewpoint of $SU(6)_{sf}$ symmetry. In the color singlet six-quark system, when all the quarks are in the same orbit, the $SU(6)_{sf}$ representation is restricted to [33] symmetric 490-dim. representation. By analyzing the [33] components of the two baryon spin-flavor wave functions, one determines the dominant state at the short distances. When the [33] component is a minor contribution in a two-baryon system, then such a system has a strong repulsion at short distances. This is the case that the state is almost forbidden by the Pauli principle.

Such a almost-Pauli forbidden state is seen for $N\Sigma$ with $(S, I) = (0, 1/2)$ and $(1, 3/2)$ systems.

In some cases, there are more than one Pauli-allowed [33] states with different flavor representations. Then for the BB coupled channel system, there are two independent Pauli-allowed states. In the case of the strangeness -2, $S = 0$ and $I = 0$ system with $\Lambda\Lambda$, $N\Xi$ and so on, we find that the Pauli-allowed components of $\Lambda\Lambda$ and $N\Xi$ are orthogonal so that the $SU(6)_{sf}$ invariant interaction does not induce the transition. This explains from the symmetry viewpoint that the $N\Xi$ conversion to $\Lambda\Lambda$ in the ${}^1S_0$ channel is suppressed. A similar suppression is also predicted in the $N\Lambda - N\Sigma$ conversion in the ${}^3S_1$ channel.

Acknowledgments

It is a great honor for me to acknowledge late Prof. Akito Arima for his continuous encouragement during my entire career in physics. This paper is based on the research carried out with Akito, Ref. 14, where we formulated the quark shell model wave functions for the six-quark ($B = 2$) systems. This approach happened to be very fruitful and successful in revealing properties of the short-range baryon-baryon interactions.

I also acknowledge Drs. Tetsuo Hatsuda, Emiko Hiyama, and Takumi Doi for useful discussions. This work was supported by Grants-in-Aid for Scientific Research No. JP20K03959, and No. JP21H00132.

Appendix A. Matrix Elements of the Spin-Flavor Exchange Operator

We follow the Nijmegen (de Swart) prescription for the $SU(3)_f$ wave functions of the octet and decuplet baryons.[30] In order to obtain the $SU(6)_{sf}$ eigenstates, it is convenient to use the matrix elements of the exchange operator, which interchanges the spin-flavor part of the quark wave functions between the baryons.

For $\mathcal{S} = -2$, $S = 0$ and $I = 0$ states, the matrix elements of the exchange operator P^{sf}_{36} of quarks 3 and 6, assuming that the 1st (2nd) baryon is formed by quarks 1, 2 and 3 (4, 5 and 6).

$$\langle P^{sf}_{36} \rangle = \frac{1}{27} \begin{pmatrix} 0 & 0 & 3\sqrt{3} & 3\sqrt{6} \\ 0 & -3 & -4\sqrt{3} & 2\sqrt{6} \\ 3\sqrt{3} & -4\sqrt{3} & 2 & -\sqrt{2} \\ 3\sqrt{6} & 2\sqrt{6} & -\sqrt{2} & 1 \end{pmatrix} \begin{matrix} \Lambda\Lambda \\ N\Xi_S \\ \Sigma\Sigma \\ \Sigma^*\Sigma^*. \end{matrix} \tag{A.1}$$

For $\mathcal{S} = -1$, $S = 0$ and $I = 1/2$ states, we have

$$\langle P^{sf}_{36} \rangle = \frac{1}{27} \begin{pmatrix} 0 & -3 & 6\sqrt{2} \\ -3 & 8 & 2\sqrt{2} \\ 6\sqrt{2} & 2\sqrt{2} & 1 \end{pmatrix} \begin{matrix} N\Lambda_S \\ N\Sigma_S \\ \Delta\Sigma^*_S. \end{matrix} \tag{A.2}$$

For $\mathcal{S} = -1$, $S = 1$ and $I = 1/2$ states, we have

$$\langle P_{36}^{sf} \rangle = \frac{1}{27} \begin{pmatrix} 0 & 1 & 2\sqrt{2} & 4\sqrt{2} & 2\sqrt{10} \\ 1 & 0 & 2\sqrt{2} & 4\sqrt{2} & -2\sqrt{10} \\ 2\sqrt{2} & 2\sqrt{2} & 7 & -4 & 0 \\ 4\sqrt{2} & 4\sqrt{2} & -4 & 1 & 0 \\ 2\sqrt{10} & -2\sqrt{10} & 0 & 0 & 1 \end{pmatrix} \begin{matrix} N\Lambda_A \\ N\Sigma_A \\ N\Sigma_A^* \\ \Delta\Sigma_A \\ \Delta\Sigma_A^*. \end{matrix} \tag{A.3}$$

For the non-strange $(S, I) = (0, 1)$ and $(1, 0)$ states, we have

$$\langle P_{36}^{sf} \rangle = \frac{1}{27} \begin{pmatrix} -1 & 4\sqrt{5} \\ 4\sqrt{5} & 1 \end{pmatrix} \begin{matrix} NN(I=1) \\ \Delta\Delta(I=1) \end{matrix} \tag{A.4}$$

$$\langle P_{36}^{sf} \rangle = \frac{1}{27} \begin{pmatrix} -1 & 4\sqrt{5} \\ 4\sqrt{5} & 1 \end{pmatrix} \begin{matrix} NN(I=0) \\ \Delta\Delta(I=0). \end{matrix} \tag{A.5}$$

By diagonalizing the above exchange operator matrix elements, we obtain the irreducible representations of the $SU(6)_{sf}$ symmetry. The eigenvalues are $-1/3$ for the [33] representation and $1/3$ for [51].

References

1. M. Oka, K. Shimizu, and K. Yazaki, Hyperon — Nucleon and hyperon-hyperon interaction in a quark model, *Nucl. Phys. A* **464**, 700–716 (1987).
2. T. A. Rijken, M. M. Nagels, and Y. Yamamoto, Baryon-Baryon interactions $S = 0, -1, -2, -3, -4$, *Few Body Syst.* **54**, 801–806 (2013).
3. A. Gal, E. V. Hungerford, and D. J. Millener, Strangeness in nuclear physics, *Rev. Mod. Phys.* **88**, 035004 (2016).
4. E. Hiyama and K. Nakazawa, Structure of $S = -2$ hypernuclei and hyperon-hyperon interactions, *Ann. Rev. Nucl. Part. Sci.* **68**, 131–159 (2018).
5. W. Weise, Equation of state and strangeness in neutron stars — role of hyperon-nuclear three-body forces -, *EPJ Web Conf.* **271**, 06003 (2022).
6. J. Haidenbauer and U. G. Meißner, Jülich hyperon-nucleon model revisited, *Phys. Rev. C* **72**, 044005 (2005).
7. T. A. Rijken, M. M. Nagels, and Y. Yamamoto, Baryon-baryon interactions: Nijmegen extended-soft-core models, *Prog. Theor. Phys. Suppl.* **185**, 14–71 (2010). M. M. Nagels, T. A. Rijken, and Y. Yamamoto, Extended-soft-core baryon-baryon model ESC16. II. Hyperon-nucleon interactions, *Phys. Rev. C* **99**, 044003 (2019).
8. S. Petschauer, J. Haidenbauer, N. Kaiser, U. G. Meißner, and W. Weise, Hyperon-nuclear interactions from SU(3) chiral effective field theory, *Front. in Phys.* **8**, 12 (2020).
9. N. Ishii, S. Aoki, and T. Hatsuda, The nuclear force from lattice QCD, *Phys. Rev. Lett.* **99**, 022001 (2007). T. Inoue *et al.* [HAL QCD], Baryon-Baryon interactions in the flavor SU(3) limit from full QCD simulations on the lattice, *Prog. Theor. Phys.* **124**, 591–603 (2010).
10. S. Gongyo *et al.* [HAL QCD], Most strange Dibaryon from lattice QCD, *Phys. Rev. Lett.* **120**(21), 212001 (2018). T. Iritani *et al.* [HAL QCD], $N\Omega$ dibaryon from lattice QCD near the physical point, *Phys. Lett. B* **792**, 284–289 (2019).

11. K. Miwa *et al.*, Recent progress and future prospects of hyperon nucleon scattering experiment, *EPJ Web Conf.* **271**, 04001 (2022).
12. L. Fabbietti, V. Mantovani Sarti, and O. Vazquez Doce, Study of the strong interaction among hadrons with correlations at the LHC, *Ann. Rev. Nucl. Part. Sci.* **71**, 377–402 (2021).
13. M. Oka and K. Yazaki, Nuclear force in a quark model, *Phys. Lett. B* **90**, 41–44 (1980). M. Oka, K. Shimizu, and K. Yazaki, Quark cluster model of baryon baryon interaction, *Prog. Theor. Phys. Suppl.* **137**, 1–20 (2000).
14. S. Ohta, A. Arima, K. Yazaki, and M. Oka, A shell model study of six quark system, *Phys. Lett. B* **119**, 35–38 (1982) [erratum: *Phys. Lett. B* **123**, 477–477 (1983)].
15. M. Harvey, Effective nuclear forces in the quark model with Delta and hidden color channel coupling, *Nucl. Phys. A* **352**, 326–342 (1981).
16. A. Faessler, F. Fernandez, G. Lubeck, and K. Shimizu, The quark model and the nature of the repulsive core of the nucleon nucleon interaction, *Phys. Lett. B* **112**, 201–205 (1982).
17. R. L. Jaffe, Perhaps a stable dihyperon, *Phys. Rev. Lett.* **38**, 195–198 (1977) [erratum: *Phys. Rev. Lett.* **38**, 617 (1977)].
18. J. K. Ahn [J-PARC E42], Search for the H-dibaryon near $\Lambda\Lambda$ and $\Xi^- p$ thresholds at J-PARC, *PoS* **Hadron2017**, 124 (2018).
19. T. Inoue *et al.* [HAL QCD], Bound H-dibaryon in flavor SU(3) limit of lattice QCD, *Phys. Rev. Lett.* **106**, 162002 (2011).
20. M. Oka, K. Shimizu, and K. Yazaki, The dihyperon state in the quark cluster model, *Phys. Lett. B* **130**, 365–368 (1983). S. Takeuchi and M. Oka, Can the H particle survive instantons? *Phys. Rev. Lett.* **66**, 1271–1274 (1991). T. Sakai, K. Shimizu, and K. Yazaki, H dibaryon, *Prog. Theor. Phys. Suppl.* **137**, 121–145 (2000).
21. M. Oka, Flavor octet dibaryons in the quark model, *Phys. Rev. D* **38**, 298 (1988).
22. K. Sasaki *et al.* [HAL QCD], $\Lambda\Lambda$ and NΞ interactions from lattice QCD near the physical point, *Nucl. Phys. A* **998**, 121737 (2020). E. Hiyama, K. Sasaki, T. Miyamoto, T. Doi, T. Hatsuda, Y. Yamamoto, and T. A. Rijken, Possible lightest Ξ hypernucleus with modern ΞN interactions, *Phys. Rev. Lett.* **124**(9), 092501 (2020).
23. K. Nakazawa *et al.*, The first evidence of a deeply bound state of $\Xi^- - {}^{14}N$ system, *PTEP* **2015**(3), 033D02 (2015). S. H. Hayakawa *et al.* [J-PARC E07], Observation of coulomb-assisted nuclear bound state of $\Xi - {}^{14}$N system, *Phys. Rev. Lett.* **126**, 062501 (2021).
24. T. Inoue [HAL QCD], Strange nuclear physics from QCD on lattice, *AIP Conf. Proc.* **2130**, 020002 (2019).
25. H. Nemura [HALQCD], Hyperon-nucleon interaction from lattice QCD at $(m_\pi, m_K) \approx (146, 525)$ MeV, *AIP Conf. Proc.* **2130**, 040005 (2019).
26. T. Nanamura *et al.* [J-PARC E40], Measurement of differential cross sections for $\Sigma+p$ elastic scattering in the momentum range 0.44–0.80 GeV/c, *PTEP* **2022**(9), 093D01 (2022).
27. E. Friedman and A. Gal, In-medium nuclear interactions of low-energy hadrons, *Phys. Rept.* **452**, 89–153 (2007).
28. A. De Rujula, H. Georgi, and S. L. Glashow, Hadron masses in a Gauge theory, *Phys. Rev. D* **12**, 147–162 (1975).
29. M. Oka and S. Takeuchi, Instanton induced quark quark interactions in two Baryon systems, *Phys. Rev. Lett.* **63**, 1780–1783 (1989). E. V. Shuryak and J. L. Rosner, Instantons and Baryon mass splittings, *Phys. Lett. B* **218**, 72–74 (1989).
30. J. J. de Swart, The Octet model and its Clebsch-Gordan coefficients, *Rev. Mod. Phys.* **35**, 916–939 (1963) [erratum: *Rev. Mod. Phys.* **37**, 326–326 (1965)].

https://doi.org/10.1142/9789819813353_0014

Resemblance Between the Quantum Many-Body System and the Many-Human-Being System

Takaharu Otsuka
The University of Tokyo,
Hongo, Bunkyo-ku, Tokyo 113-0033, Japan
RIKEN Nishina Center,
Hirosawa, Wako-shi, Saitama, Japan
otsuka@phys.s.u-tokyo.ac.jp

Some social and economical systems involving many human beings may show resemblance to many-particle quantum systems, in the mechanism to determine the most stable form of the system. The driving force and the monotonic force are introduced, and their interplay is discussed. The monotonic force changes the environment for the system in both quantum and human systems, and the driving force puts this system into particular forms (or goals) of a given mode (or movement). The environments with various "orders" can affect the finally chosen form, although the monotonic force has no direct relation to such forms. These discussions are presented in simple terms so that no knowledge of physics is required.

1. Introduction

As remembered well by many friends, colleagues and students, the activities of Prof. Akito Arima were spread over a wide variety of objects and areas, far beyond the territory of nuclear physics. So he might be interested in possible similarities between the phenomena of physical systems and those of human beings. Expecting this without firm evidences, I here try to present my idea on this intriguing subject.

The subject is the similarities between the basic mechanism for determining the shape of atomic nuclei and that for determining the behaviors of a group of people. Although I have been working on physics for some decades, I have no expertise in social sciences or humanities. So, I just state my idea without referring to any related literatures or presentations possibly made earlier. I must apologize for this at this stage. In this sense, this article cannot be a paper in the strict sense, but may still be a paper in a loose sense or something a little more than an essay.

2. Collective Behaviors of Human Systems

The ideas presented here originate in the basic mechanism of quantum many-body (or many-particle) systems, but even its simplest explanation may not be helpful enough for those who are from outside the field of physics but are still interested in the present subject. So, let me start with certain relevant aspects of collective behaviors of human systems.

Human systems can be motivated by incentives, particularly by those involving economical benefits such as jobs, goods, *etc.* A similar situation appears in quantum systems like atomic nuclei as assemblies of many protons and neutrons, where the incentives are, of course, different, and mean the ways to increase binding energies: more binding energy implies higher stability against breaking-up. The changes due to such incentives are called "modes" in this article for both natural and human systems. The mode of a human system can be interpreted as a collective movement of the individuals in the system. There can be a force driving this mode, and such force is called the "driving" force in this article.

However, we have seen examples that the incentives do not necessarily result in significant changes of the behavior of the people. For instance, a new product may not be accepted despite its lower price or superb quality. Here is an example, although it may not be the best one. The artificial (factory-made) meat is not popular among customers, no matter how its safety and/or harmlessness is authorized. In human societies, the resistance arises against almost anything, because of conservative emotions, traditional ethics, religious text or customs, or even simple fears based on bad memories from the past. Due to one or some of these, something does not happen, like the artificial meat not being a major boost. The circumstances created naïvely are generally not favorable to something new. Another possible example can be the element Cadmium (Cd) in Japan. This element was contained in a hazardous chemical waste by which many people suffered up to some deaths. Because of such negative image of this element, other useful materials involving Cadmium have been avoided, although chemical properties vary depending on other substances. Thus, the resistance can arise almost independently of the nature of the mode, and in many cases, what matters for materializing modes may be not only how to push the incentives, but also how to handle actual circumstances particularly resistance.

3. Similar Situations in Quantum Systems in Plain Words

Similar situations arise in the quantum world: we consider an object composed of a number of particles. The mode here can be, for instance, the formation of the shape of this object, with its form being sphere, ellipsoid, triangle, rod, or something more peculiar. We keep this example in mind in this article. (The behavior of many particles resulting in a form is called a collective motion in physics.) Figure 1 displays a schematic picture of this mode. Left panel of Fig. 1 (i) depicts a basically random configuration of particles, bound by featureless back-ground-type forces. By activating a driving force between particles, a substantial effect (downward blue arrow) is produced resulting in various forms of the system, like A or B (ellipsoids of different shapes), as shown in middle and right panels of Fig. 1 (i). The actions of the driving force are intuitively shown in parts also there by wavy lines.

The driving force thus gives a certain stability to the system in form A or B. The amount of the stability generally varies depending on the form, and is indeed different between the forms A and B. (Technically speaking, the driving force

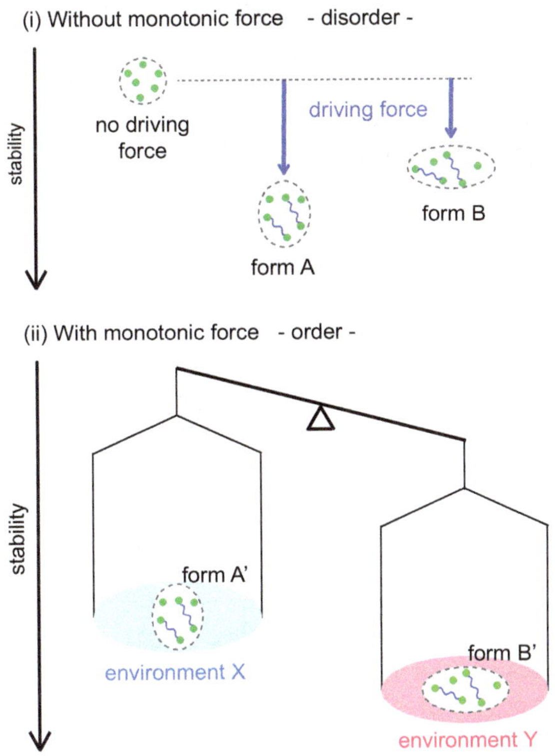

Fig. 1. Schematic pictures of many-particle systems (i) without monotonic force and (ii) with monotonic force. The environment in panel (i) remains disorder, but the driving force provides stabilities, ending up with form A or B. The disorder is altered to some order in panel (ii), and environments X and Y appear with different effects of the monotonic force.

changes binding energies, with the larger binding energy the more stability gain.) In Fig. 1 (i), the system is settled in the form A, because it gains more stability than the same system in the form B. (Technically, the form B appears as an excited state of the same system.) We see that what determines the most stable form of the mode is the driving force. This is the traditional view. However, a different picture emerges from recent studies.

In the situations shown in Fig. 1 (i), the particles are moving in a certain environment. This environment is common for all forms, inclusive A and B. The driving force is then activated, as the sole mechanism to differentiate forms A, B, *etc.* This idea seems to have been well shared.

Here, the environment has generally no reason to favor a particular form of the mode, and can even prevent the driving force from developing certain forms of the mode. This preventive effect can be seen as a resistance against the mode. Regardless of being resistance or not, the present environment can be characterized as a "disorder" for this mode.

In some systems, however, a certain component of the forces, different from the driving force, can vary the environment. Note that this component of the forces is not activated in the theoretical modeling shown in Fig. 1 (i), and the environment is of disorder. In many cases, once this component is turned on, such disordered environment is altered to another one with a certain "order" implying weakened resistance to the development of the mode. This "ordered" environment gives more stability to the system. This component of the forces varying the environment is called "monotonic force" hereafter (this is named monopole force in quantum physics). The monotonic force is a part of the entire forces, but differs from the driving force.

We now present some more details of the monotonic force. First, we would like to emphasize once more that the monotonic force does not directly induce any form of the mode. It can only change the environment. Secondly, the effect of the monotonic force is uniform for particles belonging to a certain subgroup. The name "monotonic" originates in this uniformity. Consequently, its total effect grows in proportion to the number of such particles. This proportionality is not common at all; the effects of the other forces, including the driving force, are not uniform. Due to cancellations among individual effects, the net effects of the other forces do not keep growing with the number of the particles.

When the monotonic force is turned on in the present modeling, the stability is changed, as stated above. This change varies among different environments. Figure 1 (ii) illustrates this: the form A in the disordered environment gains more stability if it is placed in environment X created by including the monotonic force. At the same time, the form A may be slightly changed to form A', but this is a minor thing in most cases. Likewise, the form B' appears in environment Y, with increased stability compared to the form B in the disordered environment.

Figure 1 (ii) also indicates that the increment of the stability in the environment Y is larger than that in the environment X. The original stability sequence between the forms A and B in Fig. 1 (i) is then inverted. This inversion is due to the monotonic force.

We note that if the driving force is absent, neither form A nor B appears, and the system will be chaotic. Thus, both the driving and monotonic forces contribute, and the interplay between them can be essential in many cases. In earlier expectations, the focus was placed just on the driving force, however.

In summary of the present discussion, the actual character of the system at its highest stability is determined by the interplay between (i) the driving force shifting the system into a certain form of the mode and (ii) the monotonic force which is neutral to the mode but can substantially affect environments.

This sort of phenomena occur in quantum systems, as exemplified in later sections. In many cases, the relation between the form and the environment is positively coupled, as referred to as the positive feedback in the (quantum mechanical version) of the self-organization.

4. Possible Relation to Human Communities

It was mentioned in Sec. 2 that the incentives in human communities generally encounter resistance powers. It was suggested in Sec. 3 that the resistance also appears in quantum many-particle systems but can be changed by monotonic forces. It is an intriguing question whether any mechanism like monotonic force exists in human communities or not.

The crucial features of the monotonic force in the quantum system are the uniformity of its effect for relevant participants, i.e., particles, and the neutrality to the outcome (form) of the mode. The uniformity and neutrality can be related, and certainly coexist. They are not the same thing, however. In moving to the human system, the concept of the monotonic force, of course, has to be extended, but the neutrality and uniformity remain as key basic concepts.

In systems comprising many human beings, the incentives illuminate directions to proceed, and it can be called driving force. Of course, the driving force goes beyond simple incentives, for instance, to those in political or social agenda.

Once a driving force gives an impact to the human system, such as a community big or small, a market, a fan group, *etc.*, one of the standard reactions is resistance as stated already. A plausible manner to encounter or soften the resistance power is a "force" uniformly acting on the individuals, like the monotonic force discussed in Sec. 3. So, we shall call this possible mechanism a monotonic force also for human systems. Note that the monotonic force is not a force in a naïve sense, but is a mechanism.

We now have the driving force and the monotonic force for human beings. The monotonic force does not directly favor any outcome of the driving force, but changes the environment instead. The environment here affect the degree of reaction, for instance, rejection or acceptance, of the relevant human community. Such environments are exemplified, as mentioned above, by traditional feelings, common senses, religious practices, fears from the past, etc., and in modern society, the so-called public opinion is important as well. The reactions of the individuals certainly vary, but the variation may occur around a common base line, to which the environment matters in the present discussion. For the same driving force, the outcome can differ for different environments.

The monotonic force, if exists, must be able to shift the mentality or behavior of almost everyone in the relevant community towards the same or similar directions, which results in a certain change of the reaction of the community as a whole, to the driving force. With varying monotonic effects, different environments are created. Thus, some environments can be considered theoretically. It is noted that the contents of the monotonic force and the driving force may not be fully uncovered, and these forces may be recognized only in terms of their functions or effects, for human systems. Even in such cases, we keep referring them by "forces".

There are many cases with a wide variety, including the size of the community. Among them, a typical example is found for the war between nations. The nations have some objectives and logics for wars, in most cases. The incentive is clear here: to win the war (in some sense) and accomplish the objectives. This is the driving force here. The relevant environments include the spirit of people, which is not necessarily suitable for the war. This is certainly a state of disorder for the war. Suppose that a war was unavoidable, what was needed for a victory must be to align the emotion of the people to a favorable direction, besides the practical capability of the war. This process works better if the goal of the war is set to be of high dignity and sufficient nobility, as characterized by ethical, religious, racial, cultural or other factors. Thus, the actual situation, like a victory or a defeat, can be crucially affected by as to how the monotonic force works. There are many examples that the war was forced to be stopped or abandoned due to the change of the environment, e.g., the decline of the spirit.

We can find a recent example in the case of the global warming. During the past ten years, the environment (in the present sense) has changed dramatically: the negative attitude against the CO_2 emission has grown, giving different balance for energy processing sectors. For instance, the coal- or LNG-burning power stations were proud of large advantages before this change, over other means of the electricity generation (like form A over B in Fig. 1 (i)), because of lower cost and higher manageability. The fear for worsening global warming has changed the environment (in the present sense), creating strong resistance to anything emitting CO_2. The basic policy has changed, like the transition from Fig. 1 (i) to Fig. 1 (ii). The actual situation may be more complex due to other factors, like international conflict. As this change occurred in a global scale shared by billions of people, this is a good example of a monotonic force. The neutrality may look less obvious, but it is mentioned that the anti-global-warming movement does not point to any particular alternatives of the electricity generation, at least in principle. The present-day dominance of solar panels and wind mills can be replaced by nicer ways. A similar change is proceeding for the car production.

Down to retail shops and its customers, recent management seems to mean that goods are frequently changed according to the sales statistics. Many goods disappear, despite their high quality. This may be interpreted as the management policy dominated by the incentives, i.e., the driving force towards short-term profit. However, the customers are often disappointed by missing favorite goods, and raise the resistance to such shops: fewer visits to such shops. This is one of the monotonic-force effects, which may not be so explicit but would emerge slowly and widely as the environmental phenomenon.

Thus, there must be many cases with all sorts of communities and groups. The environments for many of them have nothing to do with the driving forces, but can play decisive roles as to where the actual situations are settled. Because the driving forces are quite visible but the monotonic forces are rather invisible, it may be rather difficult to identify the monotonic forces for actual cases. I would like to stress that this is an intriguing subject, especially quantification of the two forces and their interplay. Concrete examples often involve substantial political, commercial or social sensitive issues, which may lead to misinterpretations or objections irrelevant to the basic theme of this article. The author does not know how to handle them. So, no further concrete examples are presented here.

Some monotonic forces may be accepted by the whole community. As the ideas behind them must be supported by almost everyone, they are like democracy, equality, patriotism, etc., and are expected to be the strongest monotonic forces. Assuming that this argument makes sense, the democracy will be the ultimate goal, apart from the intermediate passage. It is also noted that the driving force tends to sharply lead the community through a narrow window, whereas the monotonic force would yield more diversities, which modern free societies already observe. In authoritarian states, there can be a fake monotonic-force-type effect imposed externally to the people, not emerging from the people.

Along this line, the recent technological development of AI assistance to human intellectual activities can become a monotonic force. This development might cause problematical situations if the AI assistance is ill guided intentionally or unintentionally. Note that as the monotonic forces are rather invisible, its negative effects can be detected rather late.

The present idea seems to have certain relations to the existing theories and views. Among them, "The Wealth of Nations" and "The Theory of Moral Sentiments" by Adam Smith may have some direct relations. As can be imagined, the former seems to be related to the driving force, while the latter to the monotonic force.

As an extreme case, one can imagine the special case of no driving force and rather substantial monotonic forces like customs based on unknown or forgotten reasons. Such a community may suffer chaotic situations. One can find quite a few examples, unfortunately.

In the quantum physics, the driving force and the monotonic force are contained in the same mathematical/physical machinery, called the interaction between particles. If one attempts to describe the interplay of these two "forces" for human systems in some objective or mathematical ways, it will be a great challenge with incredible excitements.

5. Background of the Idea — Shape of the Atomic Nucleus

This section and next section are optional, and can be skipped for grasping the main message of this article. They present the background of the idea of driving and monotonic forces.

Atomic nuclei are composed of many particles, called protons and neutrons. Protons and neutrons are called nucleons altogether. The simplest nucleus is the one for hydrogen (element symbol, H), consisting of one proton (hydrogen-1, ^{1}H). Heavy nuclei comprise many nucleons, over 200; for instance the nucleus for uranium (element symbol U) comprises 92 protons and 146 neutrons (uranium-238, ^{238}U). Each nucleon has a radius about 10^{-15} m. As nucleons are almost sticking together

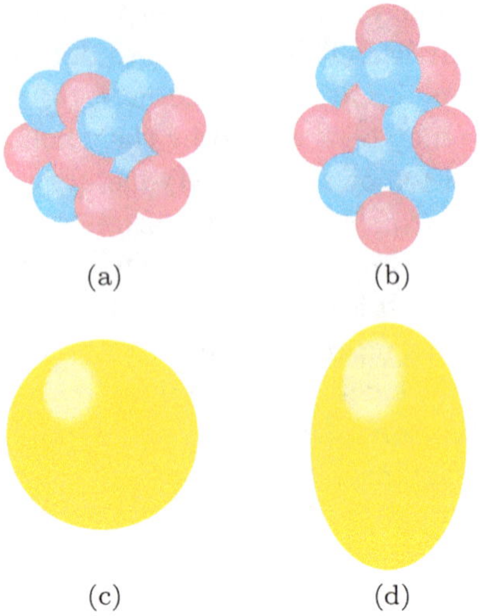

Fig. 2. Graphical image of an atomic nucleus as an assembly of protons (blues spheres) and neutrons (pink spheres). The nucleus is **a** in a spherical shape and **b** in a deformed shape (ellipsoid). **c** (**d**) More realistic view of **a** (**b**) with quantum mechanical consideration.

in a nucleus thanks to nuclear forces, the nucleus looks like the objects shown schematically in panels **a, b** of Fig. 2.

In panel **a** of Fig. 2, the nucleus is in a "spherical" shape. Because of the particle-wave character of the quantum mechanics, the actual nuclear surface is smoothed out from the rocky surface shown in Fig. 2**a** to the smooth spherical surface in panel **c**. The surface can thus be defined rather well. Some nuclei, for instance those of isotopes helium-4, oxygen-16, and calcium-40, are typical examples of "spherical nuclei".

In panel **b** of Fig. 2, the nucleus is in a deformed shape prolonged in the vertical direction: while deformation to various shapes is possible in principle, the deformation to the ellipsoids exemplified in panel **b** is discussed here. Again, due to the particle-wave nature of the quantum mechanics, the actual nuclear surface is smoothed out from that in Fig. 2**b** to the ellipsoidal picture in panel **d**.

An effective surface tension works on the nuclear surface, making the surface area as small as possible. This process results in a sphere, as well known. The situation shown in Fig. 2**c** thus emerges.

In many other nuclei, however, the shape of surface is deformed from a sphere (panel **c**) to an ellipsoid (panel **d**). This occurs due to a certain part of nuclear forces acting between nucleons. This part provides the nucleus with additional substantial binding energy or stability, like forms A and B compared to "no driving force" in Fig. 1 (i). The amount of this effect varies for different actual ellipsoidal shapes, as schematically shown by different positions of forms A and B in Fig. 1 (i). Recognizing the shape deformation as the mode, this part of nuclear forces corresponds to the *driving force*.

The nuclear shape changes from nucleus to nucleus, because the driving force acts between nucleons and the numbers of them vary from nucleus to nucleus. Traditionally, it has been supposed that the nuclear shape is determined predominantly by the driving force: which shape is favored by this force for a given nucleus. This picture of the driving force has remained for seven decades.

6. Self-organization in the Shapes of Atomic Nuclei

Nucleons in an atomic nucleus are moving around in some orbitals as schematically shown in Fig. 3. This image is a classical picture, and the precise one based on the quantum mechanics is somewhat different. But it is common between the two pictures that nucleons are making orbital motions. The orbital motions of such nucleons are not independent among themselves, and are indeed correlated due to the nuclear forces acting between nucleons. The surface is created by many nucleons in such orbitals, and the deformed shape appears when proper correlations are produced by the driving force mentioned above. (Technically this is one of the so-called Jahn-Teller effects.)

Each orbital is assumed to have a prefixed energy, and is qualitatively like the orbitals of electrons in atoms. The driving force acts on nucleons in such orbitals, and can shift a nucleon from an orbital to another. Intuitively speaking, this shift occurs more often if the energies of the orbitals are close, but the shift is hindered if these energies are far from each other. Thus, the correlation depends not only on the driving force but also the patterns of the energies of these orbitals, which is nothing but the environment. Traditionally, the energies of the orbitals are considered to be fixed from the beginning, being constant. The environment is then considered to be of *disorder*. If so, all that matters for the shape is how the driving force acts on nucleons. This is the bottomline of the traditional picture.

A recent study altered this view.[1] The energies of the orbitals are not fixed, but do change by a particular part of nuclear forces, different from the driving force.

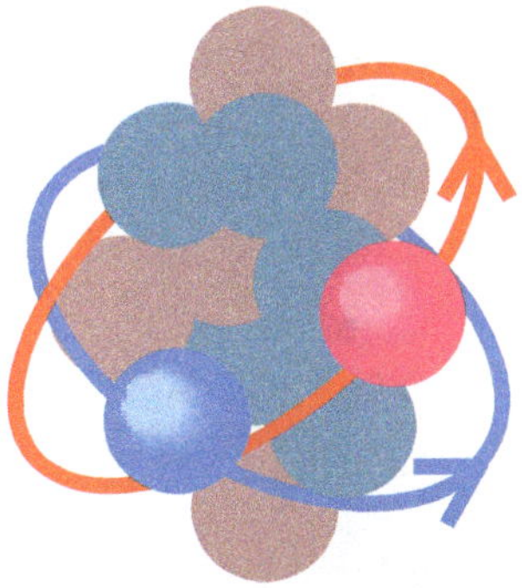

Fig. 3. Graphical image of orbital motions of nucleons in an atomic nucleus. Red and blue circles indicate protons and neutrons, respective, including dark-colored ones.

This part is nothing but the monotonic force, and was referred to as the monopole force in the terminology of quantum physics. The monotonic force raises or lowers the energies of the orbitals, giving different environments, like the environments X and Y in Fig. 1 (ii). But it does not directly favor any shape. The driving force then works in individual environments, and yields different shapes, like the forms A' and B' in Fig. 1 (ii). (These may somewhat differ from the shapes obtained without the monotonic force, which correspond to the form A and B in Fig. 1 (i).) The monotonic force creates certain "order" in each environment. Note that the "order" varies from an environment to another, for each of which the driving force produces its own shape. Among such shapes, the one with the maximum stability gain (the lowest ground-state energy) survives, with the others remain unstable (become excited states). This indirect mechanism determines the actual shapes in many nuclei, and can be understood as a quantum version of the self-organization[2] (see Ref. 1 for concrete arguments). The interplay between the driving force and the monotonic force outlined here is the background of the present proposition.

7. Summary

The essential point in the above discussion is as follows. There are two types of forces (or effects) with different characters: one drives the system into a specific mode, as denoted by the driving force. The mode here generally refers to a collective phenomenon involving many constituents, and the shape of an atomic nucleus is an example in the quantum world. A certain resistance generally exists against the mode development. The other force is to control the resistance, or the environment, without a direct or explicit relation to the mode, and is called the monotonic force. (The monotonic force is denoted the monopole force in the quantum-physics case.) The monotonic force does not create any mode, being neutral. However, it can change the disorder in the original environment to certain orders in various environments. Such orders facilitate the system to gain extra stabilities, to varying extents depending on the forms of the mode. Thus, the monotonic force, through the variation of the environment, can be another crucial factor in determining which form of the mode gains the maximum stability (i.e., ground state for physical systems). Obviously, in many systems, only the form of the maximum stability matters, which may not be the one most favored by the driving force in some sense. Applications of this idea to human systems with social, economical, political or cultural agendas are of great interest.

Acknowledgments

The author thanks Dr. Y. Utsuno for useful comment. The life-long support and encouragement from Prof. Akito Arima even for the crudest ideas like this work is heartily appreciated.

References

1. T. Otsuka, Y. Tsunoda, T. Abe, N. Shimizu, and P. Van Duppen, Underlying structure of collective bands and self-organization in quantum systems, *Phys. Rev. Lett.* **123**, 222502 (2019).
2. Wikipedia, https://en.wikipedia.org/wiki/Self-organization.

https://doi.org/10.1142/9789819813353_0015

Nuclear Mass Table with the Deformed Relativistic Continuum Density Functional Theory

C. Pan, K. Y. Zhang, S. Q. Zhang, and J. Meng*
State Key Laboratory of Nuclear Physics and Technology,
School of Physics, Peking University, Beijing 100871, China
**mengj@pku.edu.cn*

Nuclear mass is one of the most fundamental observables for atomic nucleus, and is very important for nuclear physics, particle physics, astrophysics, cosmology, etc. Although dramatic progress for the mass measurement has been achieved, most masses for nuclei far away from the stability valley still rely on the model predictions. In this paper, the relativistic density functional PC-PK1, one of the most successful density functionals, is presented. The predictive power of PC-PK1 is demonstrated with masses later measured in the regions of $50 \leq Z \leq 91$ and $102 \leq Z \leq 116$. The deformed relativistic Hartree-Bogoliubov theory in continuum (DRHBc), designed for deformed exotic nuclei, is introduced. Finally, the achievement and progress of a high-precision DRHBc mass table with PC-PK1 are reviewed.

1. Introduction

In 1969, by examining single-particle spectra in heavy nuclei, Arima, Harvey, and Shimizu,[1] as well as Hecht and Adler[2] introduced the concept of the pseudospin symmetry. This pseudospin symmetry has led to successful interpretations for many phenomena in nuclei and has attracted worldwide attentions.[3–8] However, the origin of the pseudospin symmetry had been a mystery for almost 30 years. In 1997, Ginocchio pointed out that the pseudospin symmetry is a relativistic symmetry of the Dirac Hamiltonian and the pseudo-orbital angular momentum is nothing but the orbital angular momentum of the lower component of the Dirac spinor.[9] In collaboration with Arima, one author of this review and others revealed that the pseudospin symmetry is determined by the derivative of the difference between the scalar and vector potentials.[10] The pseudospin symmetry was found to be a good approximation for normal nuclei and to become much better for exotic nuclei with a highly diffuse potential.[10–12] Following this line, the spin symmetry in anti-nucleon spectrum was found.[13–15]

Apart from the unexpected success on the origin of the pseudospin symmetry,[15,16] the relativistic density functional theory has proven to be very successful in describing a variety of properties for all nuclei in the nuclear chart.[17–25] The relativistic density functional theory naturally includes nuclear magnetism,[26] which is essential to describe the nuclear magnetic moments[27–31] and nuclear rotations.[21,32–43] The relativistic density functional theory has turned out to successfully describe both the ground-state and excited-state properties in nuclei,[24] including the nuclear mass.

In this paper, the efforts for a nuclear mass table based on the deformed relativistic continuum density functional theory with the modern functional PC-PK1 are focused on. In Section 2, the experimental and theoretical progress on the study of nuclear mass is reviewed. The relativistic density functional PC-PK1 and its predictive power for nuclear mass are presented in Section 3. The deformed relativistic Hartree-Bogoliubov theory in continuum (DRHBc) is briefly introduced in Section 4. The achievement and progress of the DRHBc mass table with PC-PK1 are reviewed in Section 5. Finally, a brief summary and perspective are given in Section 6.

2. Nuclear Mass

Nuclear mass is one of the most fundamental quantities for atomic nucleus, and is very important in various fields such as nuclear physics, particle physics, astrophysics, and cosmology. It plays an important role in the understanding of the interaction between nucleons,[44] the evolution of shell structures,[45] and the location of drip lines.[46] It also provides the basic information for nuclear reactions and the fundamental inputs for the investigation on the origin of elements.[47]

The precise measurement of nuclear mass has been an important frontier in nuclear physics. New generations of radioactive ion beam (RIB) facilities running or under construction worldwide produce more and more nuclei far from the stability valley and extend our knowledge of nuclear physics from stable nuclei to exotic ones.[48] The existence of more than 3300 nuclides has been confirmed[49,50] and the masses of about 2500 among them have been measured.[51–53] Although the proton drip line of neptunium ($Z = 93$) has been reached,[54] the neutron drip line is known only up to neon ($Z = 10$).[55] Most exotic nuclei on the neutron-rich side are beyond the experimental capability in the foreseeable future. Therefore, a nuclear mass table based on a theoretical model with reliable predictive power is highly desired.

A lot of efforts have been made to predict nuclear masses and explore great unknowns of the nuclear landscape. Precise descriptions of the nuclear masses have been achieved with various macroscopic-microscopic models.[56–59] The finite-range droplet model (FRDM) has an accuracy of 560 keV with 38 parameters adjusted to 2149 experimental masses.[59] The Weizsäcker-Skyrme 4 (WS4) model achieved an accuracy of 298 keV with 18 parameters fitted to 2353 experimental masses.[57] Recently, machine learning methods have been employed to further improve the descriptions of nuclear mass. For examples, the radial basis function approach,[60] the Bayesian neural network approach,[61] and the kernel ridge regression approach,[62] have achieved the accuracies around 200 keV, which can be reduced to less than 150 keV after including odd-even effects.[63–65]

Based on the nonrelativistic density functional theory, the nuclear mass models of Skyrme[46,66–69] and Gogny[70–72] Hartree-Fock-Bogoliubov (HFB) types have been developed. The Skyrme HFB-27* achieved an accuracy of 0.512 MeV with 24 parameters fitted to 2353 experimental masses.[69] The Gogny D1M achieved an accuracy of 0.798 MeV with 14 parameters fitted to 2149 experimental masses.[71]

The relativistic density functional theory has been applied to describe the nuclear ground-state properties including the mass.[21,73–82] In 1999, based on the functional NL3 and the BCS approach for pairing correlations, 1315 even-even nuclei with $10 \leq Z \leq 98$ were calculated, and the rms deviation of mass from available data is 2.6 MeV.[73] In 2005, based on the functional TMA and the state-dependent BCS approach, 6969 nuclei with $8 \leq Z \leq 100$ were calculated, and the rms deviation of mass from available data is 2.1 MeV.[74] In 2013, based on the functional PC-PK1 and the BCS approach, the nuclei with $Z \geq 8$ were calculated, and the rms deviation of mass from available data is 1.42 MeV.[21] The functional PC-PK1 has achieved an accuracy of 1.14 MeV for the 575 even-even nuclei by considering the dynamic correlation energies.[76,79,82]

The proper description of the exotic nuclei far from the stability valley is of crucial importance in determining drip-line location and understanding nucleosynthesis in astrophysics. In exotic nuclei, the Fermi surface is very close to the continuum threshold and the pairing interaction can scatter nucleons from bound states to resonant ones in the continuum. The density could become more diffuse due to the coupling to the continuum, and the position of the drip line might be influenced.

The relativistic continuum Hartree-Bogoliubov (RCHB) theory[83,84] solves the relativistic Hartree-Bogoliubov (RHB) equations in coordinate space, and has achieved great successes in the studies of both stable and exotic nuclei.[10,11,83,85–92] Based on the RCHB theory with PC-PK1, the first nuclear mass table including continuum effects has been constructed and the continuum effects on the limits of the nuclear landscape have been presented.[81] It is demonstrated that the continuum effects are crucial for drip-line locations and 9035 nuclei with $8 \leq Z \leq 120$ are predicted to be bound, which remarkably extends the existing nuclear landscapes. Due to the assumption of spherical symmetry in the RCHB theory, the accuracy of the RCHB mass table on the mass description for the nuclei far away from magic numbers is less satisfactory.

3. Predictive Power of PC-PK1

Among the existing relativistic density functionals, PC-PK1[93] provides one of the best descriptions for nuclear masses.[76,77,79,94–96] There are 9 parameters in the functional PC-PK1, which are fixed by 60 binding energies and 17 charge radii of selected spherical nuclei. PC-PK1 provides a good description for the isospin dependence of binding energy along either the isotopic or the isotonic chain, which makes its application reliable for exotic nuclei. With PC-PK1, by including the dynamic correlation, the description of nuclear mass for deformed nuclei can be improved. In comparison, in many density functionals, the deviation of the calculated mass

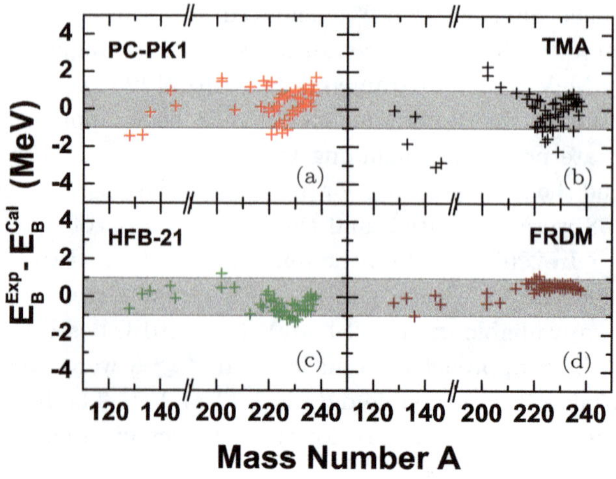

Fig. 1. Differences between the experimental data[98] and the binding energies calculated with the functionals PC-PK1[93] and TMA[74] as well as those from the mass models HFB-21[99] and FRDM.[100] Taken from Ref. 94.

from the data is isospin dependent, which is difficult to be improved by the dynamic correlation.[93,97]

The predictive power of PC-PK1 is shown in Fig. 1. The 53 newly measured data[98] in the element range of 50 to 91 are employed to examine the predictive power of PC-PK1. The results calculated with TMA[74] as well as those from the mass models HFB-21[99] and FRDM[100] are also shown for comparison. It is found that the PC-PK1 result can reproduce the experimental data quite well and the corresponding rms deviation is 0.859 MeV. Both PC-PK1 and TMA results underestimate the binding energies for nuclei near $N = 126$ and overestimate them for nuclei near $N = 82$. Moreover, the deviations for nuclei ^{144}La and ^{146}Ce in TMA are even as large as 3 MeV. This indicates that PC-PK1 is more robust in the predictions of nuclear masses, especially for the neutron-rich nuclei. The nonrelativistic mass model HFB-21[99] and the macroscopic-microscopic mass model FRDM[100] also give good agreement with the data. However, it should be noted that there are more than 20 parameters in HFB-21 including the phenomenological corrections, and some of the present data were also included in the fitting of its parameters, while PC-PK1 contains much fewer parameters, and is determined by fitting to only 60 binding energies and 17 charge radii of selected spherical nuclei.

The predictive power for PC-PK1 is also demonstrated in Fig. 2 for superheavy nuclei. The 10 measured data and 46 empirical data available in the latest "Atomic Mass Evaluation" (AME2020)[53] for $102 \leq Z \leq 120$ are employed, to examine the predictive power of PC-PK1 with the deformed relativistic Hartree-Bogoliubov theory in continuum. The results from the mass models WS4[57] and FRDM(2012)[59] are also shown for comparison. It is found that the PC-PK1 result can reproduce the 10 measured data quite well with the rms deviation of 0.635 MeV, and after

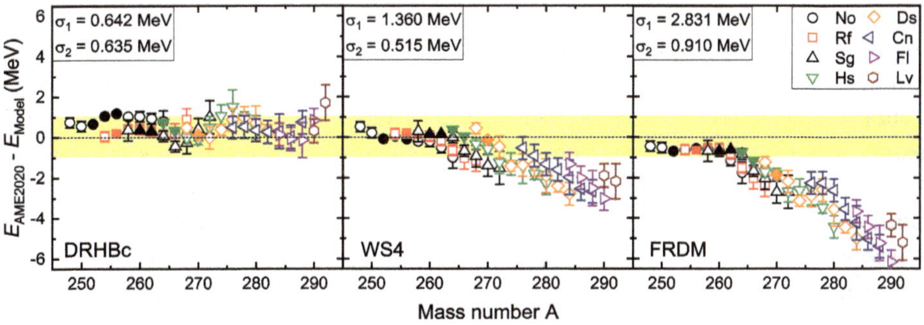

Fig. 2. Mass difference between the AME2020 data[53] and (a) the DRHBc calculation with PC-PK1, (b) the WS4 mass model,[57] and (c) the FRDM(2012) mass model[59] for superheavy even-even nuclei with data available ($102 \leq Z \leq 116$). $\sigma_1(\sigma_2)$ represents the root-mean-square deviation from the available 56 (10) measured and empirical (measured only) masses denoted by filled and open (filled) symbols. Taken from Ref. 96.

including the 46 empirical data, the rms deviation changes slightly to 0.642 MeV. The rms deviations from the 10 measured data for WS4 and FRDM are 0.525 MeV and 0.910 MeV by FRDM, and after including the 46 empirical data, the corresponding rms deviations increase to 1.360 and 2.831 MeV, respectively. The empirical data in "Atomic Mass Evaluation" are estimated from the trends in mass surface, together with all available experimental information.[52] The accuracy of these empirical data is usually proved by subsequent experiments.[101,102] This indicates that PC-PK1 provides not only a good description for the isospin dependence of the nuclear masses along the isotopic chains, but also a consistent accuracy with increasing Z. These features are in stark contrast to the macroscopic-microscopic models, and demonstrate clearly the predictive power of PC-PK1.

Apart from the nuclear mass, the density functional PC-PK1 has also been successfully applied for other nuclear ground-state properties and various excited-state properties, including low-lying excited states,[103] magnetic rotation,[37] antimagnetic rotation,[36] chiral rotation modes,[104] fission barrier,[105] and superheavy nuclei,[106] etc. In Ref. 107, based on PC-PK1, the time-dependent relativistic density functional theory was developed in a three-dimensional coordinate space without any symmetry restrictions, which could serve as an effective approach for the studies of the nuclear dynamical processes, e.g., the dynamics of chiral nuclei.[108]

4. DRHBc Theory

Except for doubly-magic nuclei, most nuclei in the nuclear chart deviate from the spherical shape, and it is important to include the deformation degree of freedom in the study of nuclear mass and determination of drip-line location. Therefore, simultaneously treating the deformation, the pairing correlation and the continuum effects is essential for a high-precision description of exotic nuclei.

To provide a proper description of deformed exotic nuclei, the deformed relativistic Hartree-Bogoliubov theory in continuum was developed.[109,110] From the Lagrangian density of the functional, the energy density functional for the nuclear system can be constructed under the mean-field and no-sea approximations. By minimizing the energy density functional, the Dirac equations for nucleons are obtained. The pairing correlations are treated by the Bogoliubov transformation. A density-dependent zero-range pairing force is adopted. The axially deformed densities and potentials are expanded in terms of the Legendre polynomials. The DRHBc theory provides a unified and self-consistent treatment of both the mean field and the pairing correlation and describes the exotic nuclei properly by solving the equations in the Dirac Woods-Saxon basis.[111]

The DRHBc theory has been developed for the functional with density-dependent meson-nucleon couplings[112] and point-coupling[113] version. Based on the DRHBc theory, the blocking effect of odd-mass and odd-odd nuclei is incorporated,[114] the rotational excitation for the exotic nuclei is explored with the angular momentum projection,[115,116] and the dynamic correlation energies are calculated by a two-dimensional collective Hamiltonian.[117]

Inheriting the advantages of the RCHB theory and including the deformation degrees of freedom, the DRHBc theory has been successfully applied to study the deformed halos in magnesium isotopes, and an interesting shape decoupling between the core and the halo is predicted in ^{42,44}Mg.[109,110] Other applications include resolving the puzzles concerning the radius and configuration of valence neutrons in ^{22}C,[118] investigating the shell evolution of carbon isotopes and neutron halos in ^{15,19,22}C,[119] exploring particles in the classically forbidden regions for magnesium isotopes,[120] and interpreting the neutron halos in ^{17}B[121] and ^{19}B.[122]

5. DRHBc Mass Table

The success of the RCHB mass table in extending the limit of nuclear existence, as well as the advantages of the DRHBc theory for deformed exotic nuclei and the predictive power of PC-PK1 for nuclear mass, stimulate the construction of an upgraded mass table including simultaneously the deformation and continuum effects by the DRHBc theory with PC-PK1. In the DRHBc calculations for all nuclei in the nuclear chart, the challenges include justifying a unified numerical setting, locating the ground-state deformation, and blocking the correct orbit(s) for odd-mass and odd-odd nuclei, etc.

In order to provide a unified description for all nuclei in the nuclear chart with the DRHBc theory, the DRHBc Mass Table Collaboration was established. The strategy and techniques for the nuclear mass table by the DRHBc theory with point-coupling density functionals are presented in Ref. 113. During the construction of the DRHBc mass table, several important and interesting results have been reported.[96,123–127] The predictive power of the DRHBc theory with PC-PK1 for the masses of superheavy nuclei has been illustrated by comparing with the latest

AME2020 data as shown in Fig. 2, and an interesting peninsula consisting of bound superheavy nuclei beyond the two-neutron drip line has been predicted.[96] Similar peninsulas of stability have been found in the region of medium-heavy nuclei,[125] and the effects of higher order deformations such as β_4 and β_6 on this phenomenon have been studied.[126] In Ref. 123, the dependence on the Legendre expansion of nuclear potentials and densities is investigated for light and heavy nuclei. By comparing the neutron drip lines predicted by the DRHBc theory and the RCHB theory, the deformation effects on the location of neutron drip line have been studied from O to Ca isotopes in Ref. 124. In Ref. 127, the bubble structures with shape coexistence in even-even isotopes of Hf to Hg are investigated.

The DRHBc mass table for even-even nuclei from $Z = 8$ to $Z = 120$ is given in Ref. 128. 2583 even-even nuclei are predicted to be bound by the DRHBc theory with the relativistic density functional PC-PK1. Figure 3 shows the nuclear landscape for even-even nuclei from O ($Z = 8$) to $Z = 120$ explored by the DRHBc theory with PC-PK1, where the squares represent bound nuclei. Among these 2583 bound even-even nuclei, the masses of 637 nuclei have been measured experimentally.[53] The binding energy differences $E_{\text{b}}^{\text{exp}} - E_{\text{b+rot}}^{\text{cal}}$ for these measured nuclei scaled by colors are plotted, where $E_{\text{b+rot}}^{\text{cal}}$ is the calculated binding energy including the rotational correction energy, and $E_{\text{b}}^{\text{exp}}$ is the experimental one. It is found that the calculated binding energies agree well with the data, and the rms deviation for these 637 nuclei is $\sigma = 1.518$ MeV.

In Fig. 3, the nucleon drip lines predicted by different mass tables are shown for comparison. It is found that compared with other results, the DRHBc calculations with PC-PK1 generally predict a more extended neutron drip line with a few exceptions. This is mainly because of the proper treatment of the continuum in the DRHBc theory and the adopted density functional.

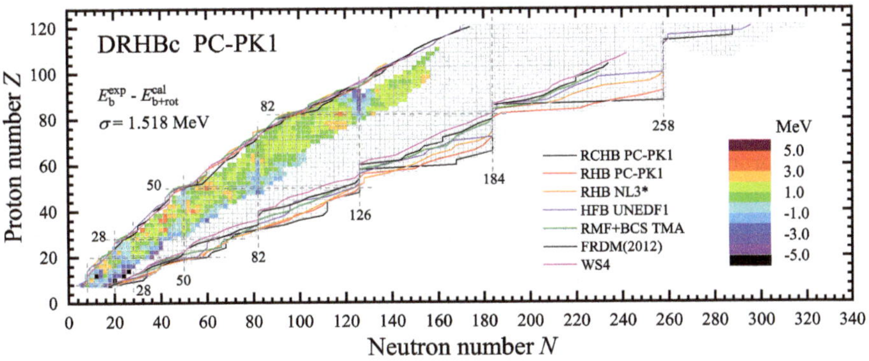

Fig. 3. 2583 bound even-even nuclei from O ($Z = 8$) to $Z = 120$ predicted by the DRHBc theory with PC-PK1. For the 637 even-even nuclei with mass measured, the binding energy differences between the data[53] and the DRHBc calculations (with rotational correction energy included) are scaled by colors. The nucleon drip lines predicted by different mass tables, RCHB with PC-PK1,[81] RHB with PC-PK1[82] and with NL3*,[77] HFB with UNEDF1,[46] RMF+BCS with TMA,[74] FRDM (2012),[59] and WS4,[57] are plotted for comparison. Taken from Ref. 128.

The successful exploration of the even-even nuclei in the nuclear chart by using the DRHBc theory with PC-PK1 demonstrates the importance of the simultaneous consideration of the deformation and continuum effects. The next step is to construct a DRHBc mass table for odd-mass and odd-odd nuclei, and the related work is in progress.[129]

6. Summary

Nuclear mass is one of the most fundamental quantities for atomic nucleus, and is very important for various fields such as nuclear physics, particle physics, astrophysics, and cosmology. In this paper, the relativistic point-coupling density functional PC-PK1 as well as its successes in the descriptions of ground-state and excited-state properties are briefly introduced, and the predictive power of PC-PK1 is illustrated for heavy nuclei and superheavy nuclei. The deformed relativistic Hartree-Bogoliubov theory in continuum is introduced, which can provide a proper description for the deformed exotic nuclei. The DRHBc Mass Table Collaboration, devoted to construct a high-precision nuclear mass table based on the DRHBc theory with PC-PK1, has made progress on the DRHBc mass table for even-even nuclei. The future work on odd-mass and odd-odd nuclei is in progress.

Acknowledgments

This work is dedicated to the memory of Professor Akito Arima who had made seminal works in nuclear physics and promotion in scientific physics research and collaboration worldwide. With great honors, we have collaborated with Professor Akito Arima on pseudospin symmetry and magnetic moments. The successful collaborations have deepened our understanding of the related fields. Professor Akito Arima had contributed a great deal in promoting nuclear physics in China and the collaboration between Chinese and Japanese institutes. He is well remembered and appreciated in China.

Helpful discussions with members of the DRHBc Mass Table Collaboration are gratefully appreciated. This work was partly supported by the National Natural Science Foundation of China (Grants No. 11935003, No. 11875075, No. 11975031, No. 12141501, and No. 12070131001), the National Key R&D Program of China (Contracts No. 2017YFE0116700, and No. 2018YFA0404400), and High-performance Computing Platform of Peking University.

References

1. A. Arima, M. Harvey, and K. Shimizu, Pseudo LS coupling and pseudo SU_3 coupling schemes, *Phys. Lett. B* **30**(8), 517–522 (1969). ISSN 0370-2693. doi: https://doi.org/10.1016/0370-2693(69)90443-2. URL https://www.sciencedirect.com/science/article/pii/0370269369904432.

2. K. T. Hecht and A. Adler, Generalized seniority for favored $j \neq 0$ pairs in mixed configurations, *Nucl. Phys. A* **137**(1), 129–143 (1969). ISSN 0375-9474. doi: https://doi.org/10.1016/0375-9474(69)90077-3. URL https://www.sciencedirect.com/science/article/pii/0375947469900773.
3. J. Dudek, W. Nazarewicz, Z. Szymanski, and G. A. Leander, Abundance and systematics of nuclear superdeformed states; relation to the pseudospin and pseudo-SU(3) symmetries, *Phys. Rev. Lett.* **59**, 1405–1408 (Sep, 1987). doi: 10.1103/PhysRevLett.59.1405. URL https://link.aps.org/doi/10.1103/PhysRevLett.59.1405.
4. W. Nazarewicz, P. J. Twin, P. Fallon, and J. D. Garrett, Natural-parity states in superdeformed bands and pseudo SU(3) symmetry at extreme conditions, *Phys. Rev. Lett.* **64**, 1654–1657 (Apr, 1990). doi: 10.1103/PhysRevLett.64.1654. URL https://link.aps.org/doi/10.1103/PhysRevLett.64.1654.
5. J. Y. Zeng, J. Meng, C. S. Wu, E. G. Zhao, Z. Xing, and X. Q. Chen, Spin determination and quantized alignment in the superdeformed bands in ^{152}Dy, ^{151}Tb, and ^{150}Gd, *Phys. Rev. C* **44**, R1745–R1748 (Nov, 1991). doi: 10.1103/PhysRevC.44.R1745. URL https://link.aps.org/doi/10.1103/PhysRevC.44.R1745.
6. F. S. Stephens, M. A. Deleplanque, J. E. Draper, R. M. Diamond, A. O. Macchiavelli, C. W. Beausang, W. Korten, W. H. Kelly, F. Azaiez, J. A. Becker, E. A. Henry, S. W. Yates, M. J. Brinkman, A. Kuhnert, and J. A. Cizewski, Pseudospin symmetry and quantized alignment in nuclei, *Phys. Rev. Lett.* **65**, 301–304 (Jul, 1990). doi: 10.1103/PhysRevLett.65.301. URL https://link.aps.org/doi/10.1103/PhysRevLett.65.301.
7. Q. Xu, S. J. Zhu, J. H. Hamilton, A. V. Ramayya, J. K. Hwang, B. Qi, J. Meng, J. Peng, Y. X. Luo, J. O. Rasmussen, I. Y. Lee, S. H. Liu, K. Li, J. G. Wang, H. B. Ding, L. Gu, E. Y. Yeoh, and W. C. Ma, Identification of pseudospin partner bands in ^{108}Tc, *Phys. Rev. C* **78**, 064301 (Dec, 2008). doi: 10.1103/PhysRevC.78.064301. URL https://link.aps.org/doi/10.1103/PhysRevC.78.064301.
8. W. Hua, X. H. Zhou, Y. H. Zhang, Y. Zheng, M. L. Liu, F. Ma, S. Guo, L. Ma, S. T. Wang, N. T. Zhang, Y. D. Fang, X. G. Lei, Y. X. Guo, M. Oshima, Y. Toh, M. Koizumi, Y. Hatsukawa, B. Qi, S. Q. Zhang, J. Meng, and M. Sugawara, Properties of the rotational bands in the transitional nucleus ^{189}Pt, *Phys. Rev. C* **80** 034303 (Sep, 2009). doi: 10.1103/PhysRevC.80.034303. URL https://link.aps.org/doi/10.1103/PhysRevC.80.034303.
9. J. N. Ginocchio, Pseudospin as a relativistic symmetry, *Phys. Rev. Lett.* **78**, 436–439 (Jan, 1997). doi: 10.1103/PhysRevLett.78.436. URL https://link.aps.org/doi/10.1103/PhysRevLett.78.436.
10. J. Meng, K. Sugawara-Tanabe, S. Yamaji, P. Ring, and A. Arima, Pseudospin symmetry in relativistic mean field theory, *Phys. Rev. C* **58**, R628–R631 (Aug, 1998). doi: 10.1103/PhysRevC.58.R628. URL https://link.aps.org/doi/10.1103/PhysRevC.58.R628.
11. J. Meng, K. Sugawara-Tanabe, S. Yamaji, and A. Arima, Pseudospin symmetry in Zr and Sn isotopes from the proton drip line to the neutron drip line, *Phys. Rev. C* **59**, 154–163 (Jan, 1999). doi: 10.1103/PhysRevC.59.154. URL https://link.aps.org/doi/10.1103/PhysRevC.59.154.
12. T.-S. Chen, H.-F. Lü, J. Meng, S.-Q. Zhang, and S.-G. Zhou, Pseudospin symmetry in relativistic framework with harmonic oscillator potential and Woods-Saxon potential, *Chin. Phys. Lett.* **20**(3), 358–361 (Feb, 2003). doi: 10.1088/0256-307x/20/3/312. URL https://doi.org/10.1088/0256-307x/20/3/312.
13. S.-G. Zhou, J. Meng, and P. Ring, Spin symmetry in the antinucleon spectrum, *Phys. Rev. Lett.* **91**, 262501 (Dec, 2003). doi: 10.1103/PhysRevLett.91.262501. URL https://link.aps.org/doi/10.1103/PhysRevLett.91.262501.

14. X. T. He, S. G. Zhou, J. Meng, E. G. Zhao, and W. Scheid, Test of spin symmetry in anti-nucleon spectra, *Eur. Phys. J. A* **28**(3), 265–269 (jun, 2006). doi: 10.1140/epja/i2006-10066-0.
15. H. Liang, J. Meng, and S.-G. Zhou, Hidden pseudospin and spin symmetries and their origins in atomic nuclei, *Phys. Rep.* **570**, 1–84 (2015). ISSN 0370-1573. doi: 10.1016/j.physrep.2014.12.005. URL http://www.sciencedirect.com/science/article/pii/S0370157315000502.
16. J. N. Ginocchio, Relativistic symmetries in nuclei and hadrons, *Phys. Rep.* **414**(4), 165–261 (2005). ISSN 0370-1573. doi: https://doi.org/10.1016/j.physrep.2005.04.003. URL https://www.sciencedirect.com/science/article/pii/S0370157305001808.
17. P. Ring, Relativistic mean field theory in finite nuclei, *Prog. Part. Nucl. Phys.* **37**, 193–263 (1996). ISSN 0146-6410. doi: 10.1016/0146-6410(96)00054-3. URL http://www.sciencedirect.com/science/article/pii/0146641096000543.
18. D. Vretenar, A. V. Afanasjev, G. A. Lalazissis, and R. Ring, Relativistic Hartree-Bogoliubov theory: static and dynamic aspects of exotic nuclear structure, *Phys. Rep.* **409**(3), 101–259 (2005). ISSN 0370-1573. doi: 10.1016/j.physrep.2004.10.001. URL http://www.sciencedirect.com/science/article/pii/S0370157304004545.
19. J. Meng, H. Toki, S. G. Zhou, S. Q. Zhang, W. H. Long, and L. S. Geng, Relativistic continuum Hartree Bogoliubov theory for ground-state properties of exotic nuclei, *Prog. Part. Nucl. Phys.* **57**(2), 470–563 (2006). ISSN 0146-6410. doi: 10.1016/j.ppnp.2005.06.001. URL http://www.sciencedirect.com/science/article/pii/S014664100500075X.
20. T. Nikšić, D. Vretenar, and P. Ring, Relativistic nuclear energy density functionals: Mean-field and beyond, *Prog. Part. Nucl. Phys.* **66**(3), 519–548 (2011). ISSN 0146-6410. doi: 10.1016/j.ppnp.2011.01.055. URL http://www.sciencedirect.com/science/article/pii/S0146641011000561.
21. J. Meng, J. Peng, S.-Q. Zhang, and P.-W. Zhao, Progress on tilted axis cranking covariant density functional theory for nuclear magnetic and antimagnetic rotation, *Front. Phys.* **8**(1), 55–79 (Feb, 2013). ISSN 2095-0470. doi: 10.1007/s11467-013-0287-y. URL https://doi.org/10.1007/s11467-013-0287-y.
22. J. Meng and S.-G. Zhou, Halos in medium-heavy and heavy nuclei with covariant density functional theory in continuum, *J. Phys. G* **42**(9), 093101 (Jul, 2015). doi: 10.1088/0954-3899/42/9/093101. URL https://doi.org/10.1088%2F0954-3899%2F42%2F9%2F093101.
23. S.-G. Zhou, Multidimensionally constrained covariant density functional theories — nuclear shapes and potential energy surfaces, *Phys. Scr.* **91**(6), 063008 (may, 2016). doi: 10.1088/0031-8949/91/6/063008. URL https://doi.org/10.1088/0031-8949/91/6/063008.
24. J. Meng, ed., *Relativistic Density Functional for Nuclear Structure*. World Scientific (2016). doi: 10.1142/9872. URL https://www.worldscientific.com/doi/abs/10.1142/9872.
25. S. Shen, H. Liang, W. H. Long, J. Meng, and P. Ring, Towards an ab initio covariant density functional theory for nuclear structure, *Prog. Part. Nucl. Phys.* **109**, 103713 (2019). ISSN 0146-6410. doi: https://doi.org/10.1016/j.ppnp.2019.103713. URL https://www.sciencedirect.com/science/article/pii/S0146641019300481.
26. W. Koepf and P. Ring, A relativistic description of rotating nuclei: The yrast line of ^{20}Ne, *Nucl. Phys. A* **493**(1), 61–82 (1989). ISSN 0375-9474. doi: 10.1016/0375-9474(89)90532-0. URL http://www.sciencedirect.com/science/article/pii/0375947489905320.

27. J. M. Yao, H. Chen, and J. Meng, Time-odd triaxial relativistic mean field approach for nuclear magnetic moments, *Phys. Rev. C* **74**, 024307 (Aug, 2006). doi: 10.1103/PhysRevC.74.024307. URL https://link.aps.org/doi/10.1103/PhysRevC.74.024307.
28. A. Arima, A short history of nuclear magnetic moments and GT transitions, *Sci. China Phys. Mech. Astron.* **54**(2), 188–193 (Feb, 2011). ISSN 1869-1927. doi: 10.1007/s11433-010-4224-6. URL https://doi.org/10.1007/s11433-010-4224-6.
29. J. Li, J. Meng, P. Ring, J. Yao, and A. Arima, Relativistic description of second-order correction to nuclear magnetic moments with point-coupling residual interaction, *Sci. China Phys. Mech. Astron.* **54**(2), 204–209 (Feb, 2011). ISSN 1869-1927. doi: 10.1007/s11433-010-4215-7. URL https://doi.org/10.1007/s11433-010-4215-7.
30. J. Li, J. M. Yao, J. Meng, and A. Arima, One-pion exchange current corrections for nuclear magnetic moments in relativistic mean field theory, *Prog. Theoret. Phys.* **125**(6), 1185–1192 (06, 2011). ISSN 0033-068X. doi: 10.1143/PTP.125.1185. URL https://doi.org/10.1143/PTP.125.1185.
31. J. Li and J. Meng, Nuclear magnetic moments in covariant density functional theory, *Front. Phys.* **13**, 132109 (2018). ISSN 2095-0470. doi: 10.1007/s11467-018-0842-7. URL https://doi.org/10.1007/s11467-018-0842-7.
32. J. König and P. Ring, Identical bands in superdeformed nuclei: A relativistic description, *Phys. Rev. Lett.* **71**, 3079–3082 (Nov, 1993). doi: 10.1103/PhysRevLett.71.3079. URL https://link.aps.org/doi/10.1103/PhysRevLett.71.3079.
33. A. V. Afanasjev, R. Ring, and J. König, Cranked relativistic Hartree-Bogoliubov theory: formalism and application to the superdeformed bands in the $A \sim 190$ region, *Nucl. Phys. A* **676**(1), 196–244 (2000). ISSN 0375-9474. doi: 10.1016/S0375-9474(00)00187-1. URL http://www.sciencedirect.com/science/article/pii/S0375947400001871.
34. A. V. Afanasjev and P. Ring, Time-odd mean fields in the rotating frame: Microscopic nature of nuclear magnetism, *Phys. Rev. C* **62**, 031302 (Aug, 2000). doi: 10.1103/PhysRevC.62.031302. URL https://link.aps.org/doi/10.1103/PhysRevC.62.031302.
35. A. V. Afanasjev and H. Abusara, Time-odd mean fields in covariant density functional theory: Rotating systems, *Phys. Rev. C* **82**, 034329 (Sep, 2010). doi: 10.1103/PhysRevC.82.034329. URL https://link.aps.org/doi/10.1103/PhysRevC.82.034329.
36. P. W. Zhao, J. Peng, H. Z. Liang, P. Ring, and J. Meng, Antimagnetic Rotation Band in Nuclei: A Microscopic Description, *Phys. Rev. Lett.* **107**, 122501 (Sep, 2011). doi: 10.1103/PhysRevLett.107.122501. URL https://link.aps.org/doi/10.1103/PhysRevLett.107.122501.
37. P. W. Zhao, S. Q. Zhang, J. Peng, H. Z. Liang, R. Ring, and J. Meng, Novel structure for magnetic rotation bands in ^{60}Ni, *Phys. Lett. B* **699**(3), 181–186 (2011). ISSN 0370-2693. doi: 10.1016/j.physletb.2011.03.068. URL http://www.sciencedirect.com/science/article/pii/S0370269311003534.
38. P. W. Zhao, J. Peng, H. Z. Liang, P. Ring, and J. Meng, Covariant density functional theory for antimagnetic rotation, *Phys. Rev. C* **85**, 054310 (May, 2012). doi: 10.1103/PhysRevC.85.054310. URL https://link.aps.org/doi/10.1103/PhysRevC.85.054310.
39. P. W. Zhao, N. Itagaki, and J. Meng, Rod-shaped nuclei at extreme spin and isospin, *Phys. Rev. Lett.* **115**, 022501 (Jul, 2015). doi: 10.1103/PhysRevLett.115.022501. URL https://link.aps.org/doi/10.1103/PhysRevLett.115.022501.
40. Y. K. Wang, Yrast band of ^{109}Ag described by tilted axis cranking covariant density functional theory with a separable pairing force, *Phys. Rev. C* **96**, 054324 (Nov, 2017). doi: 10.1103/PhysRevC.96.054324. URL https://link.aps.org/doi/10.1103/PhysRevC.96.054324.

41. Y. K. Wang, Magnetic rotations in ^{198}Pb and ^{199}Pb within covariant density functional theory with pairing correlations, *Phys. Rev. C* **97**, 064321 (Jun, 2018). doi: 10.1103/PhysRevC.97.064321. URL https://link.aps.org/doi/10.1103/PhysRevC.97.064321.
42. Z.-X. Ren, S.-Q. Zhang, P.-W. Zhao, N. Itagaki, J. A. Maruhn, and J. Meng, Stability of the linear chain structure for ^{12}C in covariant density functional theory on a 3D lattice, *Sci. China Phys. Mech. Astron.* **62**(11), 112062 (Jun, 2019). ISSN 1869-1927. doi: 10.1007/s11433-019-9412-3. URL https://doi.org/10.1007/s11433-019-9412-3.
43. Z. X. Ren, P. W. Zhao, S. Q. Zhang, and J. Meng, Toroidal states in ^{28}Si with covariant density functional theory in 3D lattice space, *Nucl. Phys. A* **996**, 121696 (2020). ISSN 0375-9474. doi: https://doi.org/10.1016/j.nuclphysa.2020.121696. URL https://www.sciencedirect.com/science/article/pii/S0375947420300063.
44. F. Wienholtz, D. Beck, K. Blaum, C. Borgmann, M. Breitenfeldt, R. B. Cakirli, S. George, F. Herfurth, J. D. Holt, M. Kowalska, S. Kreim, D. Lunney, V. Manea, J. Menéndez, D. Neidherr, M. Rosenbusch, L. Schweikhard, A. Schwenk, J. Simonis, J. Stanja, R. N. Wolf, and K. Zuber, Masses of exotic calcium isotopes pin down nuclear forces, *Nature* **498**, 346–349 (2013). doi: 10.1038/nature12226. URL https://doi.org/10.1038/nature12226.
45. E. M. Ramirez, D. Ackermann, K. Blaum, M. Block, C. Droese, C. E. Düllmann, M. Dworschak, M. Eibach, S. Eliseev, E. Haettner, F. Herfurth, F. P. Heßberger, S. Hofmann, J. Ketelaer, G. Marx, M. Mazzocco, D. Nesterenko, Y. N. Novikov, W. R. Plaß, D. Rodríguez, C. Scheidenberger, L. Schweikhard, P. G. Thirolf, and C. Weber, Direct mapping of nuclear shell effects in the heaviest elements, *Science* **337**(6099), 1207–1210 (2012). doi: 10.1126/science.1225636. URL https://www.science.org/doi/abs/10.1126/science.1225636.
46. J. Erler, N. Birge, M. Kortelainen, W. Nazarewicz, E. Olsen, A. M. Perhac, and M. Stoitsov, The limits of the nuclear landscape, *Nature* **486**, 509–512 (2012). doi: 10.1038/nature11188. URL https://doi.org/10.1038/nature11188.
47. T. Kajino, W. Aoki, A. B. Balantekin, R. Diehl, M. A. Famiano, and G. J. Mathews, Current status of r-process nucleosynthesis, *Prog. Part. Nucl. Phys.* **107**, 109–166 (2019). ISSN 0146-6410. doi: https://doi.org/10.1016/j.ppnp.2019.02.008. URL https://www.sciencedirect.com/science/article/pii/S0146641019300201.
48. T. Baumann, A. Spyrou, and M. Thoennessen, Nuclear structure experiments along the neutron drip line, *Rep. Prog. Phys.* **75**(3), 036301 (feb, 2012). doi: 10.1088/0034-4885/75/3/036301. URL https://doi.org/10.1088/0034-4885/75/3/036301.
49. M. Thoennessen, *The Discovery of Isotopes*. Springer, New York (2016).
50. Discovery of Nuclides Project. URL https://people.nscl.msu.edu/~thoennes/isotopes/ (2021).
51. F. G. Kondev, M. Wang, W. J. Huang, S. Naimi, and G. Audi, The NUBASE2020 evaluation of nuclear physics properties, *Chin. Phys. C* **45**(3), 030001 (Mar, 2021). doi: 10.1088/1674-1137/abddae. URL https://doi.org/10.1088/1674-1137/abddae.
52. W. J. Huang, M. Wang, F. G. Kondev, G. Audi, and S. Naimi, The AME 2020 atomic mass evaluation (I). evaluation of input data, and adjustment procedures, *Chin. Phys. C* **45**(3), 030002 (Mar, 2021). doi: 10.1088/1674-1137/abddb0. URL https://doi.org/10.1088/1674-1137/abddb0.
53. M. Wang, W. J. Huang, F. G. Kondev, G. Audi, and S. Naimi, The AME 2020 atomic mass evaluation (II). tables, graphs and references, *Chin. Phys. C* **45**(3), 030003 (Mar, 2021). doi: 10.1088/1674-1137/abddaf. URL https://doi.org/10.1088/1674-1137/abddaf.

54. Z. Y. Zhang, Z. G. Gan, H. B. Yang, L. Ma, M. H. Huang, C. L. Yang, M. M. Zhang, Y. L. Tian, Y. S. Wang, M. D. Sun, H. Y. Lu, W. Q. Zhang, H. B. Zhou, X. Wang, C. G. Wu, L. M. Duan, W. X. Huang, Z. Liu, Z. Z. Ren, S. G. Zhou, X. H. Zhou, H. S. Xu, Y. S. Tsyganov, A. A. Voinov, and A. N. Polyakov, New isotope ^{220}Np: Probing the robustness of the $n = 126$ shell closure in neptunium, *Phys. Rev. Lett.* **122**, 192503 (May, 2019). doi: 10.1103/PhysRevLett.122.192503. URL https://link.aps.org/doi/10.1103/PhysRevLett.122.192503.
55. D. S. Ahn, N. Fukuda, H. Geissel, N. Inabe, N. Iwasa, T. Kubo, K. Kusaka, D. J. Morrissey, D. Murai, T. Nakamura, M. Ohtake, H. Otsu, H. Sato, B. M. Sherrill, Y. Shimizu, H. Suzuki, H. Takeda, O. B. Tarasov, H. Ueno, Y. Yanagisawa, and K. Yoshida, Location of the neutron dripline at fluorine and neon, *Phys. Rev. Lett.* **123**, 212501 (Nov, 2019). doi: 10.1103/PhysRevLett.123.212501. URL https://link.aps.org/doi/10.1103/PhysRevLett.123.212501.
56. Y. Aboussir, J. M. Pearson, A. K. Dutta, and F. Tondeur, Nuclear mass formula via an approximation to the Hartree–Fock method, *At. Data Nucl. Data Tables* **61**(1), 127–176 (1995). ISSN 0092-640X. doi: https://doi.org/10.1016/S0092-640X(95)90014-4. URL https://www.sciencedirect.com/science/article/pii/S0092640X95900144.
57. N. Wang, M. Liu, X. Wu, and J. Meng, Surface diffuseness correction in global mass formula, *Phys. Lett. B* **734**, 215–219 (2014). ISSN 0370-2693. doi: https://doi.org/10.1016/j.physletb.2014.05.049. URL https://www.sciencedirect.com/science/article/pii/S037026931400358X.
58. H. Zhang, J. Dong, N. Ma, G. Royer, J. Li, and H. Zhang, An improved nuclear mass formula with a unified prescription for the shell and pairing corrections, *Nucl. Phys. A* **929**, 38–53 (2014). ISSN 0375-9474. doi: https://doi.org/10.1016/j.nuclphysa.2014.05.019. URL https://www.sciencedirect.com/science/article/pii/S0375947414001547.
59. P. Möller, A. J. Sierk, T. Ichikawa, and H. Sagawa, Nuclear ground-state masses and deformations: FRDM(2012), *At. Data Nucl. Data Tables* **109-110**, 1–204 (2016). ISSN 0092-640X. doi: https://doi.org/10.1016/j.adt.2015.10.002. URL https://www.sciencedirect.com/science/article/pii/S0092640X1600005X.
60. Z. M. Niu, Z. L. Zhu, Y. F. Niu, B. H. Sun, T. H. Heng, and J. Y. Guo, Radial basis function approach in nuclear mass predictions, *Phys. Rev. C* **88**, 024325 (Aug, 2013). doi: 10.1103/PhysRevC.88.024325. URL https://link.aps.org/doi/10.1103/PhysRevC.88.024325.
61. Z. M. Niu and H. Z. Liang, Nuclear mass predictions based on Bayesian neural network approach with pairing and shell effects, *Phys. Lett. B* **778**, 48–53 (2018). ISSN 0370-2693. doi: https://doi.org/10.1016/j.physletb.2018.01.002. URL https://www.sciencedirect.com/science/article/pii/S0370269318300091.
62. X. H. Wu and P. W. Zhao, Predicting nuclear masses with the kernel ridge regression, *Phys. Rev. C* **101**, 051301 (May, 2020). doi: 10.1103/PhysRevC.101.051301. URL https://link.aps.org/doi/10.1103/PhysRevC.101.051301.
63. Z. M. Niu, B. H. Sun, H. Z. Liang, Y. F. Niu, and J. Y. Guo, Improved radial basis function approach with odd-even corrections, *Phys. Rev. C* **94**, 054315 (Nov, 2016). doi: 10.1103/PhysRevC.94.054315. URL https://link.aps.org/doi/10.1103/PhysRevC.94.054315.
64. Z. Niu, H. Liang, B. Sun, Y. Niu, J. Guo, and J. Meng, High precision nuclear mass predictions towards a hundred kilo-electron-volt accuracy, *Sci. Bull.* **63**(12), 759–764 (2018). ISSN 2095-9273. doi: https://doi.org/10.1016/j.scib.2018.05.009. URL https://www.sciencedirect.com/science/article/pii/S2095927318302135.

65. X. H. Wu, L. H. Guo, and P. W. Zhao, Nuclear masses in extended kernel ridge regression with odd-even effects, *Phys. Lett. B* **819**, 136387 (2021). ISSN 0370-2693. doi: https://doi.org/10.1016/j.physletb.2021.136387. URL https://www.sciencedirect.com/science/article/pii/S0370269321003270.
66. M. Samyn, S. Goriely, P.-H. Heenen, J. M. Pearson, and F. Tondeur, A hartree-fock-bogoliubov mass formula, *Nucl. Phys. A* **700**(1), 142–156 (2002). ISSN 0375-9474. doi: https://doi.org/10.1016/S0375-9474(01)01316-1. URL https://www.sciencedirect.com/science/article/pii/S0375947401013161.
67. M. V. Stoitsov, J. Dobaczewski, W. Nazarewicz, S. Pittel, and D. J. Dean, Systematic study of deformed nuclei at the drip lines and beyond, *Phys. Rev. C* **68**, 054312 (Nov, 2003). doi: 10.1103/PhysRevC.68.054312. URL https://link.aps.org/doi/10.1103/PhysRevC.68.054312.
68. S. Goriely, N. Chamel, and J. M. Pearson, Skyrme-hartree-fock-bogoliubov nuclear mass formulas: Crossing the 0.6 mev accuracy threshold with microscopically deduced pairing, *Phys. Rev. Lett.* **102**, 152503 (Apr, 2009). doi: 10.1103/PhysRevLett.102.152503. URL https://link.aps.org/doi/10.1103/PhysRevLett.102.152503.
69. S. Goriely, N. Chamel, and J. M. Pearson, Further explorations of Skyrme-Hartree-Fock-Bogoliubov mass formulas. XIII. The 2012 atomic mass evaluation and the symmetry coefficient, *Phys. Rev. C* **88**, 024308 (Aug, 2013). doi: 10.1103/PhysRevC.88.024308. URL https://link.aps.org/doi/10.1103/PhysRevC.88.024308.
70. S. Hilaire and M. Girod, Relativistic mean-field mass models, *Eur. Phys. J. A* **33**, 237–241 (2016). doi: 10.1140/epja/i2007-10450-2. URL https://doi.org/10.1140/epja/i2007-10450-2.
71. S. Goriely, S. Hilaire, M. Girod, and S. Péru, First gogny-hartree-fock-bogoliubov nuclear mass model, *Phys. Rev. Lett.* **102**, 242501 (Jun, 2009). doi: 10.1103/PhysRevLett.102.242501. URL https://link.aps.org/doi/10.1103/PhysRevLett.102.242501.
72. J. P. Delaroche, M. Girod, J. Libert, H. Goutte, S. Hilaire, S. Péru, N. Pillet, and G. F. Bertsch, Structure of even-even nuclei using a mapped collective Hamiltonian and the D1S Gogny interaction, *Phys. Rev. C* **81**, 014303 (Jan, 2010). doi: 10.1103/PhysRevC.81.014303. URL https://link.aps.org/doi/10.1103/PhysRevC.81.014303.
73. G. A. Lalazissis, S. Raman, and P. Ring, Ground-state properties of even–even nuclei in the relativistic mean-field theory, *At. Data Nucl. Data Tables* **71**(1), 1–40 (1999). ISSN 0092-640X. doi: https://doi.org/10.1006/adnd.1998.0795. URL https://www.sciencedirect.com/science/article/pii/S0092640X98907951.
74. L. Geng, H. Toki, and J. Meng, Masses, deformations and charge radii — nuclear ground-state properties in the relativistic mean field model, *Prog. Theo. Phys.* **113**(4), 785–800 (2005). ISSN 0033-068X. doi: 10.1143/PTP.113.785. URL https://doi.org/10.1143/PTP.113.785.
75. A. V. Afanasjev, S. E. Agbemava, D. Ray, and P. Ring, Nuclear landscape in covariant density functional theory, *Phys. Lett. B* **726**(4), 680–684 (2013). ISSN 0370-2693. doi: https://doi.org/10.1016/j.physletb.2013.09.017. URL http://www.sciencedirect.com/science/article/pii/S037026931300734X.
76. Q.-S. Zhang, Z.-M. Niu, Z.-P. Li, J.-M. Yao, and J. Meng, Global dynamical correlation energies in covariant density functional theory: Cranking approximation, *Front. Phys.* **9**(4), 529–536 (Aug, 2014). ISSN 2095-0470. doi: 10.1007/s11467-014-0413-5. URL https://doi.org/10.1007/s11467-014-0413-5.
77. S. E. Agbemava, A. V. Afanasjev, D. Ray, and P. Ring, Global performance of covariant energy density functionals: Ground state observables of even-even nuclei and

the estimate of theoretical uncertainties, *Phys. Rev. C* **89**, 054320 (May, 2014). doi: 10.1103/PhysRevC.89.054320. URL https://link.aps.org/doi/10.1103/PhysRevC.89.054320.

78. A. V. Afanasjev, S. E. Agbemava, D. Ray, and P. Ring, Neutron drip line: Single-particle degrees of freedom and pairing properties as sources of theoretical uncertainties, *Phys. Rev. C* **91**, 014324 (Jan, 2015). doi: 10.1103/PhysRevC.91.014324. URL https://link.aps.org/doi/10.1103/PhysRevC.91.014324.
79. K. Q. Lu, Z. X. Li, Z. P. Li, J. M. Yao, and J. Meng, Global study of beyond-mean-field correlation energies in covariant energy density functional theory using a collective Hamiltonian method, *Phys. Rev. C* **91**, 027304 (Feb, 2015). doi: 10.1103/PhysRevC.91.027304. URL https://link.aps.org/doi/10.1103/PhysRevC.91.027304.
80. D. Peña-Arteaga, S. Goriely, and N. Chamel, Relativistic mean-field mass models, *Eur. Phys. J. A* **52**, 320 (2016). doi: 10.1140/epja/i2016-16320-x. URL https://doi.org/10.1140/epja/i2016-16320-x.
81. X. W. Xia, Y. Lim, P. W. Zhao, H. Z. Liang, X. Y. Qu, Y. Chen, H. Liu, L. F. Zhang, S. Q. Zhang, Y. Kim, and J. Meng, The limits of the nuclear landscape explored by the relativistic continuum Hartree-Bogoliubov theory, *At. Data Nucl. Data Tables* **121–122**, 1–215 (2018). ISSN 0092-640X. doi: 10.1016/j.adt.2017.09.001. URL http://www.sciencedirect.com/science/article/pii/S0092640X17300451.
82. Y. L. Yang, Y. K. Wang, P. W. Zhao, and Z. P. Li, Nuclear landscape in a mapped collective Hamiltonian from covariant density functional theory, *Phys. Rev. C* **104**, 054312 (Nov, 2021). doi: 10.1103/PhysRevC.104.054312. URL https://link.aps.org/doi/10.1103/PhysRevC.104.054312.
83. J. Meng and P. Ring, Relativistic Hartree-Bogoliubov description of the neutron Halo in ^{11}Li, *Phys. Rev. Lett.* **77**, 3963–3966 (Nov, 1996). doi: 10.1103/PhysRevLett.77.3963. URL https://link.aps.org/doi/10.1103/PhysRevLett.77.3963.
84. J. Meng, Relativistic continuum Hartree-Bogoliubov theory with both zero range and finite range Gogny force and their application, *Nucl. Phys. A* **635**(1), 3–42 (1998). ISSN 0375-9474. doi: 10.1016/S0375-9474(98)00178-X. URL http://www.sciencedirect.com/science/article/pii/S037594749800178X.
85. J. Meng and P. Ring, Giant Halo at the neutron drip line, *Phys. Rev. Lett.* **80**, 460–463 (Jan, 1998). doi: 10.1103/PhysRevLett.80.460. URL https://link.aps.org/doi/10.1103/PhysRevLett.80.460.
86. J. Meng, H. Toki, J. Y. Zeng, S. Q. Zhang, and S.-G. Zhou, Giant halo at the neutron drip line in Ca isotopes in relativistic continuum Hartree-Bogoliubov theory, *Phys. Rev. C* **65**, 041302 (Mar, 2002). doi: 10.1103/PhysRevC.65.041302. URL https://link.aps.org/doi/10.1103/PhysRevC.65.041302.
87. S. Zhang, J. Meng, and S. Zhou, Proton magic even-even isotopes and giant halos of Ca isotopes with relativistic continuum Hartree-Bogoliubov theory, *Sci. China G* **46**(6), 632–658 (Dec, 2003). ISSN 1672-1799. URL https://link.springer.com/article/10.1360/03yw0140.
88. J. Meng, S.-G. Zhou, and I. Tanihata, The relativistic continuum Hartree-Bogoliubov description of charge-changing cross section for C, N, O and F isotopes, *Phys. Lett. B* **532**(3), 209–214 (2002). ISSN 0370-2693. doi: 10.1016/S0370-2693(02)01574-5. URL http://www.sciencedirect.com/science/article/pii/S0370269302015745.
89. W. Zhang, J. Meng, S. Q. Zhang, L. S. Geng, and H. Toki, Magic numbers for superheavy nuclei in relativistic continuum Hartree-Bogoliubov theory, *Nucl. Phys. A* **753**(1), 106–135 (2005). ISSN 0375-9474. doi: 10.1016/j.nuclphysa.2005.02.086. URL http://www.sciencedirect.com/science/article/pii/S0375947405002423.

90. H. F. Lü, J. Meng, S. Q. Zhang, and S.-G. Zhou, Neutron halos in hypernuclei, *Eur. Phys. J. A* **17**(1), 19–24 (2003). doi: 10.1140/epja/i2002-10136-3.
91. L.-F. Zhang and X.-W. Xia, Global α-decay study based on the mass table of the relativistic continuum Hartree-Bogoliubov theory, *Chin. Phys. C* **40**(5), 054102 (May, 2016). doi: 10.1088/1674-1137/40/5/054102. URL https://doi.org/10.1088/1674-1137/40/5/054102.
92. Y. Lim, X. Xia, and Y. Kim, Proton radioactivity in relativistic continuum Hartree-Bogoliubov theory, *Phys. Rev. C* **93**, 014314 (Jan, 2016). doi: 10.1103/PhysRevC.93.014314. URL https://link.aps.org/doi/10.1103/PhysRevC.93.014314.
93. P. W. Zhao, Z. P. Li, J. M. Yao, and J. Meng, New parametrization for the nuclear covariant energy density functional with a point-coupling interaction, *Phys. Rev. C* **82**, 054319 (Nov, 2010). doi: 10.1103/PhysRevC.82.054319.
94. P. W. Zhao, L. S. Song, B. Sun, H. Geissel, and J. Meng, Crucial test for covariant density functional theory with new and accurate mass measurements from Sn to Pa, *Phys. Rev. C* **86**, 064324 (Dec, 2012). doi: 10.1103/PhysRevC.86.064324. URL https://link.aps.org/doi/10.1103/PhysRevC.86.064324.
95. A. V. Afanasjev and S. E. Agbemava, Covariant energy density functionals: Nuclear matter constraints and global ground state properties, *Phys. Rev. C* **93**, 054310 (May, 2016). doi: 10.1103/PhysRevC.93.054310. URL https://link.aps.org/doi/10.1103/PhysRevC.93.054310.
96. K. Zhang, X. He, J. Meng, C. Pan, C. Shen, C. Wang, and S. Zhang, Predictive power for superheavy nuclear mass and possible stability beyond the neutron drip line in deformed relativistic Hartree-Bogoliubov theory in continuum, *Phys. Rev. C* **104**, L021301 (Aug, 2021). doi: 10.1103/PhysRevC.104.L021301. URL https://link.aps.org/doi/10.1103/PhysRevC.104.L021301.
97. J. Meng and P. Zhao, Relativistic density functional theory in nuclear physics, *AAAPS Bulletin* **31**, 2 (2021). doi: 10.1007/s43673-021-00001-8. URL https://doi.org/10.1007/s43673-021-00001-8.
98. L. Chen, W. R. Plaß, H. Geissel, R. Knöbel, C. Kozhuharov, Y. A. Litvinov, Z. Patyk, C. Scheidenberger, K. Siegień-Iwaniuk, B. Sun, H. Weick, K. Beckert, P. Beller, F. Bosch, D. Boutin, L. Caceres, J. J. Carroll, D. M. Cullen, I. J. Cullen, B. Franzke, J. Gerl, M. Górska, G. A. Jones, A. Kishada, J. Kurcewicz, S. A. Litvinov, Z. Liu, S. Mandal, F. Montes, G. Münzenberg, F. Nolden, T. Ohtsubo, Z. Podolyák, R. Propri, S. Rigby, N. Saito, T. Saito, M. Shindo, M. Steck, P. M. Walker, S. Williams, M. Winkler, H.-J. Wollersheim, and T. Yamaguchi, New results on mass measurements of stored neutron-rich nuclides in the element range from Pt to U with the FRS-ESR facility at 360–400 MeV/u, *Nucl. Phys. A* **882**, 71–89 (2012). ISSN 0375-9474. doi: https://doi.org/10.1016/j.nuclphysa.2012.03.002. URL https://www.sciencedirect.com/science/article/pii/S0375947412001005.
99. S. Goriely, N. Chamel, and J. M. Pearson, Further explorations of Skyrme-Hartree-Fock-Bogoliubov mass formulas. XII. Stiffness and stability of neutron-star matter, *Phys. Rev. C* **82**, 035804 (Sep, 2010). doi: 10.1103/PhysRevC.82.035804. URL https://link.aps.org/doi/10.1103/PhysRevC.82.035804.
100. P. Moller, J. R. Nix, W. D. Myers, and W. J. Swiatecki, Nuclear ground-state masses and deformations, *At. Data Nucl. Data Tables* **59**(2), 185–381 (1995). ISSN 0092-640X. doi: https://doi.org/10.1006/adnd.1995.1002. URL https://www.sciencedirect.com/science/article/pii/S0092640X85710029.
101. S. Michimasa, M. Kobayashi, Y. Kiyokawa, S. Ota, R. Yokoyama, D. Nishimura, D. S. Ahn, H. Baba, G. P. A. Berg, M. Dozono, N. Fukuda, T. Furuno, E. Ideguchi, N. Inabe, T. Kawabata, S. Kawase, K. Kisamori, K. Kobayashi, T. Kubo, Y. Kubota,

C. S. Lee, M. Matsushita, H. Miya, A. Mizukami, H. Nagakura, H. Oikawa, H. Sakai, Y. Shimizu, A. Stolz, H. Suzuki, M. Takaki, H. Takeda, S. Takeuchi, H. Tokieda, T. Uesaka, K. Yako, Y. Yamaguchi, Y. Yanagisawa, K. Yoshida, and S. Shimoura, Mapping of a new deformation region around ^{62}Ti, *Phys. Rev. Lett.* **125**, 122501 (Sep, 2020). doi: 10.1103/PhysRevLett.125.122501. URL https://link.aps.org/doi/10.1103/PhysRevLett.125.122501.

102. C. Y. Fu, Y. H. Zhang, M. Wang, X. H. Zhou, Y. A. Litvinov, K. Blaum, H. S. Xu, X. Xu, P. Shuai, Y. H. Lam, R. J. Chen, X. L. Yan, X. C. Chen, J. J. He, S. Kubono, M. Z. Sun, X. L. Tu, Y. M. Xing, Q. Zeng, X. Zhou, W. L. Zhan, S. Litvinov, G. Audi, T. Uesaka, T. Yamaguchi, A. Ozawa, B. H. Sun, Y. Sun, and F. R. Xu, Mass measurements for the $T_z = -2$ fp-shell nuclei ^{40}Ti, ^{44}Cr, ^{46}Mn, ^{48}Fe, ^{50}Co, and ^{52}Ni, *Phys. Rev. C* **102**, 054311 (Nov, 2020). doi: 10.1103/PhysRevC.102.054311. URL https://link.aps.org/doi/10.1103/PhysRevC.102.054311.
103. J. M. Yao, K. Hagino, Z. P. Li, J. Meng, and P. Ring, Microscopic benchmark study of triaxiality in low-lying states of ^{76}Kr, *Phys. Rev. C* **89**, 054306 (May, 2014). doi: 10.1103/PhysRevC.89.054306. URL https://link.aps.org/doi/10.1103/PhysRevC.89.054306.
104. P. W. Zhao, Multiple chirality in nuclear rotation: A microscopic view, *Phys. Lett. B* **773**, 1–5 (2017). ISSN 0370-2693. doi: https://doi.org/10.1016/j.physletb.2017.08.001. URL https://www.sciencedirect.com/science/article/pii/S0370269317306226.
105. B.-N. Lu, E.-G. Zhao, and S.-G. Zhou, Potential energy surfaces of actinide nuclei from a multidimensional constrained covariant density functional theory: Barrier heights and saddle point shapes, *Phys. Rev. C* **85**, 011301 (Jan, 2012). doi: 10.1103/PhysRevC.85.011301. URL https://link.aps.org/doi/10.1103/PhysRevC.85.011301.
106. S. E. Agbemava, A. V. Afanasjev, T. Nakatsukasa, and P. Ring, Covariant density functional theory: Reexamining the structure of superheavy nuclei, *Phys. Rev. C* **92**, 054310 (Nov, 2015). doi: 10.1103/PhysRevC.92.054310. URL https://link.aps.org/doi/10.1103/PhysRevC.92.054310.
107. Z. X. Ren, P. W. Zhao, and J. Meng, Time-dependent covariant density functional theory in three-dimensional lattice space: Benchmark calculation for the $^{16}\mathrm{O} + ^{16}\mathrm{O}$ reaction, *Phys. Rev. C* **102**, 044603 (Oct, 2020). doi: 10.1103/PhysRevC.102.044603. URL https://link.aps.org/doi/10.1103/PhysRevC.102.044603.
108. Z. X. Ren, P. W. Zhao, and J. Meng, Dynamics of rotation in chiral nuclei, *Phys. Rev. C* **105**, L011301 (Jan, 2022). doi: 10.1103/PhysRevC.105.L011301. URL https://link.aps.org/doi/10.1103/PhysRevC.105.L011301.
109. S.-G. Zhou, J. Meng, P. Ring, and E.-G. Zhao, Neutron halo in deformed nuclei, *Phys. Rev. C* **82**, 011301 (Jul, 2010). doi: 10.1103/PhysRevC.82.011301.
110. L. Li, J. Meng, P. Ring, E.-G. Zhao, and S.-G. Zhou, Deformed relativistic Hartree-Bogoliubov theory in continuum, *Phys. Rev. C* **85**, 024312 (Feb, 2012). doi: 10.1103/PhysRevC.85.024312.
111. S.-G. Zhou, J. Meng, and P. Ring, Spherical relativistic Hartree theory in a Woods-Saxon basis, *Phys. Rev. C* **68**, 034323 (Sep, 2003). doi: 10.1103/PhysRevC.68.034323.
112. Y. Chen, L. Li, H. Liang, and J. Meng, Density-dependent deformed relativistic Hartree-Bogoliubov theory in continuum, *Phys. Rev. C* **85**, 067301 (Jun, 2012). doi: 10.1103/PhysRevC.85.067301. URL https://link.aps.org/doi/10.1103/PhysRevC.85.067301.
113. K. Zhang, M.-K. Cheoun, Y.-B. Choi, P. S. Chong, J. Dong, L. Geng, E. Ha, X. He, C. Heo, M. C. Ho, E. J. In, S. Kim, Y. Kim, C.-H. Lee, J. Lee, Z. Li, T. Luo, J. Meng, M.-H. Mun, Z. Niu, C. Pan, P. Papakonstantinou, X. Shang, C. Shen, G. Shen, W. Sun, X.-X. Sun, C. K. Tam, Thaivayongnou, C. Wang, S. H. Wong,

X. Xia, Y. Yan, R. W.-Y. Yeung, T. C. Yiu, S. Zhang, W. Zhang, and S.-G. Zhou, Deformed relativistic Hartree-Bogoliubov theory in continuum with a point-coupling functional: Examples of even-even Nd isotopes, *Phys. Rev. C* **102**, 024314 (Aug, 2020). doi: 10.1103/PhysRevC.102.024314. URL https://link.aps.org/doi/10.1103/PhysRevC.102.024314.

114. L.-L. Li, J. Meng, P. Ring, E.-G. Zhao, and S.-G. Zhou, Odd systems in deformed relativistic Hartree Bogoliubov theory in continuum, *Chin. Phys. Lett.* **29**(4), 042101 (Apr, 2012). doi: 10.1088/0256-307x/29/4/042101. URL https://doi.org/10.1088%2F0256-307x%2F29%2F4%2F042101.
115. X.-X. Sun and S.-G. Zhou, Rotating deformed halo nuclei and shape decoupling effects, *Sci. Bull.* **66**(20), 2072–2078 (2021). ISSN 2095-9273. doi: https://doi.org/10.1016/j.scib.2021.07.005. URL https://www.sciencedirect.com/science/article/pii/S2095927321004746.
116. X.-X. Sun and S.-G. Zhou, Rotating deformed halo nuclei and shape decoupling effects, *arXiv.* **2103:10886** (2021). URL https://arxiv.org/abs/2103.10886.
117. W. Sun, K.-Y. Zhang, C. Pan, X.-H. Fan, S. Zhang, and Z.-P. Li, Beyond-mean-field dynamical correlations for nuclear mass table in deformed relativistic Hartree-Bogoliubov theory in continuum, *Chin. Phys. C* p. accepted (2022).
118. X.-X. Sun, J. Zhao, and S.-G. Zhou, Shrunk halo and quenched shell gap at $N = 16$ in ^{22}C: Inversion of sd states and deformation effects, *Phys. Lett. B* **785**, 530–535 (2018). ISSN 0370-2693. doi: 10.1016/j.physletb.2018.08.071. URL http://www.sciencedirect.com/science/article/pii/S0370269318306907.
119. X.-X. Sun, J. Zhao, and S.-G. Zhou, Study of ground state properties of carbon isotopes with deformed relativistic Hartree-Bogoliubov theory in continuum, *Nucl. Phys. A* **1003**, 122011 (2020). ISSN 0375-9474. doi: https://doi.org/10.1016/j.nuclphysa.2020.122011. URL https://www.sciencedirect.com/science/article/pii/S0375947420303213.
120. K. Y. Zhang, D. Y. Wang, and S. Q. Zhang, Effects of pairing, continuum, and deformation on particles in the classically forbidden regions for Mg isotopes, *Phys. Rev. C* **100**, 034312 (Sep, 2019). doi: 10.1103/PhysRevC.100.034312. URL https://link.aps.org/doi/10.1103/PhysRevC.100.034312.
121. Z. H. Yang, Y. Kubota, A. Corsi, K. Yoshida, X.-X. Sun, J. G. Li, M. Kimura, N. Michel, K. Ogata, C. X. Yuan, Q. Yuan, G. Authelet, H. Baba, C. Caesar, D. Calvet, A. Delbart, M. Dozono, J. Feng, F. Flavigny, J.-M. Gheller, J. Gibelin, A. Giganon, A. Gillibert, K. Hasegawa, T. Isobe, Y. Kanaya, S. Kawakami, D. Kim, Y. Kiyokawa, M. Kobayashi, N. Kobayashi, T. Kobayashi, Y. Kondo, Z. Korkulu, S. Koyama, V. Lapoux, Y. Maeda, F. M. Marqués, T. Motobayashi, T. Miyazaki, T. Nakamura, N. Nakatsuka, Y. Nishio, A. Obertelli, A. Ohkura, N. A. Orr, S. Ota, H. Otsu, T. Ozaki, V. Panin, S. Paschalis, E. C. Pollacco, S. Reichert, J.-Y. Roussé, A. T. Saito, S. Sakaguchi, M. Sako, C. Santamaria, M. Sasano, H. Sato, M. Shikata, Y. Shimizu, Y. Shindo, L. Stuhl, T. Sumikama, Y. L. Sun, M. Tabata, Y. Togano, J. Tsubota, F. R. Xu, J. Yasuda, K. Yoneda, J. Zenihiro, S.-G. Zhou, W. Zuo, and T. Uesaka, Quasifree neutron knockout reaction reveals a small *s*-orbital component in the borromean nucleus ^{17}B, *Phys. Rev. Lett.* **126**, 082501 (Feb, 2021). doi: 10.1103/PhysRevLett.126.082501. URL https://link.aps.org/doi/10.1103/PhysRevLett.126.082501.
122. X.-X. Sun, Deformed two-neutron halo in ^{19}B, *Phys. Rev. C* **103**, 054315 (May, 2021). doi: 10.1103/PhysRevC.103.054315. URL https://link.aps.org/doi/10.1103/PhysRevC.103.054315.

123. C. Pan, K. Zhang, and S. Zhang, Multipole expansion of densities in the deformed relativistic Hartree-Bogoliubov theory in continuum, *Int. J. Mod. Phys. E* **28**(8), 1950082 (2019). doi: https://doi.org/10.1142/S0218301319500824. URL https://doi.org/10.1142/S0218301319500824.
124. E. J. In, P. Papakonstantinou, Y. Kim, and S.-W. Hong, Neutron drip line in the deformed relativistic Hartree-Bogoliubov theory in continuum: Oxygen to calcium, *Int. J. Mod. Phys. E* **30**(02), 2150009 (2021). doi: 10.1142/S0218301321500099. URL https://doi.org/10.1142/S0218301321500099.
125. C. Pan, K. Y. Zhang, P. S. Chong, C. Heo, M. C. Ho, J. Lee, Z. P. Li, W. Sun, C. K. Tam, S. H. Wong, R. W.-Y. Yeung, T. C. Yiu, and S. Q. Zhang, Possible bound nuclei beyond the two-neutron drip line in the $50 \leq z \leq 70$ region, *Phys. Rev. C* **104**, 024331 (Aug, 2021). doi: 10.1103/PhysRevC.104.024331. URL https://link.aps.org/doi/10.1103/PhysRevC.104.024331.
126. X. He, C. Wang, K. Zhang, and C. Shen, Possible existence of bound nuclei beyond neutron drip lines driven by deformation, *Chin. Phys. C* **45**, 101001 (2021). doi: 10.1088/1674-1137/ac1b99. URL https://doi.org/10.1088/1674-1137/ac1b99.
127. Y.-B. Choi, C.-H. Lee, M.-H. Mun, and Y. Kim, Bubble nuclei with shape coexistence in even-even isotopes of Hf to Hg, *Phys. Rev. C* **105**, 024306 (Feb, 2022). doi: 10.1103/PhysRevC.105.024306. URL https://link.aps.org/doi/10.1103/PhysRevC.105.024306.
128. K. Zhang, M.-K. Cheoun, Y.-B. Choi, P. S. Chong, J. Dong, Z. Dong, X. Du, L. Geng, E. Ha, X.-T. He, C. Heo, M. C. Ho, E. J. In, S. Kim, Y. Kim, C.-H. Lee, J. Lee, H. Li, Z. Li, T. Luo, J. Meng, M.-H. Mun, Z. Niu, C. Pan, P. Papakonstantinou, X. Shang, C. Shen, G. Shen, W. Sun, X.-X. Sun, C. K. Tam, Thaivayongnou, C. Wang, X. Wang, S. H. Wong, J. Wu, X. Wu, X. Xia, Y. Yan, R. W.-Y. Yeung, T. C. Yiu, S. Zhang, W. Zhang, X. Zhang, Q. Zhao, and S.-G. Zhou, Nuclear mass table in deformed relativistic Hartree-Bogoliubov theory in continuum, I: Even-even nuclei, *Atom. Data Nucl. Data Tabl.* **144**, 101488 (2022). ISSN 0092-640X. doi: https://doi.org/10.1016/j.adt.2022.101488. URL https://www.sciencedirect.com/science/article/pii/S0092640X22000018.
129. DRHBc Mass Table Collaboration, Deformed relativistic Hartree-Bogoliubov theory in continuum with a point-coupling functional: Examples of odd-mass Nd isotopes, *In preparation* (2022).

https://doi.org/10.1142/9789819813353_0016

Quests of Spin-Isospin Responses: A Tribute to Akito Arima

Hideyuki Sakai*
RIKEN Nishina Center,
2-1, Hirosawa, Wako, Saitama 351-0198, Japan
hidesakai@riken.jp

Decades of scientific interaction with Dr. Akito Arima on the study of spin-isospin responses are described. The episode of resolving the so-called quenching puzzle of the Gamow-Teller (GT) strengths is a pleasant recollection with Akito. Akito's interest in the spin-isospin response in the high excitation region led to the first implementation of the exothermic charge exchange reaction as a probe to explore the highly excited spin-isospin response. Our experimental efforts to explore the spin-isospin responses are presented by quoting our papers in connection with Akito's interest.

1. Prologue of GT Quenching Puzzle

Professor Akito Arima was imprisoned at the "International Conference on Spin Excitation in Nuclei" held in Telluride, Colorado, USA, March 25–27, 1982.[1] Dr. E. Vogt, Director of the TRIUMF in Canada, summarized the meeting with a portrait of Akito on the screen, saying with humor "Akito's alleged interpretation in terms of a conventional nuclear theory is outdated. It's time to accept a modern interpretation based on quarks". The cartoon Fig. 1 by Dr. Nguyen Ding Dang is a reproduction of the situation drawn at a later date. I had the opportunity to attend this international conference and was surprised by the scene.

In the 1970s, a Gamow-Teller giant spin resonance state of the nucleus was discovered.[2] The problem is that the magnitude of the resonance state (sum of transition strengths) observed in the experiment[3] was only about half of the value expected by quantum theory.[4] In quantum theory, the total sum of the transition strengths should conserve, and its whereabouts became a big puzzle as the "missing transition strength problem (quenching problem)". To explain this, A. Bohr and B. Mottelson,[5] for example, proposed that this quenching problem can be solved by considering the spin-excited state of the nucleon itself, called a delta (Δ) particle, which reflects the quark structure of the nucleon. Many nuclear physics researchers welcomed this interpretation as a long-awaited quark effect. Such was the backdrop of the timing of the international conference.

Akito gave a talk on the first day of this conference.[6] He made a different claim from that of Bohr and Mottelson. From the detailed analysis of the magnetic

*Professor Emeritus, The University of Tokyo.

Fig. 1. Akito jailed by the Δ Force.

moment in terms of the shell model,[7] he concluded that the delta particle effect is a maximum of 10% even if quarks are involved, and predicted that the missing transition strengths should be found in the highly excited region beyond the resonance state. After the talk, he faced severe criticism from the floor, a situation similar to being surrounded by enemies on all sides. Such an anomalous atmosphere probably influenced Vogt's summary and led to the illustration in the cartoon. However, Akito never gave up his view and persisted on his theory wherever he went around the world.

2. Quenching Problem of GT Strengths

Spin-isospin interactions create various interesting responses in nuclei such as the GT β-decay, the GT giant resonance (GTGR), the isovector-spin monopole giant resonance (IVSM), the double GT giant resonance (DGTGR), and so forth, depending on the momentum and excitation energy of the nuclear system. Those spin responses often play an essential role in understanding astrophysical phenomena. Note that the GT operator is defined as $\hat{O}_{\pm}^{\rm GT} = \sum_k t_{\pm}(k)\sigma_{\mu}(k)$, where $t_{\pm}$ is the raising or lowering isospin operator and σ_{μ} the μ-component spin operator, respectively, and it induces the transitions with $\Delta L = 0$ and $\Delta S = \Delta T = 1$, and consequently excites $J^{\pi} = 1^{+}$ states for an even-even target. A recent review may be found in Ref. 8.

The GT quenching phenomenon mentioned above refers to the model- independent GT sum rule value, $3(N - Z)$, which was first explicitly written by

Gaarde *et al.*,[9] often called the Ikeda sum rule. The quenching factor characterizes the degree of quenching

$$Q = \frac{S^{-}_{\mathrm{exp}}(\mathrm{GT}^{-}) - S^{+}_{\mathrm{exp}}(\mathrm{GT}^{+})}{3(N-Z)}, \tag{1}$$

where $S^{\mp}_{\mathrm{exp}}(\mathrm{GT}^{\mp})$ are the sum of the experimentally obtained $\mathrm{GT}^{\mp}$ strength $B(\mathrm{GT}^{\mp})$, respectively. The Q values were compiled by Gaarde,[9] where $B(\mathrm{GT}^{-})$ values were integrated up to just above the GTGR and theoretically estimated $B(\mathrm{GT}^{+})$ values were used, respectively, and were found to be surprisingly small, with $Q \approx 0.5(50\%)$ for medium and heavy nuclei. Note that Q should be unity in nucleonic degrees of freedom. This was recognized as the aforementioned quenching problem. In order to explain the origin of the quenching, two theoretical models were proposed. One was the Δ-hole (Δh) mixing effects where the GT strengths are shifted to the Δh region around 300 MeV excitation energies.[5] The other one strongly argued by Akito is due to the second order configuration mixing effects where the GT strengths are moved into the two-particle-two-hole state (2p2h) excitation energies, $\lesssim$50 MeV.[6,7] The Akito's assertion was well demonstrated by Bertsch and Hamamoto.[10]

The $B(\mathrm{GT})$ distributions are obtained experimentally employing various charge-exchange (CE) reactions such as the (p,n), (n,p), $(d,{}^2\mathrm{He})$, $({}^3\mathrm{He},t)$, $(t,{}^3\mathrm{He})$, etc. Nucleon CE reactions, (p,n) and (n,p), are most extensively used due partly to its reliability as a probe.[8]

The projectile (p/n) energy choice is critical due to the maximum spin-flip process and minimum distortion effect.

Maximum spinflip process

The ratio R^2 defined as $\hat{\sigma}\mathrm{GT}/\hat{\sigma}_{\mathrm{F}}$, provides an estimate of the ratio of the spin-flip to non-spinflip central components of underlying interaction, where $\hat{\sigma}_{\mathrm{GT}}(\hat{\sigma}_{\mathrm{F}})$ is the GT(F) unit cross section.[11] The R^2 value measured in the ${}^{14}\mathrm{C}(p,n){}^{14}\mathrm{N}$ reaction at 0° at various incident energies takes the maximum of about 15 at around $\mathrm{E}_{p/n} = 300$ MeV.[12] R^2 shows a target mass dependence, for example for ${}^{90}\mathrm{Zr}(p,n){}^{90}\mathrm{Nb}$ $R^2 = 25$.[13] This indicates that the energy spectrum (cross-section) of residual ${}^{90}\mathrm{Nb}$ is dominated by the spinflip character and contaminated by only 4% of the non-spinflip character that gives rise to a background.

Minimum distortion effect

The reliability of reaction analysis based on the direct nuclear reaction model depends on the distortion effects, which depend on the incident energy. The energy dependence of the distortion can be evaluated from the energy dependence of the nucleon-nucleon (NN) total cross sections. The measured total scattering cross section nearly equal to elastic scattering becomes smallest at around 200–300 MeV.[8] Thus the single-step process is dominant, and the distortion effects should be smallest in this energy region. Consequently, the nucleon (p/n) projectile energy at around 300 MeV is the best energy for the GT response search.

In order to make full use of this advantage, we proposed to construct the (p,n) and (n,p) facilities at the Research Center for Nuclear Physics (RCNP), Osaka University, in 1980s.[14,15] The Ring Cyclotron at RCNP provided us with a high-quality 300 MeV polarized proton beam to explore the spin-isospin responses. An experimental and theoretical comprehensive review of the spin-isospin responses studied with those (p,n) and (n,p) facilities is available in Ref. 8.

The (p,n) facility consists of a beam swinger system, a 100 m long time-of-flight (TOF) tunnel,[16] and a high-efficiency neutron detector system capable of neutron spin polarization analysis (NPOL, NPOL2).[17,18] Fig. 2 shows the swinger magnets (left) and the 100 m long TOF tunnel with NPOL (right). The (n,p) facility,[19] constructed around 2000, similar to the TRIUMF (n,p) facility[20] consists of a quasi-mono-energetic neutron production target (^{7}Li), a charged-particle clearing magnet, a multi-layer (n,p) target box, and the large acceptance spectrometer (LAS).[21]

^{90}Zr(p,n) and (n,p) experiments and results

Cross sections $d\sigma/d\Omega$ were measured at $E_{p/n} = 300$ MeV for $\theta = 0° - 12°$ and ω(energy transfer) $= 0 - 70$ MeV for the ^{90}Zr(p,n) reaction[22] and for the ^{90}Zr(n,p) reaction.[23]

The multipole decomposition analyses (MDA) were applied to both (p,n) and (n,p) data $(d^2\sigma(\theta,\omega)/d\Omega d\omega)$ in a consistent manner. The DWIA calculations were performed with the computer code DW81[18] for the following J^π transfers: $1^+(\Delta L = 0)$, 0^-, 1^-, $2^-(\Delta L = 1)$, $3^+(\Delta L = 2)$, and $4^-(\Delta L = 3)$. The one-body transition densities were calculated from pure 1p1h configurations. The effective

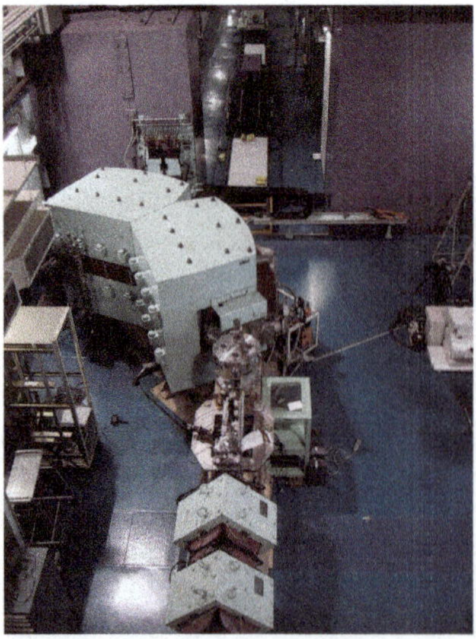

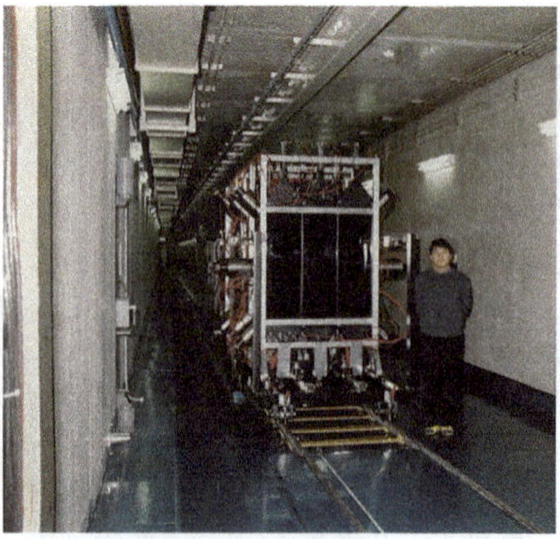

Fig. 2. (Left) Beam swinger system.[16] (Right) ToF tunnel with 100 m and neutron spin polarization analyzer NPOL,[17] and the author standing next to it.

NN interaction was taken from the t-matrix parameterization of the free NN interaction by Franey and Love at 325 MeV.[24] The derived $\Delta L = 0$ component, $d^2\sigma_{\Delta L=0}(q,\omega)/d\Omega d\omega$, of the cross sections determined in the MDA can be related to the corresponding GT strength by following a recipe known as a *proportionality ansatz*.[8]

The $\Delta L = 0$ component of the cross section, $\sigma_{\Delta L=0}(q,\omega)$, is proportional to the GT strength $B(\mathrm{GT})$,[25] the proportionality ansatz, such that

$$\sigma_{\Delta L=0}(q,\omega) = \hat{\sigma}_{\mathrm{GT}} \mathrm{F}(q,\omega) B(\mathrm{GT}), \tag{2}$$

where $\mathrm{F}(q,\omega)$ is the kinematical correction factor.[11] An advantage of Eq. (2) is that the $\hat{\sigma}_{\mathrm{GT}}$ can be calibrated by β-decay so that *absolute* $B(\mathrm{GT})$ values can be obtained from the experimentally measured $\sigma_{\Delta L=0}(q,\omega)$.

The $B(\mathrm{GT}^{\mp})$ strength against the excitation energy (ω) thus extracted are plotted in solid circles in Fig. 3.

Here, the ω is the excitation energy with respect to the *target* ground state (not residual nucleus). The obtained $B(\mathrm{GT}^-)$ strength shows a large contribution

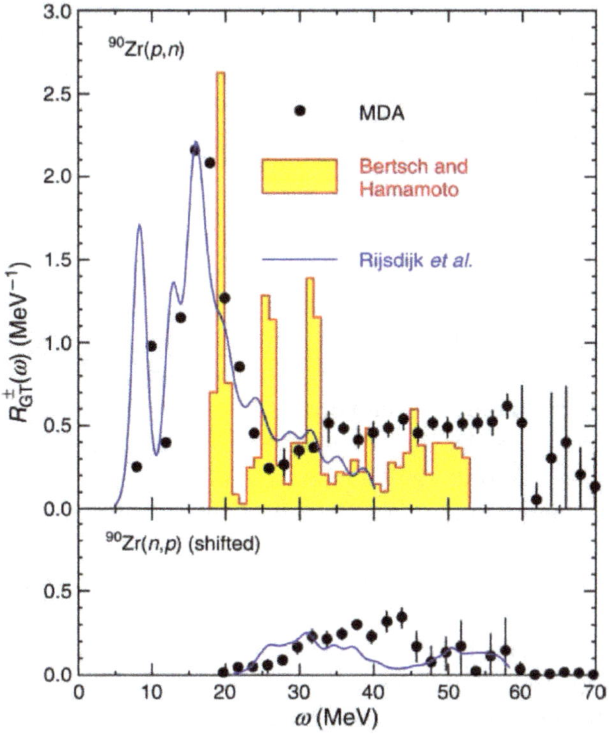

Fig. 3. Dots represent extracted $B(\mathrm{GT}^{\mp})$ distributions. The hatched histogram and solid curves represent the calculations by Bertsch-Hamamoto[10] and by Rijsdijk *et al.*,[27] respectively. Figure is taken from 8.

not only in the GTGR region, but also in the high excitation energy region up to 50 MeV.

The strengths $S_{\rm exp}^{-/+}({\rm GT}^{-/+})$ summed up to $\omega = 56\,{\rm MeV}\,(E_x \sim 50\,{\rm MeV})$ were 29.3 and 2.9, respectively. Thus the quenching factor turned out to be

$$Q = 0.88 \pm 0.07, \tag{3}$$

where the uncertainty includes those of statistical, systematic, MDA, IVSM and unit cross section $\hat{\sigma}_{\rm GT}$.[8] Note that the uncertainty was minimized by using a consistent MDA analysis in both (p,n) and (n,p) data simultaneously.[26]

Theoretical calculations have been made to reproduce the B(GT) strength distribution extending to the high-energy continuum, for example, in terms of a dressed-particle RPA (DRPA)[27] or extended-renormalized RPA[28] and obtained reasonable agreements as shown in Fig. 3.

It should be remarked that since the IVSM exists in the high-energy continuum region, IVSM strengths must be subtracted from the $d^2\sigma_{\Delta L=0}(q,\omega)/d\Omega d\omega$ spectrum. Since the IVSM strengths were only estimated theoretically, there is a model-dependent ambiguity. Experimental efforts to study IVSM resonances will be introduced later.

If the deduced $\hat{\sigma}_{\rm GT}$ value obtained with the systematic study at 300 MeV[13] in a more recent study is used, Q becomes 0.92 ± 0.07. Anyway, Q must be around 0.9. Thus obtained Q value is significantly larger than those compiled by Gaarde.[9] This fact implies that the primary quenching mechanism is due to the second-order configuration mixing of 2p2h states and hence the Δh coupling plays a minor role.

3. Epilogue of GT Quenching Puzzle

As is shown above, about 40% of the total strength is found to be widely spread in the continuum region higher than GTGR (reaching ~90% in total) in the case of the nucleus ^{90}Zr. As a result, the "missing transition strength problem" was resolved, proving that Akito's claim had been correct. In response to this, we organized the "International Symposium on New Facet of Spin Giant Resonances in Nuclei" held at the University of Tokyo in 1997. During the concluding remark of the meeting, Professor C. Glashausser of Rutgers University congratulated Akito by showing a cartoon Fig. 4 of his being released from prison.[29]

This series of events related to the quenching puzzle became a very special and unforgettable memory for me.

Sometime later, while chattering, Akito muttered, "Can't you identify the GT strengths in $\omega > 50$ MeV?". The present author immediately replied that it would be experimentally very challenging due to the momentum transfer ($q > 0.4\,{\rm fm}^{-1}$) associated with such a large energy transfer. Still, his straightforward question lingered in my mind. Soon, a novel idea to exploit an exothermic reaction to overcome the difficulty, at least partially, came up to me.

Fig. 4. Akito is released from the jail, proudly holding 2p-2h ladders in hand.

4. ICHOR Project and Search for Isovector Spin Monopole Modes

So far, the spin-isospin responses were studied mainly by using the CE reactions with *stable* beams. Note that such CE reactions lead to endothermic reactions. Generally the spin-isospin responses under consideration ($\Delta L = 0, \Delta S = 1$) are most strongly excited at small momentum transfer (preferably $q = 0$) in the reaction. However, the reaction kinematics due to the endothermic reaction is such that the excitation of a nucleus is always associated with the finite momentum transfer ($\omega < q$ space-like region)). Therefore it becomes challenging to search for spin-isospin responses in the highly excited region due to the increasing momentum transfer. This is one of the reasons why we could not identify the GT strengths beyond 50 MeV excitation energy. The exothermic CE reaction employing radioactive projectiles having a significant mass excess allows us to access an unexplored $\omega > q$ region (time-like region) owing to a large positive reaction Q-value.

Fig. 5 compares kinematical relations between q and ω at $0°$ for the ^{90}Zr(^{12}N, ^{12}C)^{90}Nb reaction at 175 MeV/u (solid line), the ^{90}Zr(p, n)^{90}Nb reaction at 300 MeV (dashed line) and the ^{90}Zr(^{3}He, t)^{90}Nb reaction at 300 MeV/u (dot dashed line). The reaction Q-values are -6.89 MeV (endothermic) and $+11.23$ MeV (exothermic) for the ^{90}Zr(p, n) and ^{90}Zr(^{12}N, ^{12}C) reactions, respectively. As is seen, the exothermic ^{90}Zr(^{12}N, ^{12}C) reaction reaches $q = 0$ (recoilless condition) at $\omega = 14\,\text{MeV}(E_x = 10\,\text{MeV})$, while the endothermic ^{90}Zr(p, n) reaction q increases monotonically according to an increase of ω. To make the best use of the exothermic reaction, we applied it to study the IVSM modes. In ^{90}Nb, the IVSM resonance in the β^- direction is estimated to exist at $E_x = 20 - 30$ MeV. Thus it is evident that the kinematics of the exothermic ^{90}Zr(^{12}N, ^{12}C)^{90}Nb reaction has the advantage to achieve $q = 0.1 - 0.2\,\text{fm}^{-1}$, while that of the endothermic ^{90}Zr(p, n)^{90}Nb reaction $q = 0.2 - 0.3\,\text{fm}^{-1}$.

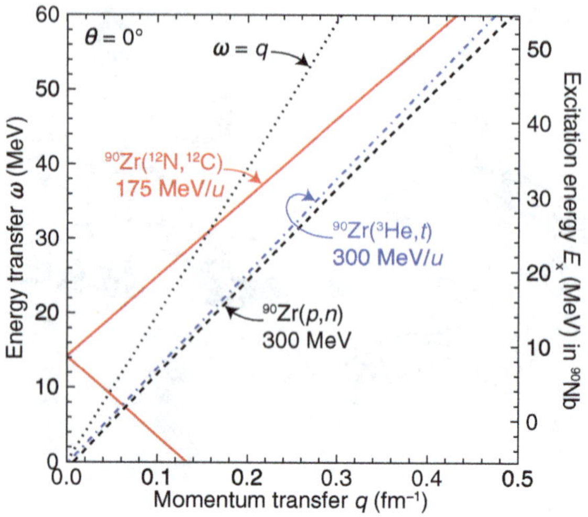

Fig. 5. q and ω kinematical relation for the representative reactions ^{90}Zr(p,n) reaction at 300 MeV and at 0° and ^{90}Zr(^{12}N, ^{12}C) reaction at 175 MeV/u.

4.1. *SHARAQ Spectrometer*

To realize the exothermic CE reaction with radioactive beams, the ICHOR (Isospin-spin responses in CHarge-exchange exOthermic Reactions) project[a] was initiated in 2005. For that purpose, a magnetic spectrometer SHARAQ was designed and constructed in collaborations with CNS and RI Beam Factory (RIBF), and installed in a RIBF experimental hall.[30] SHARAQ was intended to be used in missing mass (missing energy) spectroscopy experiments with exotic secondary RI beams.

The SHARAQ spectrometer can analyze particles with magnetic rigidity of 1.8–6.8 Tm, corresponding to an energy of 40–440 MeV/u for A/Z = 2 particles.[30] It was designed for momentum and angular resolutions of $\delta p/p \sim 1/15,000$ and $\delta\theta \sim 1$ mrad, respectively. The momentum acceptance is ±1%. The solid angle is 4.8(2.7) msr in an achromatic beam mode(in dispersion matching mode). The SHARAQ spectrometer consists of three quadrupole (Q1, Q2, Q3) and two dipole (D1, D2) magnets. Q1 and Q2 are the superconducting doublet quadrupole magnets. SHARAQ can rotate around its target position in the range from −2° to +15°.

RI beams, produced from primary heavy-ion beams in BigRIPS,[31] are transported to the SHARAQ target position through the BigRIPS-SHARAQ beamline. The beamline and the spectrometer are designed to fulfill the complete dispersion matching conditions, namely both in lateral and angular dispersions simultaneously, which is crucial to achieving the high spatial and angular resolving powers of the spectrometer because of the sizeable intrinsic momentum spread of RI beams.

[a]Supported in part by the Grant-in-Aid for Specially Promoted Research of MEXT Japan. No.17002003. Spokesperson; H. Sakai. ICHOR means a ginkgo tree in Japanese, and its leaf is used as the symbol mark logo of the University of Tokyo.

Fig. 6. The SHARAQ spectrometer with collaborators.

Fig. 6 is the photograph taken on the memorial day of the first beam. The D2 magnet weighing about 300 tons is painted blue.

5. IVSM Modes Studied via RI Beams with SHARAQ

The bulk properties of nuclei such as the incompressibilities of nuclear matter are often studied by the giant resonances at high excitation energies. The characteristics of the isovector giant monopole resonance (IVGMR) and the IVSM resonance (IVSMR) provide unique insight into the bulk properties of nuclei. The IVGMR and the IVSMR are both breathing modes in which the proton and neutron density distributions oscillate out of phase. Significant studies have been made on the IVSMR, and important information has been obtained, see for examples in Refs. 32, 33.

In a microscopic picture, IVSMR is a coherent $2\hbar\omega$ 1p-1h transition with $\Delta L = 0$ and $\Delta S = \Delta T = 1$, driven by an operator $\hat{O}_{\pm}^{\mathrm{IVSM}} = \sum_k t_{\pm}(k)\sigma_{\mu}(k)r(k)^2$.[34] Thus, the IVSM operator has the same selection rule as that of GT, but, because of having an r^2 term, there is a difference in the number of radial nodes (n) in terms of the single-particle wave function. The GT operator induces $\Delta n = 0$ ($0\hbar\omega$) excitations while the IVSM transition is associated with $\Delta n = 1$ ($2\hbar\omega$) excitations. Both operators excite $J^{\pi} = 1^{+}$ states for even-even target, and naturally they mix. Consequently, choosing an appropriate reaction probe is essential to enhance the IVSM modes.

Since the transition density associated with the IVSM mode has a node near the nuclear surface due to the $\Delta n = 1$ character, the projectile sensitive to the surface region of the target nucleus is most suited for exciting IVSM modes. The surface absorptive projectile, namely heavy-ions, is a promising tool. It should be pointed out here that the nucleonic projectiles (p or n), being used widely for the GTGR study, penetrating deeply into the nuclear interior, tend to cancel transition densities between the inner and outer parts, leading to a substantial reduction of the IVSM cross-sections.

Some experiments searching for the IVSMR were performed before the SHARAQ construction. Evidence for the existence of the IVSMR in the β^- direction came from the (p, n) reaction at 800 MeV and $(^3\text{He}, t)$ measurements.[35–37] Note that at 800 MeV, the (p, n) reaction becomes absorptive. In the β^+ direction, encouraging results had been obtained by using the $(t, ^3\text{He})$ reaction.[38]

5.1. *Study of IVSMR in the β^- Direction via Exothermic Reaction*

The IVSMR in the β^- direction was studied with a new probe by using the exothermic $^{90}\text{Zr}(^{12}\text{N}, ^{12}\text{C})^{90}\text{Nb}$ reaction at 175 MeV/u and at 0°–3° by Noji *et al.*[39]

A 250 MeV/u, 400-pnA beam of ^{14}N impinged upon a 5-mm-thick beryllium target, and ^{12}N nuclei were selected among the various projectile fragments in the BigRIPS fragment separator. The ^{12}N beam with a rate of 1.8 Mpps and purity of 92% was transported to the ^{90}Zr reaction target using the *dispersion-matching* technique. The ^{90}Zr target (99.4% isotopic enrichment) was 154 mg/cm^2 thick.

The ^{12}C ejectiles were analyzed by the SHARAQ spectrometer, and their trajectories were measured with two cathode-readout drift chambers (CRDCs) placed at the focal plane manufactured in collaboration with GANIL, France.[40] The excitation energy in ^{90}Nb was obtained in a missing-mass calculation over the range $0 \leq \text{E}_x \leq 70$ MeV with a resolution of 8 MeV. Scattering angles were measured over the range $0° \leq \theta_{\text{cm}} \leq 3°$ with a resolution of 0.6°(FWHM). ^{12}N beam particles can decay in flight (β-decay half-life of 11.0 ms) into ^{12}C near the target and contribute to the background in the data, and its contribution was roughly two orders of magnitude larger than that of the charge-exchange reaction products. These huge in-flight β-decay background events were eliminated by measuring the energy loss of particles with two 1-mm-thick plastic scintillators which sandwiched the ^{90}Zr target, resulting in the suppression of a factor of about 10^3.

The obtained excitation-energy spectrum for the $0° - 1°$ is shown in Fig. 7. Two broad peaks are clearly observed at around $\text{E}_x \approx 10$ MeV and $\text{E}_x = 20$–40 MeV in the spectrum. The lower peak is easily identified as the GTGR (see Fig. 3). As to the broad peak observed at $\text{E}_x = 20$–40 MeV, the excitation-energy spectrum was compared with those of the (p, n) reaction at 795[35] and 200 MeV,[41] and the $(^3\text{He}, t)$ reaction at 300 MeV/u,[36] as shown in Fig. 7(a). The previous data were smeared to match the resolution of the present data and scaled such that the GTGR peaks coincide. Although general feature of the broad peak nicely agrees with those of Ref. 35, 36, an enhancement at around $\text{E}_x = 20$–30 MeV is observed.

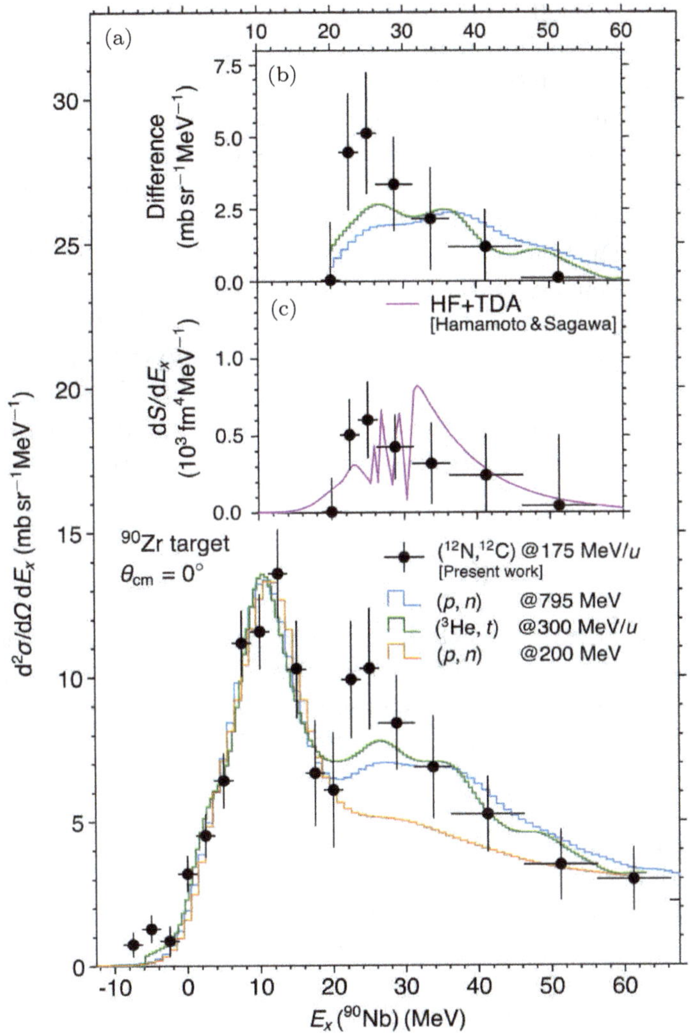

Fig. 7. (a) Comparison of the double-differential cross-section spectra with past experiments. (b) Differences of the spectra with respect to that of the 200-MeV (p, n) spectrum. (c) The isovector spin monopole strength distribution in ^{90}Zr from the experimental data compared with theoretical predictions. Figure is taken from Ref. 39.

Since the excitation of the IVSMR is strongly reduced in the (p,n) reaction at 200 MeV, due to the cancellation of the inner and surface components of the transitions amplitudes as discussed above, it was used to subtract from the (^{12}N, ^{12}C), the (p,n) spectrum at 795 MeV and the (^{3}He, t) spectrum at 300 MeV/u, as shown in Fig. 7(b). The excitation of the IVSMR is enhanced in the excitation-energy region below 30 MeV relative to the spectra from these other two reactions, indicating that the (^{12}N, ^{12}C) reaction is in-deed a powerful tool for the investigation of the IVSMR.

The extracted strength distribution for the IVSMR, which was obtained by dividing the cross-section difference by the excitation-energy-dependent distorted-wave Born approximation (DWBA) calculation is shown in Fig. 7(c). DWBA calculations were performed using the microscopic double-folding code FOLD/DWHI.[42]

The experimental data exhaust $90 \pm 54\%$ of the non-energy-weighted sum rule of the normal-mode formalism.[43] The data are compared with the theoretical strength distribution in the Hartree-Fock (HF) plus Tamm-Dancoff approximation (TDA) using the SGII Skyrme interaction for the IVSMR in ^{90}Nb.[34] The theoretical calculations do reasonably well in describing the data.

It is interesting to compare the R^2 values between the ^{90}Zr(^{12}N, ^{12}C)^{90}Nb reaction at 175 MeV/u and the ^{90}Zr$(p, n)^{90}$Nb reaction at 300 MeV. They are 54 ± 22[39] and 24 ± 2,[13] respectively. This indicates that the (^{12}N, ^{12}C) reaction tends to induce the spin modes more than two times stronger compared to the (p, n) reaction.

Thus it is clear that the exothermic (^{12}N, ^{12}C) reaction is the best suitable probe for studying IVSMR in the β^- direction.

5.2. *Study of IVSMR in the β^+ Direction via (t,^{3}He) Reaction*

In order to clearly identify the IVSMR in the β^+ direction, double-differential cross sections of the ^{90}Zr and ^{208}Pb(t, ^{3}He) reactions at 300 MeV/u were measured by Miki *et al.*[44] The GT and IVSM components are distinguished by comparing the obtained (t, ^{3}He) spectra with previously measured (n, p) spectra.[23,45]

The reaction Q-values are -2.26 and -4.98 MeV for the ^{90}Zr and ^{208}Pb targets, respectively. As such, they are endothermic reactions. The reaction kinematics of the (t, ^{3}He) reaction at 300 MeV/u is very similar to that of (p, n) reaction at 300 MeV, see Fig. 5. Since the excitation energy of IVSMR in the β^+ direction is estimated to be lower by 10 MeV compared to that of the β^- direction, the IVSMR is expected to appear at $E_x = 10 - 20$ MeV corresponding to the momentum transfer $q = 0.1 - 0.2\,\text{fm}^{-1}$. Thus one might guess that both the (t, ^{3}He) and (n, p) spectra should be very similar. Nevertheless, this is not the case. Because of the surface absorptive nature of the composite nucleus ^{3}He, the IVSMR in the β^+ direction will be enhanced as was the case in the β^- direction.

The experiment was performed at RIBF with the SHARAQ spectrometer. The primary 320 MeV/u ^{4}He beam with an intensity approximately 300 pnA bombarded onto a ^{9}Be production target with a thickness of 4 cm. From a variety of resulting fragments, tritons at 300 MeV/u were selected in the fragment separator BigRIPS. The secondary triton beam with an intensity of 10^7 pps and a purity of >99% was transported with an *achromatic-mode* to the secondary target. The beam spot at the target was 3 mm in diameter. It had an energy (angular) spread of 1.8 MeV (0.4°) at FWHM. The secondary targets included the isotopically enriched ^{90}Zr (99.36%) and ^{208}Pb (98.4%) targets with thicknesses of 308 and 400 mg/cm^2, respectively. The reaction products from the secondary target were momentum analyzed by the SHARAQ magnetic spectrometer, and detected by two CRDCs installed at the focal plane of the spectrometer. This was the first physics experiment with the SHARAQ spectrometer.

The double-differential cross sections were determined for excitation energies of $0 \leq E_x \leq 40\,\text{MeV}$ and for scattering angles of $0° \leq \theta \leq 4°$. In order to extract the monopole component, MDA similar to the $^{90}\text{Zr}(p, n)$ data analysis were performed using the J^π transfers: $1^+(\Delta L = 0), 0^-, 1^-, 2^-(\Delta L = 1), 3^+(\Delta L = 2), 4^-(\Delta L = 3)$, and $5^+(\Delta L = 4)$.

The obtained monopole cross section contains both the GT and the IVSM components. In order to distinguish the IVSM transitions from the GT transitions, the monopole cross sections of the present $(t, {}^3\text{He})$ and the previous (n, p) spectra[23,45] were compared. If the (n, p) cross section is normalized so that it contains the same amount of the GT cross section as the $(t, {}^3\text{He})$ spectra, the IVSMR should appear as an enhancement in the $(t, {}^3\text{He})$ spectra compared with the normalized (n, p) spectra.

The results are shown in the top panels of Fig. 8. At low excitation energies up to 5 MeV, the $(t, {}^3\text{He})$ cross sections were similar to the normalized (n, p) cross sections. This implies that the monopole cross section in this region is dominated by GT transitions. On the other hand, at higher excitation energies, significant enhancements were observed in the $(t, {}^3\text{He})$ spectra for both ^{90}Zr and ^{208}Pb targets. These enhancements can be attributed to the stronger excitation of the IVSMR in the $(t, {}^3\text{He})$ reaction.

The present experimental data can also provide a test of theoretical calculations. The bottom panels in Fig. 8 represent comparisons between the experimental

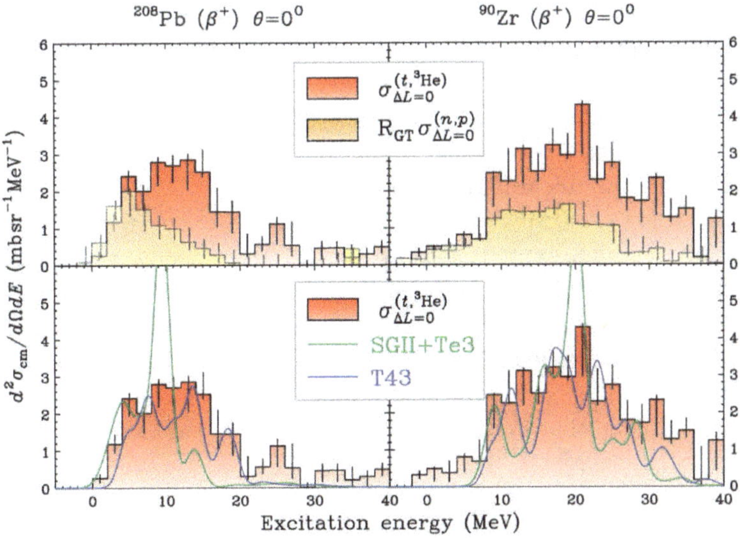

Fig. 8. [Top] Comparison of the monopole cross section of the $(t, {}^3\text{He})$ and (n, p) spectra at 0°. The (n, p) spectra were taken from Refs. 23, 45. [Bottom] Comparison of the $(t, {}^3\text{He})$ monopole cross sections of the experimental data and the DWBA+HF-RPA calculations. Only statistical errors are shown. Figure is taken from Ref. 44.

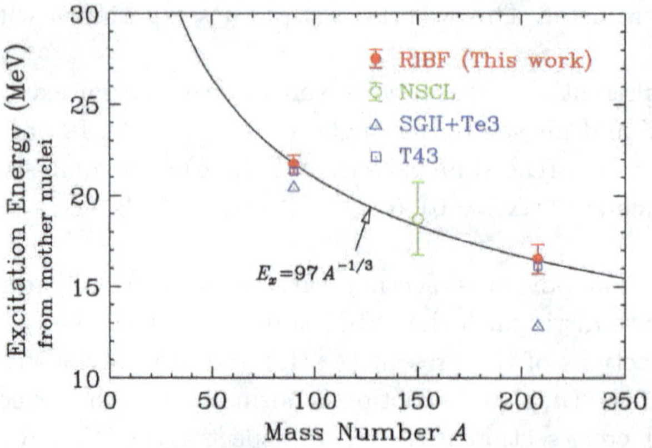

Fig. 9. Systematics of the average energy of the monopole cross section distributions (see text). The NSCL result[46] was obtained with an extrapolation to $q = 0$. Figure is taken from Ref. 44.

data and theoretical calculations on the $(t, {}^3\text{He})$ monopole cross section. For the theoretical calculations, see Ref. 44.

Since the calculations reproduced well experimental data, we concluded that IVSMR was discovered in the present $(t, {}^3\text{He})$ spectra.

It is interesting to derive the mean excitation energy of the monopole cross sections as a function of the mass number A. It is plotted in Fig. 9. The excitation energy, here, is measured from the ground state of the mother nucleus. The filled circles represent the average energies of the $\Delta L = 0$ cross sections in our $(t, {}^3\text{He})$ spectra. They are $21.7 \pm 0.5(\text{stat})$ MeV for ^{90}Zr and $16.5 \pm 0.8(\text{stat})$ MeV for ^{208}Pb. The data are fitted with a simple $A^{-1/3}$ dependence. The obtained function is $\text{E}_x = 97A^{-1/3}$ MeV, which is shown as a solid curve in Fig. 8. This is larger than the typical $2\hbar\omega$ value of $80A^{-1/3}$ MeV,[47] which is consistent with the repulsive nature of the spin-isospin part of the residual interaction.

In summary, the double-differential cross sections of the ^{208}Pb and ${}^{90}\text{Zr}(t, {}^3\text{He})$ reactions at 300 MeV/u using the SHARAQ spectrometer were measured. The monopole cross sections extracted from our $(t, {}^3\text{He})$ spectra were found to significantly exceed those of the previous (n, p) spectra. The absorptive nature of the $(t, {}^3\text{He})$ reaction allowed us to identify these enhancements as the IVSMRs (β^+). The $(t, {}^3\text{He})$ monopole cross sections were well reproduced by the theoretical calculations.

Thus, we concluded that IVSMR in the β^+ direction was established for the first time using the high energy $(t, {}^3\text{He})$ reaction.

6. Summary

Our study of spin isospin physics inspired by Akito is briefly presented in this article, along with our experimental measurements and facility constructions.

In the 1980s, the so-called 'quenching problem', missing strength of about 50% from the spin sum-rule value $3(N-Z)$, was one of the topics in spin-isospin physics. Two different theoretical interpretations were made as to the origin of this quenching. One was due to the Δ-hole mixing, which was highlighted at that time, and the other due to the 2p2h second-order configuration mixing, which Akito strongly asserted. We successfully identified ~40% of the GT strengths in the high continuum region beyond GTGR by the ^{90}Zr(p,n) reaction at 300 MeV, which solved the quenching problem and at the same time proved Akito's assertion.

In order to extend our spin-isospin response study to the high excitation region, we introduced the exothermic CE reaction with a radioactive projectile. Under the ICHOR program, the SHARAQ magnetic spectrometer was designed and constructed to realize the exothermic CE reaction with a radioactive beam. It was, for the first time, demonstrated to excite the IVSMR in the β^- direction, which is located $\omega = 20 - 40$ MeV employing the exothermic ^{90}Zr(^{12}N, ^{12}C)^{90}Nb reaction at 175 MeV/u with SHARAQ in the dispersion-matching mode operation. The observed IVSMR turns out to exhaust $90 \pm 54\%$ of the non-energy-weighted sum rule of normal-mode formalism. Thus, we established the exothermic CE reaction as a new probe to explore the spin-isospin responses in the high excitation-energy region.

The IVSMR in the β^+ direction that had not been established was for the first time confirmed by using the $(t, ^3\text{He})$ reaction at 300 MeV/u with SHARAQ in the achromatic mode operation. The measured mean excitation energies of the IVSMR for ^{90}Zr and ^{208}Pb targets provided us the target mass number dependence. It was found to be $\text{E}_x = 97A^{-1/3}$ MeV showing the repulsive nature of underlying effective interactions. Thus, it was concluded that the IVSMR in the β^+ direction was discovered.

The SHARAQ spectrometer has been used for various researches with radioactive beams. For example, an exotic multi-neutron state, namely the tetra-neutron n^4 as a resonance state, was discovered by the exothermic double CE reaction ^{4}He(^{8}He, ^{8}Be)n^4.[48] The projectile ^{8}He changes to ejectile ^{8}Be by the double GT transitions, while the target ^{4}He to the residual n^4 by the double spin-dipole transitions. The significant mass excess of ^{8}He about +31.6 MeV is the key to compensating the large negative mass difference between ^{4}He and $4n$ system of about 29.8 MeV, which realized the almost recoilless condition of about 10 MeV/c (0.05 fm^{-1}) for the $4n$ system. n^4 is observed as the resonance at excitation energy 0.83 ± 0.65 (stat.) ± 1.25 (sys.) MeV with the width of 2.6 MeV (FWHM).

7. In Closing

Akito always continued his research enthusiastically, even as he grew older, and even when he was overwhelmed by other duties. Once, Akito, who was at that time the president of RIKEN, invited me to come over for a meeting with him, and I hurried to his office wondering what it would be for. When I arrived, he said,

"I am going to the airport now, and you will come with me." As soon as we jumped in a taxi, he took out his transparencies and commenced a physics discussion. All the way to the airport, he continued discussing physics delightfully. I got a glimpse of Akito's genuine passion for research that transcended time restrictions. Although we approached nuclear physics differently, theoretically or experimentally, I was truly inspired by his sincerity as a researcher.

Acknowledgements

First of all, I am very thankful for all the help and support Akito gave me during my scientific career at RCNP, the University of Tokyo, and RIKEN. This article is mainly based on the experimental results presented in published articles.[8, 22, 23, 39, 44] The author is grateful to all coauthors of these articles for their excellent collaboration. Let me mention in particular, specific names. Section 2 for GT quenching physics is based on the work with Tomotsugu Wakasa and Kentaro Yako, while section 5 on the ISMR experiments with Shumpei Noji and Kenjiro Miki. Many thanks also to Kichiji Hatanaka and the late Hiroyuki Okamura for their collaboration and help. Thanks also to Tomohiro Uesaka and Susumu Shimoura for the SHARAQ construction and physics discussions. It is greatly appreciated to Shumpei Noji and Tomotsugu Wakasa for their critical reading and comments on this manuscript. Last but not least, I am grateful to the late Munetake Ichimura for enlightening discussions and collaboration in spin-isospin physics.[8]

References

1. *Spin Excitations in Nuclei*, ed. F. Petrovich *et al.*, Plenum Press, New York (1984).
2. R. R. Doering *et al.*, *Phys. Rev. Lett.* **35**, 1691 (1975).
3. C. D. Goodman *et al.*, *Phys. Rev. Lett.* **44**, 1755 (1980).
4. C. Gaarde, *Nucl. Phys. A* **396**, 127c (1983).
5. A. Bohr and B. R. Mottelson, *Phys. Lett. B* **100**, 10 (1981).
6. A. Arima, p. 7 in Ref. [1].
7. K. Shimizu, M. Ichimura, and A. Arima, *Nucl. Phys. A* **226**, 282 (1974).
8. M. Ichimura, H. Sakai, and T. Wakasa, *Prog. Part. Nucl. Phys.* **56**, 446 (2006).
9. C. Gaarde *et al.*, *Nucl. Phys. A* **334**, 248 (1980).
10. G. F. Bertsch and I. Hamamoto, *Phys. Rev. C* **26**, 1323 (1982).
11. T. N. Taddeucci *et al.*, *Nucl. Phys. A* **469**, 125 (1987).
12. J. Rapaport and E. Sugarbaker, *Annu. Rev. Nucl. Part. Sci.* **44**, 109 (1994).
13. M. Sasano, H. Sakai *et al.*, *Phys. Rev. C* **79**, 024602 (2009).
14. H. Sakai *et al.*, Workshop Reports to the Program Advisary Committee. RCNP-T-14, 1984 and RCNP-P-74, 1985. In Japanese.
15. Cyclotron Cascade Project and Its Prospects, Brochure published by RCNP, 1986, p. 90. In Japanese.
16. H. Sakai *et al.*, *Nucl. Instrum. Methods Phys. Res. A* **369**, 120 (1996).
17. H. Sakai *et al.*, *Nucl. Instrum. Methods Phys. Res. A* **320**, 479 (1992).
18. T. Wakasa, H. Sakai *et al.*, *Nucl. Instrum. Methods Phys. Res. A* **404**, 355 (1998).
19. K. Yako, H. Sakai *et al.*, *Nucl. Phys. A* **684**, 563c (2001).
20. H. Conde, *Nucl. Instrum. Methods Phys. Res. A* **292**, 121 (1990).

21. N. Matsuoka and T. Noro, RCNP Annual Report, 1987, p. 176.
22. T. Wakasa, H. Sakai *et al.*, *Phys. Rev. C* **55**, 2909 (1997).
23. K. Yako, H. Sakai *et al.*, *Phys. Lett. B* **615**, 193 (2005).
24. M. A. Franey and W. G. Love, *Phys. Rev. C* **31**, 488 (1985).
25. C. Gaarde *et al.*, *Nucl. Phys. A* **369**, 258 (1981).
26. H. Sakai and K. Yako, *Nucl. Phys. A* **731**, 94 (2004).
27. G. A. Rijsdijk, *Phys. Rev. C* **48**, 1752 (1993).
28. Nguyen Dinh Dang, A. Arima *et al.*, *Phys. Rev. Lett.* **79**, 1638 (1997).
29. *International Symposium on New Facet of Spin Giant Resonances in Nuclei*, ed. H. Sakai, H. Okamura, and T. Wakasa, World Scientific, Singapore 1998, p. 437.
30. T. Uesaka, H. Sakai *et al.*, *Nucl. Instrum. Methods Phys. Res. B* **266**, 42185 (2008).
31. T. Kubo, *Nucl. Instrum. Methods Phys. Res. B* **204**, 97 (2003).
32. M. N. Harakeh and A. van der Woude, *Giant Resonances: Fundamental High-Frequency Modes of Nuclear Excitation*, (Oxford University Press, New York, 2001).
33. G. Colò, The compression modes in atomic nuclei and their relevance for the nuclear equation of state, *Phys. Part. Nucl.* **39**, 286 (2008).
34. I. Hamamoto and H. Sagawa, *Phys. Rev. C* **62**, 024319 (2000).
35. D. L. Prout *et al.*, *Phys. Rev. C* **63**, 014603 (2000).
36. A. Brockstedt *et al.*, *Nucl. Phys. A* **530**, 571 (1991).
37. R. G. T. Zegers *et al.*, *Phys. Rev. Lett.* **90**, 202501 (2003).
38. J. Guillot *et al.*, *Phys. Rev. C* **73**, 014616 (2006).
39. S. Noji, H. Sakai *et al.*, *Phys. Rev. Lett.* **120**, 172501 (2018).
40. S. Michimasa *et al.*, *CNS-REP-80* 55 (2009); *CNS-REP-83*, 55 (2010).
41. C. Gaarde, *Nucl. Phys. A* **396**, 127 (1983).
42. J. Cook and J. A. Carr, Computer program FOLD, Florida State University (unpublished); based on F. Petrovich and D. Stanley, *Nucl. Phys. A* **275**, 487 (1977); modified as described in J. Cook, K. W. Kemper, P. V. Drumm, L. K. Fifield, M. A. C. Hotchkis, T. R. Ophel, and C. L. Woods.
43. S. Y. van der Werf, Computer program NORMOD (unpublished).
44. K. Miki, H. Sakai *et al.*, *Phys. Rev. Lett.* **108**, 262503 (2012).
45. K. J. Raywood, S. A. Long, and B. M. Spicer, *Nucl. Phys. A* **625**, 675 (1997).
46. C. J. Guess *et al.*, *Phys. Rev. C* **83**, 06318 (2011).
47. A. Bohr and B. R. Mottelson, *Nuclear Structure* (Benjamin, New York, 1969), Volumes I–II.
48. K. Kisamorii, S. Shimoura, H. Sakai *et al.*, *Phys. Rev. Lett.* **116**, 052501 (2016).

https://doi.org/10.1142/9789819813353_0017

Spin-dependent Modes in Nuclei and Astrophysical Processes

Toshio Suzuki

Department of Physics, College of Humanities and Sciences,
Nihon University, 3-25-40 Sakurajosui, Setagaya-ku,
Tokyo 156-8550, Japan
suzuki@chs.nihon-u.ac.jp

Roles of nuclear forces on spin responses in nuclei and shell evolutions are studied. Important roles of tensor and two-body spin-orbit forces on shell inversion as well as three-nucleon forces on shell evolution in neutron-rich isotopes toward driplines are pointed out. Spin modes in nuclei such as Gamow-Teller transitions are well described by new shell-model Hamiltonians, which have proper spin-isospin interactions. Electron-capture and β-decay rates in stellar environments are evaluated with the Hamiltonians, and applied to nuclear Urca processes in *sd*-shell and *sd-pf* shell and nucleosynthesis of iron-group elements in Type Ia supernovae. ν-induced reaction cross sections are updated with new shell-model Hamiltonians, and used to study synthesis of light elements in supernovae and detection of supenova ν.

1. Introduction

A recent progress in shell model caculations, which properly takes into account important roles of spin-isospin interactions, is found to lead to significant improvements in magnetic properties of nuclei[1–3] as well as proper shell evolutions toward the driplines[4–6] An important general role of the tensor interaction, that obeys a general sign rule, is pointed out.[4] The monopole terms of the neutron-proton tensor interaction are attractive between $j_>$ ($j_> = \ell + 1/2$) and $j_<$ ($j_< = \ell - 1/2$) orbits but repulsive between $j_>$ and $j_>$ or $j_<$ and $j_<$ orbits.[4]

New Hamiltonians in *p-sd* shell with proper tensor components can explain spin properties of *p*-shell nuclei such as Gamow-Teller (GT) transitions as well as magnetic moments systematically.[1,2] Attraction between proton (π) $p_{3/2}$ and neutron (ν) $p_{1/2}$ orbits weakens as the proton number in $p_{3/2}$-shell decreases from $Z = 5$ (^{13}B) to $Z = 3$ (^{11}Li). This leads to a change of effective single-particle energy of neutron $p_{1/2}$ orbit upward closer to the $1s_{1/2}$ orbit, and results in the change of the magic number from $N = 8$ to $N = 6$ in $N = 8$ isotones toward the dripline. Evidence for proton magic number at $Z = 6$ is also found in neutron-rich carbon isotopes.[7]

In Sect. 2, roles of nuclear forces on shell evolutions in carbon and oxygen isotopes are discussed. In addition to the tensor interaction, the importance of two-body spin-orbit (LS) and three-nucleon forces is pointed out. In Sect. 3, weak transition rates in stellar environments relevant to astrophysical processes in stars are evaluated with the new shell-model Hamiltonians in *sd*-shell, *pf*-shell and *sd-pf* shell, which can describe spin properties of nuclei quite well. Electron-capture and β-decay rates thus obtained are applied to study nuclear Urca processes in stars and nucleosynthesis in supernova (SN) explosions. Neutrino-nucleus reaction cross sections updated are used to study synthesis of light elements in SN explosions and detection of SN neutrinos.

2. Roles of Nuclear Forces on Shell Evolution

Roles of each component of nuclear forces, that is, tensor, two-body LS and three-body forces on shell evolution are investigated. The tensor and two-body LS components are shown to be important in the level inversion of $1s_{1/2}$ and $0d_{5/2}$ orbits in $N = 9$ isotones. The 3-body interaction is found to be important to induce repulsion between valence neutrons and determine the driplines of oxygen and carbon isotopes.

2.1. *Inversion of $1s_{1/2}$ and $0d_{5/2}$ Orbits in $N = 9$ Isotones*

We discuss level inversion between $\nu 1s_{1/2}$ and $\nu 0d_{5/2}$ orbits between ^{15}C and ^{17}O. The $1s_{1/2}$ orbit is above the $0d_{5/2}$ orbit in ^{17}O, while they are iverted in ^{15}C.[8] This inversion can be explained by the roles of tensor and two-body LS interactions. The contributions to the monopole terms from the tensor interaction are attractive for $\pi p_{1/2}$-$\nu d_{5/2}$ and vanish for $\pi p_{1/2}$-$\nu s_{1/2}$. The two-body LS interaction gives a repulsive contribution to the monopole term for $\pi p_{1/2}$-$\nu s_{1/2}$. As the number of ptotons in $p_{1/2}$ orbit decreases from ^{17}O to ^{15}C, the $\nu 0d_{5/2}$ orbit is pushed up because of less attraction from the tensor interaction, while $\nu 1s_{1/2}$ orbit is pulled down due to less repulsion from the two-body LS interaction. Thus the $\nu s_{1/2}$ orbit is lowered and comes below the $\nu 0d_{5/2}$ orbit in ^{15}C. Figure 1(a) shows how the monopole terms from tensor, two-body LS and central forces contribute to the iversion of the levels. Energy eigenstates otained by shell model calculation are shown in Fig. 1(b).

2.2. *Shell Evolution in Oxygen and Carbon Isotopes*

The ground state energy of oxygen isotopes decreases with more binding as neutrons are added to ^{16}O-core one by one, but it stops decreasing and turns out to increase at a certain point. The isotope with the minimum energy corresponds to the dripline. When the ground state energies are calculated with the G-matrix derived from fundamental nucleon-nucleon (NN) interactions,[9,10] the energy continues to decrease and the isotopes remain bound up to ^{28}O ($N = 20$). Calculated energies for oxygen isotopes obtained with the G-matrix for the CD-Bonn potential[11] are shown in Fig. 2 (dashed line). Here, the folded-diagram method with a third-order Q-box has been used, and the configuration space is taken up to $16\hbar\omega$ excitations.[10]

Repulsie interaction is needed to reproduce the observed dripline at $N = 16$, ^{24}O. The three-nucleon (3N) forces can induce repulsion among valence neutrons

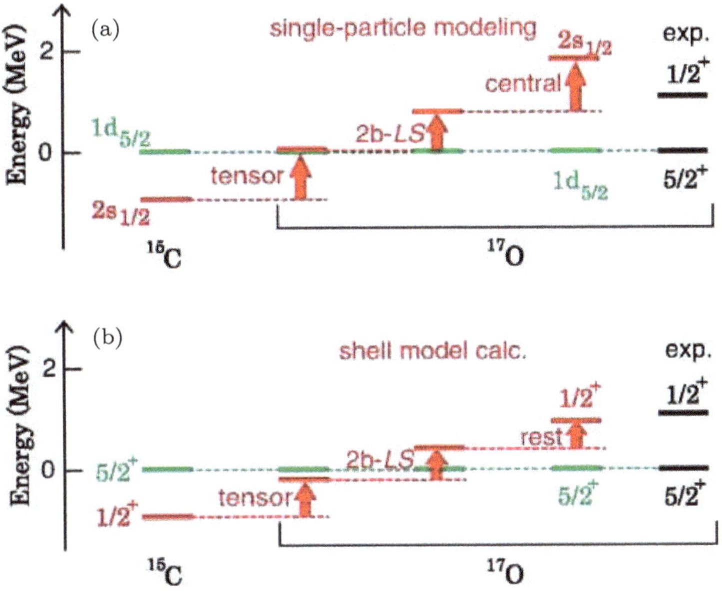

Fig. 1. Energies of the neutron $1s_{1/2}$ orbit relative to the $0d_{5/2}$ orbit obtained (a) within single-particle scheme using the monopole-based universal interaction (VMU[5]) and the M3Y two-body LS (2b-LS) force, and (b) in terms of the shell model calculation with the YSOX.[2] Experimental energy level of ^{17}O is shown at the far right. Taken from Ref. 6.

when the third nucleon is summed over all nucleons in the ^{16}O-core. Here, we adopt Fujita-Miyazawa 3N forces,[12] in which one nucleon is virtually excited to the Δ (1232 MeV) resonance and de-excited to the nucleon. The monopole terms for the induced NN interaction is obtained by using the exchange matrix elements, which prove to be repulsive.[13] The energies for oxygen isotopes with the contributions from the 3N forces are shown in Fig. 2 (solid line). The dripline for oxygen isotopes is reproduced with the inclusion of the 3N forces.

Next, we discuss shell evolution in carbon isotopes. The energies of the isotopes obtained by shell-model calculations in *p-sd* shell are shown in Fig. 3(a). The SFO Hamiltonian[1] is used except that the same G-matirix and Fujita-Miyazawa 3N forces used for the oxygen isotopes are taken for the *sd*-shell part. The repulsive contributions from the 3N forces lead to the energy minima at $N = 14$ (^{20}C) and $N = 16$ (^{22}C), and the energy at $N = 14$ is slightly lower than that at $N = 16$. However, the effects of halo structure in ^{22}C are not taken into account. The effective single-particle energy (ESP) of $\nu 1s_{1/2}$ orbit is pushed up due to the repulsive monopole term for $\nu 0d_{5/2}$-$\nu 1s_{1/2}$ as neutrons are occupied in the $0d_{5/2}$ orbit. There occurs an inversion between $\nu 0d_{5/2}$ and $\nu 1s_{1/2}$ orbits, and the $\nu 0d_{5/2}$ orbit becomes lower than the $\nu 1s_{1/2}$ orbit around $A = 17$.[6] This results in the formation of two-neutron halo in ^{22}C.[14] Because of the halo effect, the energy is lowered compared with the harmonic oscillator case, which makes the halo nucleus ^{22}C to be at the dripline.

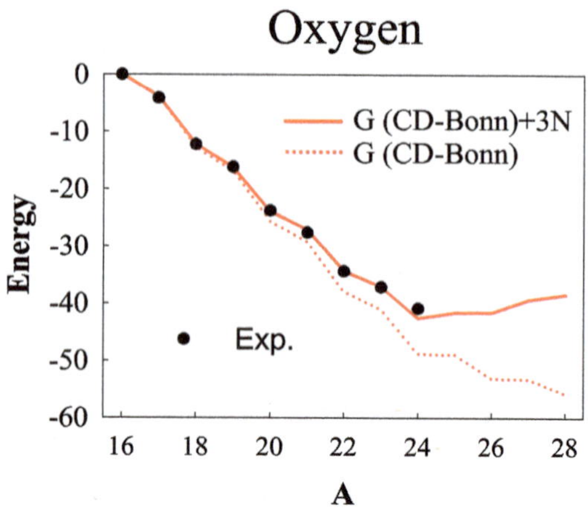

Fig. 2. Energies of oxygen isotopes obtained by shell-model calculations in *sd*-shell. Solid curve is obtained with CD-Bonn G-matrix and Fujita-Miyazawa 3N forces, while dotted curve is obtained with the G-matrix only. Experimental energies are denoted by filled circles.

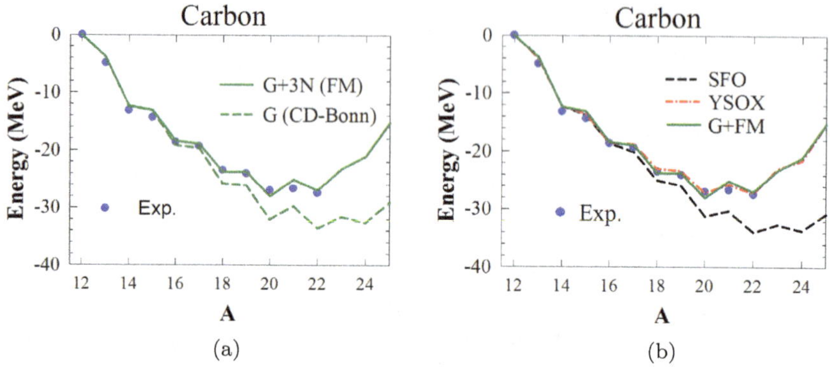

Fig. 3. Energies of carbon isotopes obtained by shell-model calculations in *p*-*sd* shell. (a) Solid curve is obtained with CD-Bonn G-matrix and Fujita-Miyazawa 3N forces in the *sd*-shell part, while dashed curve is obtained with the G-matrix only. (b) Dash-dotted and dashed curves are obtained with YSOX and SFO Hamiltonians, respectively, while solid curve is the same as the solid curve in (a). Experimental data are denoted by filled circles.

Calculated energies of the carbon isotopes obtained with SFO[1] and YSOX[2] Hamiltonians are shown in Fig. 3(b). In the *sd*-shell part of YSOX, repulsive contributions from the 3N forces are taken into account phenomenologically, while that of SFO is the Kuo's G-matrix. The YSOX reproduces the dripline at $N = 16$ though the energies of ^{20}C and ^{22}C are almost degenerate.

Calculated low-lying energy levels in ^{16}C, ^{17}C and ^{18}C obtained with and without the 3N forces are shown in Fig. 4. The 3N forces push down the 2^+ energies closer to the observed values in ^{16}C and ^{18}C. The order of the levels of the three low-lying states are well reproduced with the inclusion of the 3N forces in ^{17}C. Thus the 3N forces are found to improve the energy levels in the isotopes.

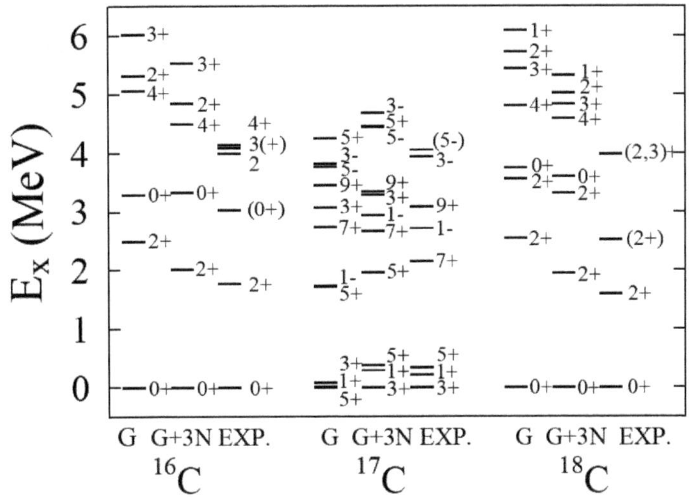

Fig. 4. Energy levels of ^{16}C, ^{17}C and ^{18}C obtained by shell-model calculations with the G-matrix and G-matrix +3N interactions. Experimental energy levels are also denoted. In case of ^{17}C, twice the spin is denoted instead of the spin.

3. Weak Rates in Stellar Environments and Astrophysical Processes in Stars

Spin dependent transitions, especially, the GT transitions give dominant contributions to the weak rates. Electro-capture and β-decay rates induced by GT transitions are evaluated with new shell-model Hamiltonians in *sd*-shell, *sd*-*pf* shell and *pf* shell, and applied to nuclear Urca processes in stars and nucleosynthesis in supernova (SN) explosions. ν-nucleus reactions on nuclei important for nucleosythesis and ν detection are evaluated with new shell-model Hamiltoninans.

3.1. *Electron-Capture and β-Decay Rates and Nuclear Urca Processes*

Electron-capture and β-decay rates in *sd* shell obtained by USDB Hamiltonian[15] in stellar environments, that is, at high temperatres ($T = 10^8$–10^{10} K) and high densities ($\rho Y_e = 10^8$–10^{10} g cm^{-3} with Y_e the electron fraction) are applied to nuclear Urca processes. The e-capture rates increase while the β-decay rates decrease as the density increases due to the increase of electron chemical potential at high densities. Both the weak rates coincide at a certain density, called a Urca density, almost independent of temperatures. Both ν and $\bar{\nu}$ are emitted at the Urca density thus taking away the energy from the star, which results in a drastic cooling of the core of the star. This mechanism, called the nuclear Urca process, occurs quite efficiently for the nuclear pairs with $A = 23$ and 25, where the transitions between the ground states are GT ones.[16,17] The weak rates for the nuclear pair, ^{23}Na -^{23}Ne, and the cooling of ONeMg core of a star with 8.8 M$_\odot$ are shown in Fig. 5.[16,18]

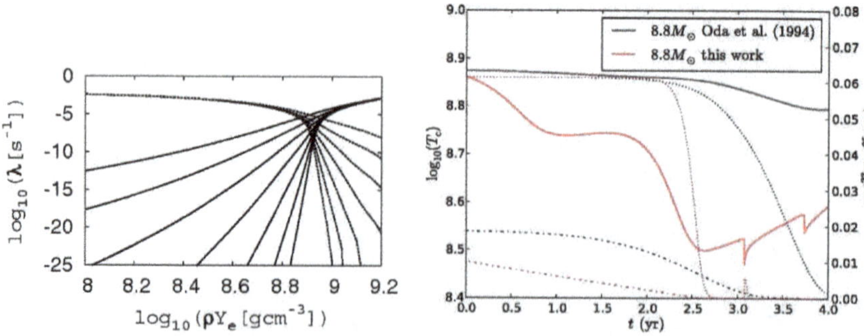

Fig. 5. (Left) Weak rates for the A = 23 Urca nuclear pair, ^{23}Na-^{23}Ne, as functions of density $\log_{10}\rho Y_e$ at temperatures in the range of $\log_{10}T = 8$ to 9.2 in steps of 0.2. Electron-capture and β-decay rates are denoted by solid and dashed lines, respectively. Taken from Ref. 16. (Right) The evolution of the central temperature T_c of an ONeMg core of the 8.8M$_\odot$ star as a function of time in units of years. The temperature drop up to t = 1 year and the temperature drop during t = 2–2.5 year are caused by the Urca processes by the A = 23 and A = 25 nuclear pairs, respectively. Changes of the mass fraction X_c of ^{25}Mg and ^{23}Na are also shown. The case with the use of Oda *et al.*'s table[19] using sparce grid is also shown, where there is no effect of the Urca process. Taken from Ref. 16.

The star with 8.8 M$_\odot$ collapses triggered by subsequent e-capture processes on ^{24}Mg and ^{20}Ne, and ends with an e-capture (EC) SN explosion. Stars with M > 9 M$_\odot$ likely to end with core-collapse (CC) SN explosions. Border of CC-SN or EC-SN is at M ~ 9 M$_\odot$, and the fate of the stars with 8–10 M$_\odot$ is sensitive to nuclear weak rates as well as their masses.

Next, we discuss a nuclear Urca process for a nuclear pair in the island of inversion.[20] Several nuclear pairs have been pointed out to be important for the cooling of neutron star crusts.[21] Here, the pair with $A = 31$ in the island of inversion, ^{31}Mg $\leftrightarrow$ ^{31}Al, is considered as it is one of the potential candidates that contribute to the cooling.[21, 22]

Neutron-rich nuclei in the island of inversion are studied by shell-model calculations with the EEdf1 interaction[6] obtained by the extended Kuo-Krenciglowa (EKK) method[23] starting from chiral EFT N^{3}LO and Fujita-Miyazawa 3N interaction.[24] Neutron ESP's between *sd*-shell and *pf*-shell orbits become much closer in the n-rich region, $Z = 10$–12, for the EEdf1 compared with conventional *sd-pf* shell Hamiltonians such as sdpf-m.[25] This results in larger admixtures of *pf*-shell components for the EEdf1: the contributions from 4p-4h excitations are found to be larger than those from 2p-2h excitations in ^{32}Mg.

Energy levels of ^{31}Mg can be well explained by the EEdf1.[24] Configurations including up to 6p-6h excitations are taken. The ground state (g.s.) of ^{31}Mg is calculated to be $1/2^+$ consistent with the experimental observation,[26] while it is predicted to be $7/2^-$ by the sdpf-m. The first excited state is predicted to be $3/2^+$, very close to the ground state $1/2^+$. As the g.s. of ^{31}Al is $5/2^+$, GT transition between $3/2^+$ state in ^{31}Mg and $5/2^+_{g.s.}$ in ^{31}Al gives the main contribution to the e-capture and β-decay rates for the $A = 31$ pair.

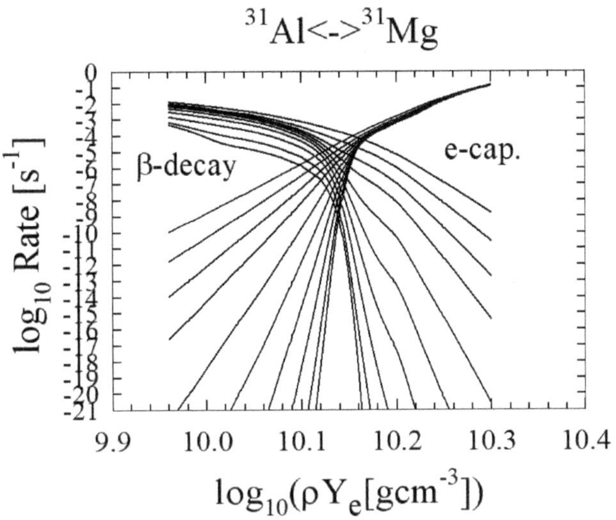

Fig. 6. Electron-capture and β-decay rates for the nuclear pair, ^{31}Al-^{31}Mg, as functions of density $\log_{10}(\rho Y_e)$ for various temperatures. The rates are evaluated with the EEdf1 interaction obtained by the EKK method.[24]

The weak rates in stellar environments obtained with the EEdf1 are shown in Fig. 6. The GT transitions between ^{31}Mg ($3/2^+$, $1/2^+$) and ^{31}Al ($5/2^+$, $1/2^+$, $3/2^+$) are taken into account. Calculated log ft values are 5.43 and 7.40 for ^{31}Mg ($3/2^+$) $\rightarrow$ ^{31}Al ($5/2^+$) and ($3/2^+$), respectively. Those for ^{31}Mg ($1/2^+$) $\rightarrow$ ^{31}Al ($3/2^+$) and ($1/2^+$) are 5.98 and 5.78, respectively, while the experimental values are 5.59 and 6.025, respectively. Deviations of the calculated GT strengths from the experimental ones are by a factor of 0.4–1.8.

There exists a Urca density as shown in Fig. 6 since the excitation energy of the $3/2^+$ state in ^{31}Mg is as small as 0.05 MeV. If the g.s. of ^{31}Mg is taken to be $7/2^-$, there does not exist a Urca density because of non-existence of GT transitions between low-lyig states.

3.2. *Electron-Capture Rates in pf Shell and Synthesis of Iron-Group Elements*

GT strengths in Ni and Fe isotopes are described quite well with a new shell-model Hamiltonian. GXPF1J,[27] in pf shell with a universal quenching factor $f = g_A^{eff}/g_A = 0.74$. In particular, the GT strength in ^{56}Ni obtained by (p, n) reaction has two-peak structure,[28] and it can be reproduced well by the GXPF1J.[40] Calculated GT strength with the GXPF1J as well as those with KB3G[29] and KBF[30] and the experimental values are shown in Fig. 7. Both KB3G[29] and KBF[30] fail to reproduce the two-peak structure. The GXPF1J is shown to reproduce experimental e-capture rates in various pf-shell nuclei better than KB3G and RPA calculations.[31]

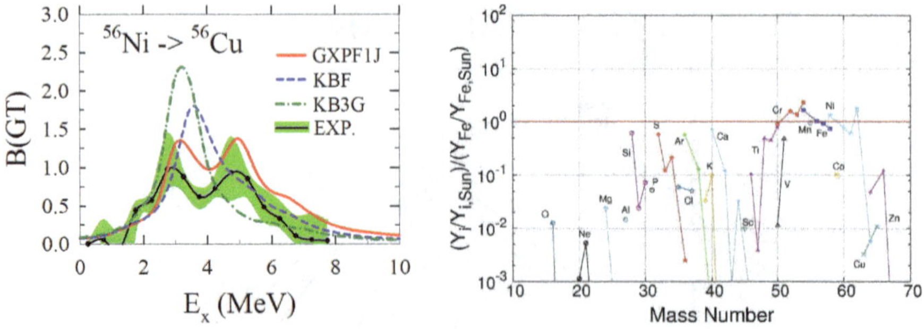

Fig. 7. (Left) The $B(GT)$ strengths obtained by shell-model calculations with GXPF1J (solid), KBF (dashed) and KB3G (dash-dotted) Hamiltonians. Experimntal data[28] are also shown. (Right) Abundances ratio relative to Fe relative to the solar abundance ratio for nuclei produced in Type Ia SN explosions for the WDD2 delayed-detonation model. The weak rates of GXPF1J are used. Taken from Ref. 32.

The GT strength is more spread for GXPF1J compared with KB3G, which leads to smaller e-capture rates for GXPF1J. As the KBF, conventionally used for stellar calculations, take into account experimental data of energies and GT strength available, the e-capture rates of KBF come close to those of GXPF1J.

Now, we discuss synthesis of iron-group elements in Type Ia SN. Here, single-degenerate 'Chandra' model is treated. Accretion of matter from non-degenerate companion to a white dwarf (WD) occurs, and a SN explosion takes place when WD mass approaches the Chandrasekhar limit. A lot of ^{56}Ni is produced during the explosion, and successive e-capture reactions produce neutron-rich (n-rich) nucleides and Y_e decreases. Decrease of e-capture rates leads to less production of n-rich nucleides and larger Y_e.

There was an over-production problem of n-rich iron-group elements compared with the solar abundances when e-capture rates of FFN[33] were used. N-rich nuclei such as ^{54}Cr and ^{58}Ni are produced several times more than the solar abundances when an explosion model W7 with fast deflagration with subsonic frame front was used.[34] Another explosion model WDD2[34] with slow deflagration and delayed detonation with supersonic shock wave is used also here. The production yields of elements are obtained with the e-capture rates of GXPF1J ($21 \leq Z \leq 32$) and KBF (other Z) for both the W7 and WDD2 explosion models. The over-production problem is now found to be suppressed within a factor of about twice as shown in Fig. 7 for the production yields for the WDD2 model.[32]

3.3. *Neutrino-Nucleus Reactions Relevant to Nucleosynthesis and ν Detection*

ν-induced reaction cross sections are evaluated by new shell-model Hamiltonians for ^{12}C,[35] ^{16}O,[36] ^{56}Fe,[37,38] ^{56}Ni[39] and ^{40}Ar[40] targets. Experimental charged-current cross sections on ^{12}C and ^{56}Fe for decay-at-rest (DAR) neutrinos are found to be well reproduced by the evaluations as the GT transition strengths are well explained by the shell model calculations. In case of ^{56}Fe, multipoles other than 1^+ and 0^+ are evaluated by RPA calculations.

Updated cross sections on ^{12}C are applied to study nucleosynthesis of light elements in SN explosions.[35,41] Neutral-current reaction ^{4}He (ν, ν' p) ^{3}H followed by (α, γ) process is important for the production of ^{7}Li. ^{11}B is produced by neutral-current reaction ^{12}C (ν, ν' p) ^{11}B as well as ^{7}Li (α, γ) ^{11}B reaction. Abundances of ^{7}Li and ^{11}B produced are found to be enhanced compared with those obtained by ν-induced cross sections of Woosley *et al.*[42]

The MSW ν oscillations occur in case of normal mass hierarchy. Charged-current reactions, ^{4}He (ν_e, e$^-$p) ^{3}He and ^{12}C (ν_e, e$^-$p) ^{11}C, increase in He/C layer due to more energetic ν_e converted from ν_μ and ν_τ. This results in an enhancement of production yield ratio N(^{7}Li)/N(^{11}B) in case of normal hierarchy. The ^{7}Li/^{11}B abundance ratio depends on mass hierarchy and the mixing angle θ_{13}. For $\sin^2 2\theta_{13} = 0.1$,[43] where the high-density resonance occurs adiabatically, the ^{7}Li/^{11}B ratio is larger for the normal hierarchy than for the inverted hierarchy. Inverted hierarchy looks to be statistically more preferred based on a Bayesian analysis of the Murchison meteorite.[44] Accelerator experiments, on the other hand, favor normal hierarchy.[45]

Next, we discuss ν-induced reactions on ^{16}O. ^{16}O is an important target as it is contained in water Cherenkov detectors. The reactions are dominantly induced by spin-dipole transitions. Charged- and neutral-current reaction cross sections are evaluated by shell-model calculations with a modified version of SFO, SFO-tls,[3] which takes full account of tensor interaction in the *p*-*sd* cross shell. Cross sections of various γ and particle emission channels are evaluated.[36] Charged-current reactions on ^{16}O can be useful for the detection of SN neutrinos. Event spectra of the reactions at Super-Kamiokande are evaluated for a future SN neutrino burst for the cases with and without the oscillation effects.[46]

We finally discuss ν capture reactions on ^{40}Ar, ^{40}Ar (ν_e, e$^-$) ^{40}K. A liquid argon detector has excellent potentiallities to detect SN neutrinos. Direct measurements of the charged-current reaction cross section on ^{40}Ar are accessible by using a liquid argon time projection chamber (TPC) detector and a spallation neutron source for neutrinos. The GT and isobaric-analog (IA) transitions are evaluated by shell model calculations.[40] The sdpf-m and GXPF1J interactions are used for the *sd* shell and *pf* shell, respectively. The monopole-based universal interaction (VMU),[5] which has the tensor force of π+ρ meson exchanges is adopted for *sd*-*pf* cross-shell part. The two-body LS interaction due to meson exchanges is also added to the cross-shell part. This interaction is referred to as SDPF-VMU-LS, and configurations with $2\hbar\omega$ excitations, $(sd)^{-2}(pf)^2$, are taken with a quenching factor $f = g_A^{eff}/g_A = 0.775$. Cumulative sum of $B(GT)$ is shown in Fig. 8. The experimental GT strength from the (p, n) reaction[47] is consistent with the shell-model calculation. Dashed curve is obtained in Ref. 48 based on β-decay rates from the mirror nucleus ^{40}Ti. Total cross sections obtained with the SDPF-VMU-LS are shown in Fig. 8. Multipoles

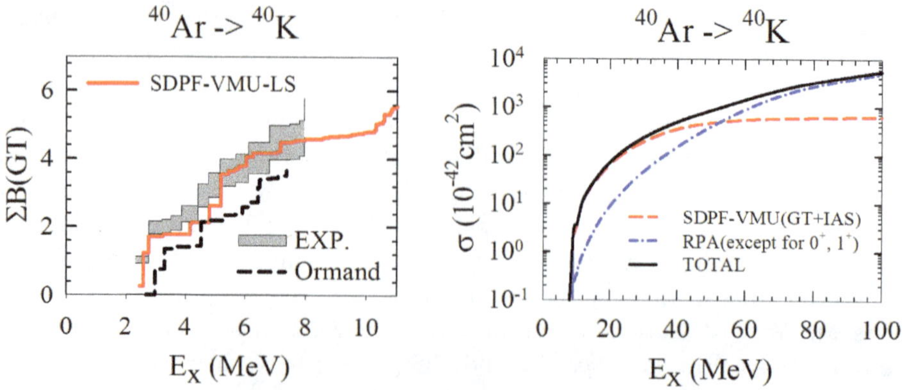

Fig. 8. (Left) Cumulative sum of GT strength for ^{40}Ar $\to$ ^{40}K up to excitation energies of ^{40}K, E_x, obtained by the shell-model calculation with SDPF-VMU-LS. The experimental data are shown by dashed area.[47] Calculated values of Ref.[48] are also shown by the dashed line. (Right) Calculated reaction cross sections for ^{40}Ar (ν_e, e$^-$) ^{40}K obtained by the shell-model calculations with SDPF-VMU-LS for the GT and IAS transitions, and by RPA calculations for the multipoles except for 0^+ and 1^+. Total cross sections including all the multipoles are shown by the solid line. Taken from Ref. 40.

other than 1^+ and 0^+ are evaluated by RPA calculations, which become important at $E_\nu \geq 50$ MeV.

4. Summary

Roles of nuclear forces on spin responses in nuclei and shell evolutions are investigated. Tensor and two-body LS forces are pointed out to be important for shell inversion of $0d_{5/2}$ and $1s_{1/2}$ orbits between ^{15}C and ^{17}O. Three-nucleon forces are shown to lead to repulsive interactions among valence neutrons, which results in proper shell evolutions in carbon and oxygen isotopes toward driplines and the determination of the driplines. The 3N forces are also shown to improve energy levels of ^{16}C, ^{17}C and ^{18}C.

Spin responses in nuclei are successfully described by shell-model calculations with new Hamiltonians, which take into account proper spin-isospin interactions, especially the tensor interaction. The GT transition strengths in *p*-shell, *sd*-shell, *sd*-*pf* shell and *pf*-shell nuclei are described quite well with the SFO, USDB, EEdf1 and GXPF1J Hamiltoninas, respectively. The weak rates are dominantly induced by spin-dependent transitions such as the GT ones, and they have been updated with the use of the new shell-model Hamiltoninans.

Electron-capture and β-decay rates in *sd* shell are applied to nuclear Urca processes in ONeMg cores of 8–10$M_\odot$ stars. The nuclear pairs with $A = 23$ and 25 are found to be important for the cooling of the ONeMg core. The weak rates in the island of inversion are obtained by an interaction derived with the EKK method in *sd*-*pf* shell. GT transitions between low-lying states in the nuclear pair with $A = 31$,

^{31}Al-^{31}Mg, are shown to lead to an Urca density, where both ν and $\bar{\nu}$ emissions become efficient for the cooling of neutron star crusts.

GT strengths in pf-shell nuclei are described quite well with the GXPF1J Hamiltonian. Electron-capture rates obtained by the GT strengths are used to study nucleosynthesis of iron-group elements in Type Ia SN explosions. An over-production problem of neutron-rich iron group isotopes compared with the solar abundances seen for the FFN weak rates is found to be suppressed within a factor of about twice.

Neutrino-nucleus reaction cross sections are also updated with the use of new shell-model Hamiltonians. Now we can explain experimental data for ν-induced reaction cross sections on ^{12}C and ^{56}Fe. Synthesis of light elements by neutrio processes in SN is discussed. Neutral-current reactions ^{12}C (ν, ν' p) ^{11}B and ^{4}He (ν, ν' p) ^{3}H followed by α capture are important to produce ^{11}B and ^{7}Li. Production yields obtained by the updated Hamitonians are found to be enhanced compared with previous calculations. In case of MSW oscillations, the yield ratio ^{7}Li/^{11}B is predicted to be enhanced for the normal hierarchy. Statitical analysis of presolar grains in a Meteorite suggests to favor inverted hierarchy, while accelerator experiments favor a normal one.

Cross sections for charged and neutral current reactions on ^{16}O in various γ and particle emission channels are updated. Event spectra for charged reactions at Super-Kamiokande are evaluated for a future SN neutrino burst with and without the oscillation effects. Charged-current reaction on ^{40}Ar is investigated with a new shell-model Hamiltonian in *sd*-pf shell, which can explain experimental GT data rather well. Total reaction cross sections are obtained by a hybrid model, where GT and IA transitions are evaluated with shell model while other multipoles are treated by RPA.

Acknowledgements

The present chapter is based on my talk in the International Conference on Simplicity, Symmetry and Beauty of Atomic Nuclei "in honor of Prof. Arima's 88 year-old Birthday" held at Shanghai during Sept. 25–29, 2018.

I would like to express my sincere thanks to Prof. Arima for inviting me to the study of spin and magnetic properties of nuclei. It started from the work concerned with my Thesis.[49] I would like to add that there was an attempt to include 3N forces in shell-model calculations in early days.[50]

The present work was supported in part by JSPS KAKENHI grant No. JP19K03855.

References

1. T. Suzuki, R. Fujimoto, and T. Otsuka, *Phys. Rev. C* **67**, 044302 (2003).
2. C. Yuan, T. Suzuki, T. Otsuka, F. Xu, and N. Tsunoda, *Phys. Rev. C* **85**, 064324 (2012).

3. T. Suzuki and T. Otsuka, *Phys. Rev. C* **78**, 061302 (2008).
4. T. Otuska, T. Suzuki, R. Fujimoto, H. Grawe, and Y. Akaishi, *Phys. Rev. Lett.* **95**, 232502 (2005).
5. T. Otsuka, T. Suzuki, M. Honma, Y. Utsuno, N. Tsunoda, K. Tsukiyama, and M. Hjorth-Jensen, *Phys. Rev. Lett.* **104**, 012501 (2010).
6. T. Otsuka, A. Gade, O. Sorlin, T. Suzuki, and Y. Utsuno, *Rev. Mod. Phys.* **92**, 015002 (2020).
7. D. T. Tran, H. J. Ong, G. Hagen, D. Morris, N. Aoi, T. Suzuki *et al.*, *Nature Communications* **9**, 1594 (2018).
8. I. Talmi and I. Unna, *Phys. Rev. Lett.* **4**, 469 (1960).
9. T. T. S. Kuo and G. E. Brown, *Nucl. Phys.* **85**, 40 (1966).
10. M. Hjorth-Jensen, T. T. S. Kuo, and E. Osnes, *Phys. Rep.* **261**, 125 (1995).
11. R. Machleidt, *Adv. Nucl. Phys.* **19**, 189 (1989).
12. J. Fujita and H. Miyazawa, *Prog. Theor. Phys.* **17**, 360 (1957).
13. T. Otsuka, T. Suzuki, J. D. Holt, A. Schwenk, and Y. Akaishi, *Phys. Rev. Lett.* **105**, 032501 (2005).
14. T. Suzuki, T. Otsuka, C. Yuan, and N. Alahari, *Phys. Lett. B* **753**, 199 (2016).
15. B. A. Brown and W. A. Richter, *Phys. Rev. C* **74**, 034315 (2006).
16. H. Toki, T. Suzuki, K. Nomoto, S. Jones, and R. Hirschi, *Phys. Rev. C* **88**, 015806 (2013).
17. T. Suzuki, H. Toki, and K. Nomoto, *Astrophys. J.* **817**, 163 (2016).
18. S. Jones, R. Hirschi, K. Nomoto *et al.*, *Astrophys. J.* **772**, 150 (2013).
19. T. Oda, M. Hino, K. Muto, M. Takahara, and K. Sato, *At. Data and Nucl. Data Tables* **56**, 231 (1994).
20. E. K. Warburton, J. A. Becker, and B. A. Brown, *Phys. Rev. C* **41**, 1147 (1990).
21. H. Schatz *et al.*, *Nature* **505**, 62 (2014).
22. W.-J. Ong *et al.*, *Phys. Rev. Lett.* **125**, 262701 (2020).
23. K. Takayanagi, *Nucl. Phys. A* **852**, 61 (2011); *ibid.* **864**, 91 (2011).
24. N. Tsunoda, T. Otsuka, N. Shimizu, M. Hjorth-Jensen, K. Takayanagi, and T. Suzuki, *Phys. Rev. C* **95**, 021304 (2017).
25. Y. Utsuno, T. Otsuka, T. Mizusaki, and M. Honma, *Phys. Rev. C* **60**, 054315 (1999).
26. J. R. Terry *et al.*, *Phys. Rev. C* **77**, 014316 (2008).
27. M. Honma, T. Otsuka, T. Mizusaki, M. Hjorth-Jensen, and B. A. Brown, *J. Phys. Conf. Ser.* **20**, 7 (2005).
28. M. Sasano *et al.*, *Phys. Rev. Lett.* **107**, 202501 (2012).
29. E. Caurier, G. Martinez-Pinedo, G. F. Nowacki, A. Poves, and A. P. Zuker, *Rev. Mod. Phys.* **77**, 427 (2005).
30. K. Langanke and G. Martinez-Pinedo, *At. Data and Nucl. Data Tables* **79**, 1 (2001).
31. A. L. Cole *et al.*, *Phys. Rev. C* **86**, 015809 (2012).
32. K. Mori, M. A. Famiano, T. Kajino, T. Suzuki *et al.*, *Astrophys. J.* **833**, 179 (2016).
33. G. M. Fuller, W. A. Fowler, and M. J. Newman, *Astrophys. J.* **252**, 715 (1982); *Astrophys. J. Suppl.* **48**, 279 (1982).
34. K. Iwamoto, F. Brachwitz, K. Nomoto *et al.*, *Astrophys. J. Suppl.* **125**, 439 (1999).
35. T. Suzuki, S. Chiba, T. Yoshida, T. Kajino, and T. Otsuka, *Phys. Rev. C* **74**, 034307 (2006).
36. T. Suzuki, S. Chiba, T. Yoshida, K. Takahashi, and H. Umeda, *Phys. Rev. C* **98**, 034613 (2018).
37. T. Suzuki and T. Kajino, *J. Phys. G* **40**, 083101 (2013).
38. N. Paar, T. Suzuki, M. Honma, T. Marketin, and D. Vretenar, *Phys. Rev. C* **84**, 047305 (2011).

39. T. Suzuki, M. Honma, K. Higashiyama *et al.*, *Phys. Rev. C* **79**, 061603 (2009).
40. T. Suzuki and M. Honma, *Phys. Rev. C* **87**, 014607 (2013).
41. T. Yoshida, T. Suzuki, S. Chiba *et al.*, *Astrophys. J* **686**, 448 (2008).
42. S. E. Woosley, D. H. Hartmann, R. D. Hoffman, and W. C. Haxton, *Astrophys. J* **356**, 272 (1990).
43. K. Abe *et al.* (T2K Collaboration), *Phys. Rev. Lett.* **107**, 041801 (2011).
44. G. J. Mathews, T. Kajino, W. Aoki, W. Fujiya, and J. B. Pitt, *Phys. Rev. D* **85**, 105023 (2012).
45. K. Abe *et al.* (T2K Collaboration), *Phys. Rev. Lett.* **121**, 171802 (2018).
46. K. Nakazato, T. Suzuki, and M. Sakuda, *Prog. Theor. Exp. Phys.* **2018**, 123E02 (2018).
47. M. Bhattacharya, C. D. Goodman, and A. Garcia, *Phys. Rev. C* **80**, 055501 (2009).
48. W. E. Ormand, P. M. Pizzochero, P. F. Bortignon, and R. A. Broglia, *Phys. Lett. B* **345**, 343 (1995).
49. A. Arima, Y. Horikawa, H. Hyuga, and T. Suzuki, *Phys. Rev. Lett.* **40**, 1001 (1978).
50. A. Arima, N. Onishi, T. Inoue, YA. Arima, N. Onishi, T. Inoue, Y. Akiyama, and T. Suzuki, *Nucl. Phys. A* **459**, 286 (1986).

https://doi.org/10.1142/9789819813353_0018

Harmonic Oscillator Wave Functions in the Nuclear Shell Model

Igal Talmi
The Weizmann Institute of Science
Rehovot, Israel

70 years ago I published a paper, *nuclear spectroscopy with harmonic oscillator wave functions*.[1] The abstract was written in German, by R. Schafroth, assistant of my thesis adviser, Wolfgang Pauli in the ETH in Zurich. Its beginning is:

Zusammenfassung: Die Beschränkung auf harmonische Zentralpotentiale ermöglicht die Entwicklung einer neuen (vom üblichen Slaterschen Verfahren verschiedenen) Methode zur Berechnung der Matrixelemente der Wechselwirkung zwischen den äusseren Nukleonen im Schalenmodell, welche insbesondere auch den Fall nichtzentraler Kräfte in einfacher Weise zu behandeln gestattet.

In English: Restriction to harmonic oscillator wave functions enables the development of a new method, different from the usual Slater expansion, for calculating matrix elements of the interaction between valence nucleons in the shell model. This method is specially useful for non-central forces.

The following is intended to mark the birth of the method. But before describing it, it is worth-while to describe the situation in nuclear research at that time. It will also give me the opportunity to mention some of the important contributions of our late distinguished colleague and friend, Akito Arima to whose memory this volume is dedicated.

Nuclear physics started by experiments of Rutherford and his allies in 1911. Their scattering experiments showed that the positive charge of the atom and most of its mass, are concentrated in a region whose diameter is smaller by many thousands than that of the atom. There were various speculations about the composition of this nucleus until the neutron was discovered by Chadwick in 1932. In the same year, Heisenberg published a paper suggesting that nuclei are composed of protons and neutrons. In any nucleus there are Z protons and N neutrons, electric charge Z times the proton charge and mass roughly proportional to A = N + Z multiplied by the proton/neutron mass. There were no accurate measurements of binding energy at that time, but it was evident that some nuclei are more stable than others. In atoms, stability of atoms of noble gases is associated with closed shells of electrons.

In the same year, Bartlett proposed a similar structure for some nuclei. He suggested that in ^{4}He there are closed shells of protons and neutrons in the $1s$ orbit and in ^{16}O there is also a closed $1p$ shell.[2]

Not many papers followed the suggestion of Bartlett. The idea of shell structure was taken up by Elsasser in a systematic and thorough way. Studying the experimental data he discovered several nuclei with extra stability.[3] Many physicists could not accept the idea that an independent particle model could give a description of a system of particles strongly interacting by short range forces. Also the magic numbers of protons or neutrons extra in stable nuclei were baffling. The lowest of them, 2, 8 and 20 could be due to the full occupation, by identical nucleons, of the lowest orbits $1s, 1p, 2s1d$ of a three dimensional harmonic oscillator. Higher magic numbers, discovered by Elsasser, 50, 82 and 126 could be due to only very strange potentials.

A devastating statement against any shell model was published in 1936 by Niels Bohr.[4] He wrote: "In the atom and in the nucleus we have indeed to do with two extreme cases of mechanical many-body problems for which a procedure of approximation resting on a combination of one-body problems, so effective for the former case, looses any validity in the latter."

During the years of World War II, nuclear physicists in major countries were busy in developing nuclear weapons. In 1948, Maria Mayer published a detailed study[5] in which she showed that nuclei whose numbers of protons/neutrons were found by Elsasser to be magic, have indeed extra stability. This paper revived the interest in the shell model. Feenberg and Hammack[6] and Nordheim[7] published papers in which they tried to find shells which reproduce the magic numbers whose reality was established by Mayer. The shells they suggested, like those of Elsasser, contained orbits characterized by the orbital angular momenta of the nucleons.

Maria G. Mayer[8] and independently, Jensen *et al.*[9] introduced a novel idea. The shell structure due to the central potential is modified by a strong spin-orbit interaction between the spins of nucleons and their orbital angular momenta

$$2\alpha(\mathbf{s}.\mathbf{l}) = \alpha[(\mathbf{s}+\mathbf{l}).(\mathbf{s}+\mathbf{l}) - \mathbf{l}.\mathbf{l} \;\; - \mathbf{s}.\mathbf{s}] \rightarrow \alpha[j(j+1) - l(l+1) - 3/4] \qquad (1)$$

The interaction (1) is equal to αl for the state with $j = l + 1/2$ and to $-\alpha(l+1)$ for the state with $j = l - 1/2$. Hence, an orbit with given $l(l > 0)$ is split into two orbits. They usually remain in the same shell but there is an important exception. The single nucleon state with the largest value of l is mostly affected. If its energy is sufficiently lowered by a negative coefficient α, it may fall into a lower shell. This situation reproduces the observed magic numbers as demonstrated by the following example. The closed orbits $1s, 1p, 2s, 1d, 2p, 1f$ contain 40 protons or neutrons but the shell may contain 10 more protons (or neutrons) with $l = 4, j = 9/2(1g_{9/2})$ from the next higher shell. Thus, the magic number 50 is obtained.

The strong spin-orbit interaction gives rise to the jj-coupling scheme. A nucleon with orbital angular momentum l has also a definite value of its total angular momentum $j, l+1/2$ or $l-1/2$. The states of a group of nucleons are obtained by couplings of the various j-values of the nucleons.

The state of the protons (neutrons) in a nucleus with magic Z (or N) nucleons is well defined. This is also the case for a nucleus with Z+1 and Z-1 protons or N+1 and N-1 neutrons. If there are more nucleons outside closed shells (valence nucleons) they can couple to form several states. Maria Mayer was well aware of this situation and she added to her model, coupling rules for ground states. The ground states of even-even nuclei should all have spin (total angular momentum) $J=0$. There are no exceptions to this rule. Ground states of odd-even (and even-odd) nuclei should have spin J equal to j of an orbit in the shell. A few exceptions to this rule were noted by Mayer. She looked for a theoretical basis of these rules in the interactions between nucleons. She considered some simple j^n configurations of identical nucleons, and "for simplicity" calculated their energy levels due to a mutual zero range δ-interaction.[10] The results for these cases agreed with her coupling rules. Also the effect of the "pairing energy" showed up. It is clear that the effective interaction between nucleons cannot have zero range. But Maria's choice is the limit of a short range approximation of the effective interaction. In it the main effect of the interaction is well within the region where the wave functions of the two nucleons do not vanish. Before that, the long range approximation was popular, in which the interaction could be replaced by a constant in the region in which all wave functions do not vanish.

Mayer thought that the exceptions to her rule on the spin of ground states of odd-even nuclei are due to the effective interaction having a finite range > 0. Her student, Dieter Kurath, wrote a paper about it[11] and I, a student of Pauli at the ETH in Zurich wrote a paper about this possibility.[12] The "bible" of spectroscopy in those days was Theory of Atomic Spectra by Condon and Shortley.[13] In the shell model, energies of states of several valence nucleons are obtained by adding to the energies of the single nucleons the interaction energies of nucleon pairs. The matrix elements of these interactions which are used in the "bible" are very simple, $\langle\psi_a\psi_b|V|\psi_c\psi_d\rangle$ where $\psi_a = R(\mathbf{r}_1)l_a m_a$, or simpler, $\langle l_1m_1, l_2m_2|V|l_3m_3, l_4m_4\rangle$.

To calculate such matrix elements the mutual interaction may be expressed as a function of $\mathbf{r}_1$ and $\mathbf{r}_2$. It was introduced by Slater for the electrostatic interaction between atomic electrons, but can be generalized for more complicated interactions. It is an expansion of the interaction in a series of Legendre polynomials. For a simple potential interaction it is

$$V(|\mathbf{r}_1 - \mathbf{r}_2|) = \sum_k \mathbf{f}_k(r_1, r_2) P_k(cos\omega) \tag{2}$$

In (2), ω is the angle between $\mathbf{r}_1$ and $\mathbf{r}_2, cos\omega = \mathbf{r}_1.\mathbf{r}_2/r_1r_2$. The energy levels of a two particle interaction are given as finite sums of the radial integrals

$$F^k = \int_0^\infty R_l^2(r1) \int_0^\infty R_{l'}^2(r_2) f_k(r_1, r_2) dr_1, dr_2 \tag{3}$$

The Slater expansion is useful for the long range approximation where $F^k \ll F^0$. In the case of the δ-potential which is the limit of the short range approximation, the situation is very different. It can be expressed as

$$\delta(\mathbf{r}_1 - \mathbf{r}_2) = \delta(r_1 - r_2) \sum_{\mathrm{k}\kappa} ((-1)^\kappa Y_{\mathrm{k}\kappa}(\vartheta, \varphi) Y_{k,-\kappa}(\vartheta, \varphi))$$

$$= \sum_k f_k(|\vec{r}_1 - \vec{r}_2|) P_\kappa(cos\omega) = \sum (2k+1) F_0 \quad (4)$$

It is clear that the δ-potential is only a limit of a short range interaction between nucleons. Still, it became clear that the short range approach agrees better with experiment than the long range approximation that was accepted before Mayers' paper[10] and even later. Racah, who established modern spectroscopy, criticized a calculation in jj -coupling of the ground state of a $j^3 T = 3/2$ configuration, "proving" it is wrong [14]. Kurath in his paper [11], pointed out that Racah's argument is based on the long range approximation, i.e. assuming that of all F^k only F^0 is non-vanishing.

Younger physicists followed Mayer's use of the δ-potential. Arima and Horie used it in their seminal paper on magnetic moments of odd-even and even-odd nuclei.[15] They used jj-coupling wave functions, slightly perturbed by nearby configurations which have non-vanishing matrix elements, of the δ-potential with the main configuration.

I was impressed by the results of Maria Mayer for the δ-potential and was looking for an expansion in which non-vanishing matrix elements of the δ-potential will have only one non-vanishing term. In the case of the short range interactions, the first terms will be dominant. In the case of the δ-potential, the operator vanishes for any $r = |\mathbf{r}_1 - \mathbf{r}_2| > 0$ and, for an attractive interaction, $V \to \infty$ so that $\int_{-\infty}^{\infty} \delta(r) dr = \int_{-\alpha}^{\alpha} \delta(r) dr = 1, a > 0$. In the following, I will try to give a short description how this idea was formulated and the activity around it in the 70 years since its inception.

A simple potential two-body interaction is a function of $\mathbf{r} = \mathbf{r}_1 - \mathbf{r}_2$. If in more complicated interactions, where the center-of-mass appears, the interaction is independent of its value. This follows from its being independent of the position in space of the interacting particles. Hence, if instead of expressing the interaction in terms of $\mathbf{r}_1$ and $\mathbf{r}_2$, it may be simpler to express the products of the single nucleon wave functions in terms of the center-of-mass and the relative coordinate. This may be very complicated for most central potentials but possible in the case of a harmonic oscillator potential. The transformation that I introduced is between $\mathbf{r}_1$ and $\mathbf{r}_2$ into the center-of-mass $\mathbf{R}$ and relative coordinate

$$\mathbf{r}_1, \mathbf{r}_2 \to R = (r_1 + \mathbf{r}_2)/2 \quad \mathbf{r} = \mathbf{r}_1 - \mathbf{r}_2 \quad \mathbf{r}_1^2 + \mathbf{r}_2^2 = 2R^2 + \frac{1}{2}\mathbf{r}^2 \quad (5)$$

The kinetic energy follows a similar transformation ($\mathbf{P} = \mathbf{p}_1 + \mathbf{p}_2, \mathbf{p} = (\mathbf{p}_1 - \mathbf{p}_2)/2$). The Hamiltonian includes a constant negative energy term which makes certain low-lying orbits bound. Thus, the transformation (5) yields a linear combination of products of oscillator wave functions of **R** and **r**. The result of the transformation which I introduced may be expressed by

$$\psi_{n1l1m1}(\mathbf{r}_1)\psi_{n2l2m2}(\mathbf{r}_2) = \sum T^{NLMnlm}_{n1l1m1n2l2m2}\Psi_{NLM}(\mathbf{R})\psi nlm(\mathbf{r}) \tag{6}$$

Harmonic oscillator wave functions have been widely used in shell model calculations. Also positions of orbits are similar to those in a harmonic oscillator potential. The radial part the wave function of a single particle in a harmonic oscillator potential, apart from normalization is $exp(-\nu r^2)r^{l+1}V_{nl}(r^2)$ where V is an associated Laguerre polynomial (properly normalized).

To obtain matrix elements of a two body interaction after the transformation, two terms contribute if and only if the wave functions of **R** are the same or proportional. What remains to do are integrations over a product of functions of **r**. The integration over angles is well known and what remains is a simple radial integration over the product of the radial components of the operator and two oscillator wave functions. These integrals may be reduced to a sum of simpler ones I_l, which are equal to

$$N_l \int_0^\infty V(r)exp(-2\nu r^2)r^{2l+2}dr \tag{7}$$

The normalization factor is equal to

$$N_l = (2^{\ell+2}\nu^{l+\frac{3}{2}})/\nu\pi(2l+1) \tag{8}$$

In a paper of Tauber and Ta Yu Wu,[17] the authors named the integrals (7) "Talmi integrals" and this was adopted by most authors.

Not everyone understood my paper.[1] Marcos Moshinsky[16] wrote: "In the Talmi method one could say that the evaluation of the matrix elements is given in terms of two steps: a) The matrix element is reduced to Slater coefficients by means of the standard methods of Racah and b) the Slater coefficients are evaluated assuming an harmonic oscillator potential, taking advantage of the relations between the wave functions of two nucleons in that potential, with the wave functions associated with the relative and centre of mass coordinates of the same two nucleons".

I wrote clearly that I introduce an expansion which is different from that of Slater and it would not have the Slater expansion as an intermediate step. In the Appendix of,[1] there are expansions of some two particle wave functions after the transformation.[5]

This misunderstanding led Moshinsky to claim credit: "In the present paper we show that by introducing a transformation bracket connecting the two nucleon wave function to the relative and centre of mass wave function, the two nucleon matrix elements are given directly in Talmi radial integrals, without the intermediate step of the Slater coefficients".

Introducing a new expansion, I thought that it is nice to relate it to an existing one. There was also a practical aspect. If the Slater expansion in some case is published, its expression in harmonic oscillator integrals could be directly carried out even by the rather simple methods of the "bible". In the shell model, to obtain energies of states it is necessary to calculate energies due to two nucleon interactions and add them to the single nucleon energies. Since the Slater expansion of the latter in simple configurations was available, Tauber and Ta You-Wu published a paper on light nuclei in 1953[17] starting with the Slater integrals and then transforming them to harmonic oscillator integrals ("Talmi integrals"). These authors published another paper in 1956[18] using the Slater expansion and then expressing it by harmonic oscillator integrals (Talmi integrals). Moshinsky may have been misled by these papers as well as by a paper by my former student Thieberger in 1956.[19] He calculated the direct and exchange terms of several configurations given by the Slater expansion, in terms of the I_l(Talmi integrals). Actually, the procedure which Moshinsky attributed to me by mistake, was seriously suggested by Ford and Konopinski.[20] They do not throw away the Slater expansion, they suggest to start from it and transform it to the I_l, "without recourse to the Talmi expansion". They do not realize that this expansion depends on the operator whose matrix element is calculated. The transformed wave functions may be used to calculate any matrix element.

In my paper,[1] I addressed this point on page 198. In view of "the vast amount of results calculated in atomic spectroscopy" all in terms of Slater F_k integrals, "if one is interested in central interactions only, it is not necessary to calculate the wave-functions of the configuration in terms of $\mathbf{R}$ and $\mathbf{r}$ as only the Slater parameters F_k are required and with the above transformation they can be directly calculated in terms of the I_l". I used this approach in some of the actual calculations of interactions. In the rest of the paper, where non-central interactions are considered, this possibility did not exist and the transformation was applied to the wave functions.

The misunderstanding or missrepresentation continued. In 1960, the University of Mexico published a volume by Brody and Moshinsky, Tables of Transformation Brackets.[21] It is written there "Talmi has shown that for this [harmonic oscillator potential] it is possible to reduce the Slater coefficients to simple one-dimensional integrals of the two-particle interaction". A few lines below, it is written "Recently one of the authors, making use of the symmetry properties of the harmon[c oscillator, was able to define and obtain an explicit expression for the transformation brackets... (Moshinsky 59)". These statements are only half true which may be worse than untrue ones. A more recent example is from 1969 in a volume by Moshinsky, which was published in the series Documents on Modern Physics. It is The Harmonic Oscillator in Modern Physics: From Atoms to Quarks.[22] In the list of references, my 1952 paper[1] is number 3 but it refers only to the definition of Talmi integral in the section on single particle states.

Not everyone missed the correct aim of my method. Balashov and Eltekov in 1959 wrote a paper in which they present my approach and introduce an important improvement.[23] I defined the coefficients of the transformation[5] in the m-scheme and to obtain states with given spin J, Clebsch-Gordan coefficients should be employed. I did that following the "bible" since I did not know yet the efficient and elegant methods of Racah. In that book, energies of states were calculated by looking at traces of matrices of the two — nucleon interaction for the various values of the total M. For example, in the j^2 configuration, the expectation value for $M = 2j-1$ is equal to the energy of the state with $J = 2j-1$. The trace of the matrix with $M = 2j-3$ is equal to the energy of that state plus the interaction energy of the $J = 2j-3$ state etc. Balashov and Eltekov defined transformation coefficients which are m-independent and kindly called them "generalized Talmi coefficients". They added tables of these coefficients for several configurations.

Marcos Moshinsky and his coworkers started a high activity in calculating coefficients of the transformation of products of two harmonic oscillator wave functions to linear combination of products of wave functions of center of mass and relative coordinates.[21, 24, 25] More precisely, they defined the coordinates

$$\mathbf{R}' = (\mathbf{r}_1 + \mathbf{r}_2)/2 \quad \mathbf{r}' = (\mathbf{r}_1 - \mathbf{r}_2)/\sqrt{2} \quad \mathbf{r}_1^2 + \mathbf{r}_2^2 = \mathbf{R}'^2 + \mathbf{r}'^2 \tag{9}$$

This definition is less physical but more symmetric as will be shown below. Where it has been used, the coordinates of (9) were denoted simply by $\mathbf{R}$ and $\mathbf{r}$. The Mexico group called the transformation coefficients "Moshinsky brackets". They even combined the last two words into "brashinskets" which did not catch up. Their activity helped very much the use of the transformation which was usually called in the literature the "Talmi-Moshinsky method".

In addition to Ref. 16 of Moshinsky, he published in 1960 with Brody and Jacob, a paper[25] in which they calculate matrix elements of some nuclear interactions in terms of Talmi integrals. A "relation between Slater and Talmi integrals is also obtained". A comprehensive table of transformation brackets was published by Brody, with an English text, in 1959.[24] Brody and Moshinsky also published in 1960, tables of transformation brackets.[21]

Moshinsky realized the group theoretical aspects of the transformation and, with Bargman wrote two papers on the group theory of the harmonic oscillator. The first in 1960[26] and the second written also in 1960, was published in 1961.[27] In both papers, N particles are occupying the same harmonic oscillator potential well. In 1964 Kaufman and Noack wrote a detailed paper on "unitary symmetry of the Talmi transformation".[28] Although aware of the papers in this field they write: "The methods used in all these calculations do not, however, give a direct insight into the mathematical structure underlying the Talmi transformation." The Hamiltonian (3) is invariant under the unitary group U_6 in 6 dimensions and "it is natural to consider the subgroup $U_2 \times SU_3$ since the six coordinates of the functions are physically grouped in 2×3 coordinates." "the Talmi transformation happens to be a finite element of U_6 which belongs to the subgroup $U_2 \times SU_3$." They explain that "Thus, once we have set up our scheme,... the Talmi transformation may be written down immediately". To make the situation clear, they use the transformation (9)... They give examples and tables of some transformation brackets.

In the following, some papers that will be listed are those in which contributions to the method are made. Just applications of it are not listed, not even my own. The first paper where the method applied deserves mentioning. In 1953, Kronheimer published a paper in which he calculated "nuclear binding energies with single-particle oscillator wave functions".[29] In 1959, Lawson and Maria Mayer wrote a paper,[30] whose title is a permutation of the title of Ref. 1 — harmonic oscillator wave functions in nuclear spectroscopy. They defined an operator **p.P**+mk **r.R** which must give zero when applied to the transformed wave functions. They used methods to calculate pairing energies in certain nuclei.

Arima and Terasawa in 1959 tried to calculate the splitting between the ^{17}O ground state ($J = 5/2$) and the first excited $J = 3/2$ state. They used the transformation and calculated the necessary brackets.[31] Smirnov in 1960 wrote a paper on the transformation for particles with different masses.[32]

A simpler method was published by Smirnov in 1962[33] for "calculating the Talmi coefficients for particles with different masses by passing to Cartesian coordinates".

An unexpected application of the transformation turned up in the kinetic theory of gases. Kumar wrote about it in 1965.[34] He defined a generating functions which helped him in calculating brackets of the transformation in the case of different masses for gas mixtures. In 1966, Kumar wrote a detailed paper where the situation is described in detail.[35]

Baranger and Davies wrote a paper in 1965 where they derive the oscillator brackets occurring in calculating matrix elements of nuclear interactions.[36] These are important for Hartree-Fock calculations. Bakri in 1966 wrote a paper on the transformation brackets for unequal masses.[37] The expression which he derived "is much simpler than that given previously by Smirnov." In 1967, Bakri published a paper[38] in which he is using group theory to prove an important property of the transformation brackets which he used in Ref. 37. They transform like simpler products and "the transformation matrix elements are the Talmi coefficients."

Chasman and Wahlborn, in 1966, wrote a paper on "transformation scheme for harmonic oscillator wave functions".[39] Among those, the transformation considered here. Their method "can be used to evaluate Moshinsky coefficients, but its flexibility makes it useful in a wider range of contexts."

In 1968 Gal published a detailed paper on the transformation for different particles "- a complete construction and new results".[40] His research was mainly in hypernuclei which involves nucleon-hyperon interactions. "Explicit expressions for the transformation matrix are obtained which show the group-theoretical structure of the transformation".

A rather fat volume by Langworthy and Podgor was issued in 1968 by the Naval Research Laboratory[41] "A Table of Central-Force Geometrical Coefficients Associated With Talmi Integrals. For Use in Calculating Nuclear Two-Body Matrix Elements Using Harmonic Oscillator Wave Functions".

Buck in 1968, wrote a paper in which he shows that there exists a simple generating function used to derive a new formula for the transformation for the most general case, particles with different masses moving in oscillator potentials with different frequencies. This paper was mimeographed as a Report of Brookhaven

National Laboratory[42] Only in 1995 it was included in a paper, with Merchant,[43] where it is shown that there is a generating function for oscillator wave functions which are simultaneous eigenstates of angular momentum. It is used to derive a new formula for the transformation brackets in the most general case.

Bayman and Kallio published in 1967 a paper on the Relative-angular-momentum-zero part of two-nucleon wave functions.[44] The transformation to relative and c.m. for a finite well, was treated by first expanding them in terms of various harmonic oscillator wave functions and then applying to each of them the Talmi-Moshinsky transformation.

In 1969, Talman (not Talmi. . .) wrote a paper with a modest title "Some properties of three-dimensional harmonic oscillator wave functions",[45] but it is rather interesting. In addition to other results, "a relatively simple expression for the Talmi bracket is obtained, and an efficient method of calculating two-particle matrix elements is described".

Chun Wa Wong published in 1970 a long paper on Generalized harmonic oscillator and generalized oscillator transformation brackets.[45] He is treating a system of oscillators with "different oscillator lengths and of separated centers of well". "It is hoped that the (transformed) wave functions will be useful in problems of nuclear reactions as well as in those of nuclear structure".

Raynal and Revai published a paper in 1970, on the hyperspherical approach to the transformation coefficients in the three-body problem.[47] In 1971, Trlifaj derived "An explicit formula for Talmi-Moshinsky transformation brackets of unequal-mass particles, . . . it is specially suited for numerical calculations."[48] In 1971, Zohni wrote "A fortran program for the computation of the generalized Talmi coefficients"[49]

In 1971, Richardson, Shapiro and Malik, derived a "Direct Reduction to Talmi Integrals" without going through the transformation, for central and tensor forces".[50] As already pointed out above, without the transformation, the wave functions of the system are not obtained. In 1971, Domany used second quantination to derive the transformation coefficients for two different particles, centered on different sites.[51] In 1972, Domany, Goshen and Katriel published an "algebraic derivation of the Talmi expansion" for particles with different masses and oscillator constants.[52] They present the derivation "for the case of two one-dimensional harmonic oscillator with different parameters. The generalization to three dimensions is trivial.

In 2021 Efros published a paper on Calculation of oscillator (Talmi-Moshinsky-Smirnov) brackets.[53] In it he mentions his earlier papers. In 1972 he wrote a paper on some properties of the Talmi coefficients.[54] Then a paper presenting a program for computing the coefficients of transformations between oscillator states,[55] and in 1973.[56]

Hasse and Waghmare published in 1972 a paper. In which they express the contributions of central, spin-orbit and tensor forces to nuclear energies in terms of Talmi integrals.[58] Koh and Goldhammer in 1978[59] carry out a similar treatment to the Reid potential. In 1979, Rashid wrote a paper "on a new expression for the

Talmi-Moshinsky harmonic oscillator computer program bracket".[60] He explains that he obtained expressions which can be easily used.

Feng and Tamura published in 1975 a paper in which they present a computer program which calculates the transformation coefficients.[61] They claim that the speed of these calculations is very much bigger than the speed available then. In 1977, Dobes published a paper on "A new expression for harmonic oscillator brackets".[62] In it, an explicit formula for general harmonic oscillator is derived which is particularly useful and convenient in numerical computations.

In 1979, Fieck published a paper on Racah algebra and Talmi transformation in the theory of multi-centre integrals of Gaussian orbitals.[63] The author is "using the Talmi transformation and the Moshinsky-Smirnov coefficients" with application to molecular physics. In 1979, Maretis published a paper on "Talmi transformation and the multicenter integrals of harmonic oscillator functions".[64] In it he evaluates those integrals by using a generalization of the Talmi transformation obtaining closed formulas.

In 1980, Tobocman wrote a paper on containing a generalized transformation for "few-body and direct interaction matrix elements".[66] A set of basis states which are products of harmonic oscillator wave functions, having a set of Jacobi coordinates as arguments is considered. He shows that the integral for evaluating the few-body matrix element can be transformed into a sum of products of 3-dimensonal integrals, "The coefficients of such an expansion are generalized Talmi-Moshinsky coefficients".

In 1981, Shlomo and Prakash derived for the transformation brackets a new class of sum rules which are quite different from the orthogonality relations.[67] These were derived by exact expression for the Wigner transform of the one body density matrix $\nu^0(\mathbf{r}, \mathbf{r}')$. These sum rules hold for brackets of the transformation between two identical particles in the same nl orbits. In 1983, Shlomo extended the derivation to particles in different orbits.[68] The derivation was at first obtained in one dimensional case but the method works also for higher dimensions. The three dimensional case, in spherical coordinates, is presented in detail. The sum rules are particularly useful in providing a check on the results of computer programs for the brackets.

Gan, Gong, Wu, and Bao wrote in 1984, a "program to calculate generalized Talmi-Moshinsky coefficients in 3-body and 4-body systems".[69] The particles they consider may have different masses and they use Jacobi coordinates following Tobocman.[66] In 1988, Bao and He published a paper on "a procedure to calculate the Talmi-Moshinsky coefficients for N-body systems with arbitrary masses by using those of its subsystems".[70] Also here they follow Tobocman.[66] The "relations between hyperspherical harmonics and generalized Talmi-Moshinsky transformations" are described in a paper, published in 1990 by Liu.[71] These are studied for the three-body and four-body systems.

A pleasant surprise was a paper, published in 1997 written by Ruan Wenying and Liu Youan, on "Talmi-Moshinsky transformation brackets for two-dimensional systems".[72] In it, an arbitrary number of particles with arbitrary masses is considered. The authors state that it will be quite useful in investigating the energy spectra and electron-electron correlations in the so-called artificial atoms. Quoting another paper of Ruan, Xie and Chen published in 1998 a paper on a generalized Talmi-Moshinsky transformation for 1D harmonic oscillator functions and interaction matrix elements.[73] They consider systems with arbitrary number of particles and arbitrary masses. They evaluate N-particle matrix elements into a sum of products of one-dimensional integrals. They consider materials in which there is high electrical conductance in one direction and very low conductance in other directions.

In the following papers, the effort seems to be directed to large scale computing. In 2001, Kamuntavicius, Kalinauskas, Barrett, Mickevicius, and Germanas published a paper on the transformation "brackets: compact expression, symmetries, sums and Fortran code".[74] The authors "present a very simple expressions and a Fortran code for a fast and precise calculation of the three-dimensional harmonic oscillator transformation brackets."

This paper was followed by a paper published in 2010, by Germanas, Kalinauskas, and Mickevicius on the "Calculation of four-particle harmonic-oscillator transformation brackets".[75] The same authors published in 2013 a paper with a "revised calculation of four-particle harmonic-oscillator transformation brackets matrix".[76]

A real pleasant surprise came with a paper by Burns on "Strong decay amplitudes using the Talmi-Moshinsky transformation and the suppressed production of the 1^{-+} hybrid"[76]. In the abstract it is explained that "the Talmi-Moshinsky transformation is applied to the strong decay and the production of mesons. . . a new class of selection rules is uncovered with implications for production of the exotic 1^{-+} hybrid in e^+e^- experiments". In the introduction it is stated that the "Talmi-Moshinsky transformation — provides a simple framework for the calculation and uncovers new selection rules sensitive only to the gluonic angular momentum of the states". The next section is on "Quark-pair creation and the Talmi-Moshinsky transformation".

The papers mentioned above do not constitute a comprehensive list. They came to my attention but I believe that there are more. Still, they show how wide and intensive research followed a simple transformation from about 70 years ago.

The author is grateful to the reviewers for their opinions.

References

1. Igal Talmi, Nuclear spectroscopy with harmonic oscillator wave functions, *Helv. Phys. Acta* **25**, 185 (1952).
2. J. N. Roberts, Nuclear structure, *Nature* **130**, 165 (1932).
3. W. M. Elsasser, Sur le principe de Pauli dans les noiyaux, *J. Phys. Radium* **4**, 549 (1933). Remarques sur la constitution des noiyaux atomiques, I, **5**, 253 (1934). Sur le principe de Pauli dans les noiyaux, II, **5**, 389 (1934). Remarques sur la constitution des noiyaux atomiques, II, **5**, 475 (1934). Sur le principe de Pauli dans les noiyaux, III, **5**, 635 (1934).
4. N. Bohr, Neutron capture and nuclear constitution, *Nature* **137**, 344 (1936).

5. Maria G. Mayer, On closed shells in nuclei, *Phys. Rev.* **74**, 235 (1948).
6. E. Feenberg and K. C. Hammack, Nuclear shell structure, *Phys. Rev.* **75**, 1877 (1949).
7. L. Nordheim, On spins, moments and shells in nuclei, *Phys. Rev.* **75**, 1894 (1949).
8. M. G. Mayer, On closed shells in nuclei, II, *Phys. Rev.* **75**, 1969 (1949).
9. O. Haxel, J. H. D. Jensen, and H. E. Suess, Zur interpretazion der ausgezeichneten Nukleonzahlen im Bau der Atomkerne, *Naturwissenschaften* **5**, 376 (1948).
10. M. G. Mayer, Nuclear configurations in the spin-coupling model, II, theoretical considerations, *Phys. Rev.* **78**, 22 (1950).
11. D. Kurath, Effect of finite range interactions in the jj-coupling shell model, *Phys. Rev.* **80**, 98 (1950).
12. I. Talmi, On the spin of the nuclear ground state in the jj-coupling scheme, *Phys. Rev.* **82**, 101 (1951).
13. E. U. Condon and G. H. Shortley, *Theory of Atomic Spectra*, Cambridge University Press (1951).
14. G. Racah, Nuclear Coupling and shell model, *Phys, Rev.* **78**, 622 (1950).
15. A. Arima and H. Horie, Magnetic moments of odd-even nuclei, *Prog. Theor. Phys.* **121**, 509 623 (1954).
16. M. Moshinsky, Transformation brackets for harmonic oscillator functions, *Nucl. Phys.* **13**, 104 (1959).
17. G. E. Tauber and Ta-You Wu, Some p- and d-shell nuclei in intermediate coupling, *Phys. Rev.* **93**, 295 (1954).
18. G. E. Tauber and Ta-You Wu, Self-consistent treatment of the independent-particle c central-field nuclear model, *Phys. Rev.* **105**, 2772 (1957).
19. R. Thieberger, Calculations in nuclear spectroscopy with the harmonic oscillator shell model, *Nucl. Phys.* **2**, 533 (1956/57).
20. K. W. Ford and E. J. Konopinski, Evaluation of Slater integrals with harmonic oscillator Wahkborn wave functions, *Nucl. Phys.* **9**, 218 (1958/59).
21. T. A. Brody, and M. Moshinsky, Tables of transformation brackets for nuclear s shell model calculations, *Monografias del Instituto del Fisica* (1960).
22. M. Moshinsky, The harmonic oscillator in modern physics, *From Atoms to Quarks, Documents on Modern Physics*, Gordon and Breach (1969).
23. V. V. Balashov and V. A. Eltekov, Method for separating the relative motion of two nucleons in oscillator potential well, *Nucl. Phys.* **16**, 423 (1960).
24. T. A. Brody, Brackets de transformacion para functiones de oscilador armonico, *Rev. Mex. de Fisica* **8**, 139 (1959).
25. T. A. Brody, G. Jakob, and M.Moshinsky, Matrix elements in nuclear shell theory, *Nucl. Phys.* **17**, 16 (1960).
26. V. Bargman and M. Moshinsky, Group theory of harmonic oscillators, (I) The collective modes, *Nucl. Phys.* **18**, 697 (1960).
27. V. Bargman and M. Moshinsky, Group theory of harmonic oscillators, (II) The integrals of motion for the quadrupole-quadrupole interaction, *Nucl. Phys.* **23**, 177 (1961).
28. B. Kaufman and C. Noack, Unitary symmetry of oscillators and the Talmi transformation, *Jour. of Math. Physics* **6**, 142 (1965).
29. E. H. Kronheimer, Calculation of nuclear binding energies with single-particle oscillator wave-functions, *Phys. Rev.* **90**, 1003 (1953).
30. R. D. Lawson and M. G. Mayer, Harmonic oscillator wave functions in nuclear spectroscopy, *Phys. Rev.* **117**, 174 (1960).
31. A. Arima and T. Terasawa, Spin-orbit splitting and tensor force. II, *Prog. of Theor. Physics* **23**, 115 (1960).

32. Yu. F. Smirnov, Talmi transformation for particles with different masses, *Nucl. Phys.* **39**, 183 (1962).
33. Yu. F. Smirnov, Talmi transformation for particles with different masses (II), *Nucl. Phys.* **39**, 346 (1962).
34. K. Kumar, Talmi transformation for unequal-mass particles and related formulas, *Jour. of Math. Physics.* **7**, 671 (1966).
35. K. Kumar, The Chapman-Enskog solution of the Bolzmann equation: a reformulation in terms of irreducible tensors and matrices, *Australian Jour. of Physics* **20**, 205 (1967).
36. M. Baranger and K. T. R. Davies, Oscillator brackets for Hartree-Fock calculations, *Nucl. Phys.* **79**, 403 (1966).
37. M. M. Bakri, A simple expression for the Talmi coefficients, *Nucl. Phys. A* **96**, 115 (1967).
38. M. M. Bakri, Linear transformation connecting mass-centre and laboratory wave- f functions of two particles, *Nucl. Phys. A* **96**, 377 (1967).
39. R. R. Chasman and S. Wahlborn, Transformation scheme for harmonic oscillator wave functions, *Nucl. Phys. A* **90**, 401 (1967).
40. A. Gal, The Talmi transformation for different particles — a complete construction and new results, *Annales of Physics* **49**, 341 (1968).
41. J. B. Langworthy and S. Podgor, A table of central-force coefficients associated with Talmi integrals, *Naval Research Lab.* NRL Report 6733.
42. B. Buck, A simple expression for the general oscillator brackets, *Brookhaven Lab. Report* BNL 12979, October 1960.
43. B. Buck and A. C. Merchant, A simple expression for the general oscillator bracket, *Nucl. Phys. A* **600**, 387 (1996).
44. B. F. Bayman and A. Kallio, Relative-angular-momentum-zero part of two-nucleon wave functions, *Phys. Rev.* **156**, 1121 (1967).
45. J. D. Talman, Some properties of three-dimentional harmonic oscillator Wave functions, *Nucl. Phys. A* **141**, 273 (1970).
46. Chun Wa Wong, Generalized harmonic oscillator shell model and generalized oscillator transformation brackets, *Nucl. Phys. A* **147**, 563 (1970).
47. J. Raynal and J. Revai, Transformation coefficients in the hyperspherical approach to the three body problem, Il *Nuovo Cimento Serie X* **68**, 612 (1970).
48. L. Trlifaj, Simple formula for the general oscillator brackets, *Phys. Rev. C* **5**, 1534 (1972).
49. O. Zohni, A fortran program for the computation of the generalized Talmi coefficients, *Computer Physics Communications* **3**, 61 (1972).
50. R. H. Richardson, J. Shapiro, and E. B. Malik, Direct reduction to Talmi integrals without use of transformation brackets, *Phys. Rev. C* **3**, 84 (1971).
51. E. Domany, An expansion of two particle harmonic oscillator wavefunction products, *J. Phys. A: Gen. Phys.* **5**, 241 (1972).
52. E. Domany, S. Goshen, and J. Katriel, Algebraic derivation of the Wahkborn generalized Talmi expansion, *Nuovo Cimento* **4**(9), 389 (1972).
53. V. D. Efros, Calculation of oscillator (Talmi-Moshinsky-Smirnov) brackets (it refers also to his earlier papers), *Comp. Phys. Comm.* **265**, 108005 (2021).
54. V. D. Efros, Some properties of the moshinsky coefficients, *Nucl. Phys. A* **202**, 180 (1973).
55. V. D. Efros, Program to calculate coefficients of transformations between three-particle hyperspherical harmonics, *Comput. Phys. Commun.* **255**, 107281 (2020).

56. V. D. Efros, Calculation of coefficients of transformations between three-particle hyperspherical harmonics, *Comput. Phys. Commun.* 261 107817 (2021).
57. R. W. Hasse and Y. R. Waghmare, Talmi integrals and effective interactions, *Phys. Lett.* **41B**, 264 (1972).
58. Y. S. Koh and P. Goldhammer, Separation of realistic nuclear forces in terms of Talmi integrals, *Phys. Rev. C* **19**, 487 (1979).
59. M. A. Rashid, A new expression for the Talmi-Moshinsky harmonic Oscillator brackets, mimeograph of the *Inter. Centre of Theor. Phys.* Trieste (1979).
60. D. H. Feng and Tamura, *Comp.Phys. Commun.* **19**, 87 (1975).
61. J. Dobes, A new expression for harmonic oscillator brackets, *J. Phys. A: Math. Gen.* **12**, 1083 (1979).
62. G. Fieck, Racah algebra and Talmi transformation in the theory of multi-centre integrals of Gaussian orbitals, *J. Phys. B: Atom. Mol. Phys.* **12**, 1063 (1979).
63. D. K. Maretis, Talmi transformation and the multicenter integrals of harmonic oscillator functions, *J. Chem. Phys.* **71**, 917 (1979).
64. W. Tobocman, A generalized Talmi-Moshinsky transformation for few-body and direct interaction matrix elements, *Nucl. Phys. A* **357**, 293 (1981).
65. M. Fabre de la Ripelle, H. Fiedeldey, G. Wiechers, Beyond the first order of the hyperspherical harmonic expansion method, *Annals of Physics*, **138**, 275–318 (1982).
66. S. Shlomo and M. Prakash, Phase space distribution of an N-dimentional harmonic oscillator, *Nucl. Phys.* **357**, 157 (1981).
67. S. Shlomo, Sum rules for harmonic-oscillator brackets, *J. Phys. A: Math. Gen.* **16**, 3463 (1983).
68. Y. P. Gan, M. Z. Gong, C. E. Wu, and C. G. Bao, Program to calculate generalized Talmi-Moshinsky coefficients of three-body and four-body systems, *Comp. Phys. Comm.* **34**, 387 (1985).
69. C. G. Bao and J. C. He, A procedure to calculate the Talmi-Moshinsky coefficients for N- systems with arbitrary masses, *J. Phys. A: Math. Gen.* **21**, 3253 (1988).
70. X. Liu, relations between spherical harmonic transformations and generalized Talmi-Moshinsky transformations, *J. Math. Phys.* **31**, 1621 (1990).
71. Ruan Wenying and Liu Youyan, Talmi-Moshinsky transformation brackets for two-dimensional systems, preprint post 1992, *Dept. App. Phys. China Univ. of Tech.*
72. Xie Wenfang and Chen Chuanyu, A generalized Talmi-Moshinsky transformation for 1D harmonic oscillator functions and interaction matrix elements, *Commun. Theor. Phys.* **30**, 419 (1998).
73. G. P. Kamuntavicius, R. K. Kalinauskas, B. R. Barrett, S. Mickevicius, and D. Germanas, The general harmonic-oscillator brackets: compact expressions, symmetries, sums, and Fortran code, *Nucl. Phys. A* **695**, 191 (2001).
74. D. Germanas, R. K. Kalinauskas, and S. Mickevicius, Calculation of four-particle harmonic-oscillator transformation brackets, *Comp. Phys. Comm.* **181**, 420 (2010).
75. S. Mickevicius, D. Germanas, R. K. Kalinauskas, Revised calculation of four-particle harmonic-oscillator transformation, *Comp. Phys. Comm.* **184**, 409 (2013).
76. T. J. Burns, Strong decay amplitudes using the Talmi-Moshinsky transformation and the suppressed production of the 1-+ hybrid, arXiv:hep-ph/0611132v2 (2006).

https://doi.org/10.1142/9789819813353_0019

Symmetries in Subatomic Systems — Searching for Six Quark Clusters in Nuclei

J. D. Vergados* and D. Strottman†

**Physics Department, University of Ioannina,*
Ioannina, Gr 451 10, Greece
† Theoretical Division, LANL, Los Alamos, NM 87544, USA
and FIAS, Goethe University, Frankfurt, Germany

The evidence of the existence of tetra-quark meson and penta-quark fermion systems has revived interest in the hunt of other multi-quark systems. We explore the possibility of six-quark cluster configurations with the two nucleon spin-isospin quantum numbers existing in the nuclear medium. The investigation of such clusters is feasible if one exploits all the symmetries involved, in particular orbital, color-spin $SU(6)_{cs}$ and isospin SU(2). We investigate whether such clusters contribute to neutrinoless double-beta decay mediated by heavy neutrinos or other exotic particles, since in conventional nuclear physics the relevant nuclear matrix elements are suppressed due to the presence of the nuclear hard core.

1. Introduction

We first met Akito when he visited Rutgers and Stony Brook Universities in 1967–68 when we were both at Stony Brook. In addition he visited Oxford for several weeks where one of us (DS) was a Research Fellow as well as numerous intervening visits at locations around the world. His ideas and guidance then have laid the foundation for much work through the decades including parts of this paper. We dedicate this paper to his memory.

Multi-quark systems are fermions containing more than three-quarks or mesons containing more than one quark and one anti-quark. The first calculations[1,2] of such systems employed the Jaffe version[3,4] of the MIT bag model. Hints of such multi-quarks have since been seen observed experimentally, the penta-quark[5–8] and the tetra-quark.[9,10] One wonders are there not multi-quark configurations in the nucleus? And what signatures might reveal their presence? In this paper we examine the possibility of the presence of six-quark clusters in the nucleus, a more complex problem. Such clusters, if present with a reasonable probability, may contribute to various processes such as neutrinoless double-beta decay mediated by heavy neutrinos or other exotic particles. The relevant nuclear matrix elements are suppressed due to the presence of the nucleon-nucleon hard core. In the presence of q^6 clusters, however, the interacting quarks are in the same hadron and one can have a contribution even in the case of a δ-function interaction.[12]

The paper is organized as follows: In section 2 we present the mathematical model describing six-quark states, in 3 we exhibit the techniques to obtain the wave functions, in section 3.3 we discuss the structure of the q^6 eigenstates, in 4 we expand the six-quark states into two clusters of the form $q^3 \times q^3$, in 5 we examine the mixing of the six-quark states with two-particle shell model states, and in 6 we consider the role of six-quark admixtures in the $0\nu\beta\beta$ decay, mediated by heavy Majorana neutrinos. Our conclusions are presented in section 7.

2. Model to Describe a Six-Quark State

Symmetries play a crucial role in reliably estimating the probability of calculating such six-quark clusters in the nucleus. Already in 1937 Wigner[13,14] introduced his supermultiplet model which uses the $SU(2)_\sigma \times SU(2)_\tau \subset SU(4)$. Wave functions described by such a labelling can naturally describe selection rules for operators that describe beta decay and most electromagnetic transitions.

Akito Arima and colleagues[15–17] developed a shell model code to describe nuclei A=18-22 based on the SU(3) model. We shall use a similar approach albeit with a generalized group structure to describe the system of six-quarks. In a previous article[18] we developed the formalism needed in the evaluation of the energy of six-quark cluster configurations that can arise in a harmonic oscillator basis of up to $2\hbar\omega$ excitations. The symmetries that were found useful for this purpose are those from the Jaffe Model: the combined colour-spin symmetry $SU_c(3) \times SU_\sigma(2) \subset SU_{cs}(6)$, and the isospin symmetry $SU_I(2)$. However, unlike the Jaffe model and the $q^4\bar{q}$ calculations in which the radial wave functions were a spherical Bessel function, here we employ three-dimensional harmonic oscillator wave functions to confine the quarks. In this sense our calculations are not unlike those using a harmonic oscillator functions to describe a three-quark system; see Ref. 19 and references therein.

Building on the work of Jahn and collaborators,[20–22] in 1953 Elliott[23–25] introduced the $SU(3)$ group to characterize states in light nuclei and showed how characteristics of rotational bands could develop. Akito Arima and colleagues[15,16] advanced this work by implementing the technology in a computer code. This effort was extended to other nuclei in the 1s0d shell[26–28] by Manakos and the Darmstadt group as well as to the 1p0f shell.[29] Necessary SU(3) recoupling coefficients and Wigner coefficients in algebraic form were calculated by Hecht[30] and Vergados.[31] Subsequently, computer codes for the recoupling coefficients were published by Akiyama and Draayer.[32,33] Nine-$(\lambda\mu)$ coefficients were defined by Millener.[34] The $SU(3) \times SU(2) \subset SU(6)$ coupling coefficients were published in Refs. 35, 36. These results are critical for the calculations described in this paper.

2.1. *Basis States*

In the following $[f] = [f_1 f_2 \cdots f_6]$ will denote the Young tableaux that labels an irreducible representation of $SU(6), SU(3)$, and and S_n.[37] The $SU(3)$ provides a convenient labeling scheme useful for many-quark states. The $SU(3)$ be they flavour, colour, or spatial, by two integers $(\lambda\mu)$. Our notation is briefly described in references.[18,35] In this notation a color singlet is $(00)_c$, a colour octet $(11)_c$; a single quark carries the $SU(3)_c$ representation $(10)_c$. Removing a single quark from a colour singlet q^6 state leaves a q^5 $(01)_c$ state.

As described in our previous paper[18] all basis states are products of four components: space, colour, spin, and isospin; the colour and spin will be combined in colour-spin wave function described by $SU(6)_{cs}$. We assume the six-quarks must be a color singlet. The isospin will be described by $SU(2)_I$ and the spatial wave functions – being three-dimensional harmonic oscillator functions – will be labelled by $S(N)$, $[f]_r$, and $SU(3)_r$, $(\lambda\mu)_r$. The $SU(6)_{cs}$ representations relevant to this work and their spin and isospin content are shown in Tables 1-4 of[38] The $SU(N)$ coefficients are usually referred to as fractional-parentage coefficients following the work of Racah.[39,40]

2.2. $0\hbar\omega$ *Excitations*

In the lowest energy basis state all six-quarks are assumed to be in the $0s$ state. The wave function is then

$$|(0s)^6[6]_r\ [f]_{cs}(00)_c\ S\ I=1\rangle. \tag{1}$$

The orbital angular momentum is zero. Since the spatial wave function is totally symmetric, the product of the colour-spin and isospin wave functions must be antisymmetric. For $I=1$ the isospin representation is necessarily$[42]_I$; from Table 1 of Ref. 38 one sees that the spin is 0 or 2.

The former is diproton-like but in the nucleus, one cannot exclude the $S=2$ possibility. Note that the six-quarks can also have isospin two or three as well as one or two. One may think of these states as having Δ admixtures. However, if one decomposes the $I=0$ six-quark wave function into two hadrons as we shall do in section 4 there will be Δ admixtures.

2.3. *One Particle,* $2\hbar\omega$ *Excitations*

These positive parity states are formed by promoting a $0s$ quark to the $1s0d$ shell. The wave function is

$$\begin{aligned}&||(0s)^5[5]_r\ [f]_{cs}(01)_c\ L_1=0\ \ S_1\ I_1\\&\times 1s/0d\ [1]_{cs}(10)_c\ L_2\ S_2\ I_2=\frac{1}{2}\frac{1}{2};\ (00)_c\ L=L_2\ S\ J\ I=1\rangle.\end{aligned} \tag{2}$$

Depending on the total isospin of the six quarks, the five-quark wave function can have $I_1=\frac{1}{2}$ or $I_1=\frac{3}{2}$. For the five-quark state with $I_1=\frac{1}{2}$, S_1 can have the values $\frac{1}{2},\frac{3}{2},\frac{5}{2}$. If the isospin, I_1, is $\frac{3}{2}$, the $SU(2)$ representation is $[41]_I$; the conjugate $SU(6)_{cs}$ is then $[2111]_{cs}$ which allows $S_1=\frac{1}{2},\frac{3}{2}$. The angular momentum of the $(1s0d)$ quark is either 0 or 2.

2.4. *Two-Particle, 2ℏω Excitations*

These states are formed by promoting two $0s$ quarks to the $0p$ shell. The wave function can be written as

$$\begin{aligned}|(0s)^4[4]_r\ [f]_{cs}(\lambda_c\mu_c)_c\ L_1 = &\ S_1\ I_1 \\ \times (0p)^2_r\ [f_r](\lambda_r\mu_r)\ [f_2]_{cs}\ (\mu_c\lambda_c)_c\ S_2\ I_2;\ (00)_c\ L\ S\ J\ I = 1\rangle. \quad (3)\end{aligned}$$

The spatial symmetry the $0p^2$ wave functions, $[f_r]$, may be either [2] or [11]. Note that with spatial symmetry included which can be either symmetric or anti-symmetric, $[f]_{cs}$ no longer determines $]f_r]$ uniquely.

For the $0s^4$ states the spatial part is symmetric so we need deal only with colour-spin and isospin. The possible SU(2) isospin representations are $[22]I = 0$, $[31]I = 1$, and $[4]I = 2$. The corresponding $SU(6)_{cs}$ representations are $[22]_{cs}, [211]_{cs}$, and $[1111]_{cs}$, respectively. The colour-spin content is shown in Table 4 of Ref. 38, see also 41.

By combining all these considerations, one produces the complete basis states for our problem. It is important that one includes all possible basis states for reasons of projecting out spurious states; this will be explained below. For $J = 0, I = 1$ there are 31 basis states of which 10 are $L = 0$, 11 are $L = 1$, and 10 are $L = 2$.

3. Calculation of the Wave Functions

The methods of calculation and relevant formulae were presented in Ref. 18. Our approach generalizes the earlier work of Jahn,[20] Elliott,[21,22] Arima,[15] and others.

3.1. *Matrix Elements of the Interaction*

In this work we shall use two different quark-quark interactions motivated by QCD and previous work. The leading order of the one-gluon exchange between two quarks is

$$\frac{\alpha_s}{b}\frac{1}{(r_{12}/b)}\boldsymbol{\lambda}_1 \cdot \boldsymbol{\lambda}_2. \quad (4)$$

The matrix elements of the SU(3) generators are

$$\langle\,(\lambda,\mu)|\ \boldsymbol{\lambda}_1.\boldsymbol{\lambda}_2|\,(\lambda,\mu)\,\rangle_c = \begin{cases} \dfrac{2}{3}, & (\lambda,\mu)_c = (2,0)_c \\ -\dfrac{4}{3}, & (\lambda,\mu)_c = (0,1)_c. \end{cases}$$

This factor must then be multiplied by the matrix elements of the radial components which may have an additional spin dependence.

The two-body matrix elements in the case of the one-gluon exchange potential are given in terms units of α_s/b, see,[38] section 10. Hadron studies seem to suggest a value of the strong coupling constant α_s to be a factor of five times larger than that employed in high energy physics.[11] We adopt a value $\alpha_s = 1$. Another possibility is an interaction employed by Ohta *et al*,[42] with an explicit spin dependence.

The QCD interaction is spin-independent and because all our states are colour singlets and the radial matrix elements of $0s^2$ are identical, all $0s^6$ states are degenerate in energy.

An additional parameter is the harmonic oscillator parameter $\hbar\omega$. We have no experimental information regarding its size, but simple estimates indicate that it must be in the range between 200 and 300 MeV, see,[38] section 3.2.

The multi-quark matrix elements can easily be calculated using the one-particle and two particle cfp's (coefficients of fractional parentage), see,[38] sections 3.3 and 11, employing wherever necessary standard Racah techniques. Additional formulae for the needed matrix elements may be found in our previous paper.[18]

3.2. *Removal of Spurious States*

Diagonalization of the interactions produce wave functions that can be used to interact with the remaining $A-2$ nucleons. However, the resulting wave functions contain components from spurious states. As first pointed out by Skyrme,[43] when converted to center-of-mass and relative coordinates, the $2\hbar\omega$ wave functions contain parts in which the center-of-mass is not in the ground state. Including such components can modify calculated transition rates and energies. Therefore we project out such spurious states. We should mention, however, there exists an alternative, more elegant method developed by Hecht.[44] For $L = 0$, $J = 0$, $I = 1$, of interest to our work, our results are presented in Table 1; for a more complete list see,[38] Table 3.

3.3. *The Structure of the q^6 Eigenstates*

We shall focus on the $I = 1$, $J = 0$ six-quark system which is most relevant below. Results for three-quark and other six-quark states, both positive and negative parity, will be published elsewhere. The calculations assumed $\alpha_s = 1$ and $b_q = 0.8$ fm. The energies of the $2\hbar\omega$ states were increased by $\Delta = 300 - -500$ MeV. The spurious states were projected out in all cases.

The QCD interaction produces relatively little mixing between the ground state and the $2\hbar\omega$ states. This is in part because the lowest lying $2\hbar\omega$ state before projection is largely spurious. After projection the ground state is 99% $0s^6$. As already

Table 1. Six-quark states in the case of $L = 0 = J = 0$, $I = 1$ relevant to our work. There are 10 basis states of which nine are $2\hbar\omega$ excitations; one combination of the states is singly spurious and one is doubly spurious.

$0s^6$	$0s^5$	$0s^4$	L	S	J	I	singly	doubly
1	2	7	0	1	1	0	1	1
1	2	7	0	0	0	1	1	1

stated the $0s^6$ $J = 1, I = 0$ state and the $0s^6$ $J = 0, I = 1$ states are degenerate using a spin-independent interaction. This degeneracy is broken by the effects of the $2\hbar\omega$ states.

The Ohta interaction was fit somewhat to experiment and one can expect that its results reflect this to some extent. The $I = 0, J = 1$ state is separated from the $I = 1, J = 0$ state by over 70 MeV. The $I = 1, J = 0$ ground state is 93% $0s^6$ for $\Delta = 500$ MeV and its energy is depressed by 71 MeV by the effects of the $2\hbar\omega$ states. The first excited $I = 1$, $J = 0$ level — which could help determine Δ — lies 423 MeV higher, although such a state could also be explained by $q^8\bar{q}^2$ states.

4. Separation into Two Baryon Clusters, $q^3 \times q^3$

To investigate the effect of three-quark clusters on electroweak transitions it is necessary to convert the six-quark states into two three-quark clusters. One can do this in two ways. The methods produce identical results.

4.1. *Separation Using $SU(6)_{cs} \subset SU(3)_c \times SU(2)_s$ Algebra*

In the first method one uses the recoupling coefficients and algebra of $SU(6)_{cs} \subset SU(3)_c \times SU(2)_s$. Here, we briefly explain the formulae required using schematic formulae.

Let K represent the variables $(\lambda\mu)S, I$ for the six-quark state and k for a single quark; the significance of the other K variables will be obvious. Then, one has

$$|q^6 K\rangle = \sum_{K_3, K_{23}, K_5} \langle q^5 K_5 \times q\ k; K|\}q^6\ K\rangle\langle q^3 K_3 \times q^2\ K_{23}; K_5|\}q^5\ K_5\rangle \\ \times |[[|q^3 K_3 \times q^2 K_{23}\rangle]^{(K_5)} \times q]^{(K)}\rangle \tag{5}$$

where

$$|[[|q^3 K_3 \times q^2 K_{23}\rangle]^{(K_5)} \times q]^{(K)}\rangle = \sum_{K'} \times q\ k\rangle]^{(K')}]^{(K)}\rangle$$
$$|q^2 K_{23} \times q\ k;\ K'\rangle = \sum \langle q^3 K' \mid q^2 K_{23} \times q\ k;\ K'\rangle |q^3 K'\rangle. \tag{6}$$

The Racah coefficient $U(K_3 K_{23} K\, k; K_5 K')$ in reality is the product of an SU(3) Racah coefficient and two SU(2) Racah coefficients, one for spin and one for isospin. The factor $\langle q^5 K_5 \times q\ \ k; K\ \ |\}q^6\ \ K\rangle$ was called a fractional parentage coefficient by Racah[39,40] and can be written as a product of $SU(6_{cs}) \subset SU(3)_c \times SU(2)_\sigma$ coefficients. Note that the six-quark cluster is a color singlet. When one separates the six-quarks into two three-quark entities, the individual three-quark cluster can be in a colour singlet or a colour octet.

Separation for the $2\hbar\omega$ states proceeds similarly. We omit the equations here.

4.2. *Combined Orbital-Color Spin-Isospin Projection*

We begin by evaluating the orbital part is of the projection. It can be shown see,[38] section 4.2, that

$$0s^6 = (0s^2)^2 0s(\mathbf{R}_1)0s(\mathbf{R}_2), 0s^5 1s = \frac{1}{3}(0s^2)^2 0s(\mathbf{R}_1)1s(\mathbf{R}_2),$$

$$0s^4(0p^2 L = 0) = \frac{1}{3\sqrt{2}}(0s^2)^2 0s(\mathbf{R}_1)1s(\mathbf{R}_2) \tag{7}$$

The part $0s^2$ refers to the internal structure of each nucleon-like component, which is of no particular interest, and for convenience it will be subsequently omitted. All other orbital configurations do not lead to two nucleon like configurations, since they are expected to occur in the ground state of the relevant $6-q$ cluster. Thus we can write a factor $\kappa_{orb} = 1, 1/3, 1/(3\sqrt{2})$ respectively for the above configurations. For all other configurations is taken to be zero.

The color spin isospin part depends, of course, on the configuration and be calculated using the techniques of section 4. Thus we can write:

$$|i\rangle = \kappa_{Orb}(i)\kappa(i)nsn's\left([2,1](00)\frac{1}{2}\times[2,1](00)\frac{1}{2}\right)_{IS}, \quad nsn's = \begin{cases} 0s0s, & i=1 \\ 0s1s, & \text{otherwise} \end{cases} \tag{8}$$

The interesting conclusion that in fact we encounter only two nucleon like configurations for given spin and isospin. Thus we find

$$\Phi(6q; J, I) = A_{0s}[J,I]\left[0s\frac{1}{2}\times 0s\frac{1}{2}\right]^{J,I} + A_{1s}[J,I]\left[0s\frac{1}{2}\times 1s\frac{1}{2}\right]^{J,I} \tag{9}$$

where

$$A_{0s}[J,I] = (c[1]\kappa[i])_{J,I},\ A_{1s} = \sum_{i=2}^{6}(c[i]\kappa_{orb}[i]\kappa[i])_{J,I}$$

where c[i] is the amplitude of the configuration i in the $\Phi(6q; J, I)$ (for the relevant values see Tables 2a) and 2b). The values of $c[i]$ will be given below.

i) $J = 0, I = 1$ case.
We find that the coefficients are:

$$c = \{0.9982, -0.0395, -0.0354, -0.0165, 0.0055, 0.0039 \text{ for } \hbar\omega = 300\text{MeV}\}$$

and

$$A_{0s}[0,1] = 0.333,\ A_{1s}[0,1] = -0.0024 \tag{10}$$

Table 2. a) The six-quark $J = 0$, $I = 1$ basis states, top six numbered lines, and b) the $J = 1$, $I = 0$ ones, the next six numbered lines. The corresponding orbital-spin-isospin coefficients $\kappa_{orb} \times \kappa$, needed to transform the shown states into two nucleon-like configurations. are shown in the last column.

$0s^6$	$[f]_{cs}$	(λ, μ)					S	L	$\kappa_{orb} \times \kappa_{NN}^{0s^6}$
1	$[2^2 1^2]$	$(0, 0)$					0	0	1/3
$0s^5 1s$	$[f]_{cs}$	(λ_1, μ_1)	$S_1 I_1$	$[f_2]$	$(\lambda_2, \mu_2$	$S_2\ I_2$	S	L	$\kappa_{orb} \times \kappa_{NN}^{0s^5 1s}$
2	$[2^2 1]$	$(0, 1)$	$\frac{1}{2}\frac{1}{2}$	[1]	$(1, 0)$	$\frac{1}{2}\frac{1}{2}$	0	0	5/27
3	$[21^3]$	$(0, 1)$	$\frac{1}{2}\frac{1}{2}$	[1]	$(1, 0)$	$\frac{1}{2}\frac{1}{2}$	0	0	$-4/27$
$0s^4 0p^2$	$[f]_{cs}$	(λ_1, μ_1)	$S_1 I_1$	$[f_2]$	(λ_2, μ_2)	$S_2\ I_2$	S	L	$\kappa_{orb} \times \kappa_{NN}^{0s^4 0p^2}$
4	$[2^2]$	(1,0)	1 0	$[1^2]$	(0,1)	1 1	0	0	$1/(9\sqrt{3})$
5	$[21^2]$	(1,0)	1 1	$[1^2]$	(0,1)	1 1	0	0	$1/(9\sqrt{2})$
6	$[21^2]$	(1,0)	0 1	[2]	(0,1)	0 0	0	0	$1/(6\sqrt{6})$
$0s^6$	$[f]_{cs}$	(λ, μ)					S	L	$\kappa_{orb} \times \kappa_{NN}^{0s^6}$
1	$[2^3]$	(0,0)					1	0	$-1/6$
$0s^5 1s$	$[f]_{cs}$	(λ_1, μ_1)	$S_1 I_1$	$[f_2]$	(λ_2, μ_2)	$S_2 I_2$	S	L	$\kappa_{orb} \times \kappa_{NN}^{0s^5 1s}$
2	$[2^2 1]$	(0,1)	$\frac{1}{2}\frac{1}{2}$	[1]	(1,0)	$\frac{1}{2}\frac{1}{2}$	1	0	5/27
3	$[2^2 1]$	(0,1)	$\frac{3}{2}\frac{1}{2}$	[1]	(1,0)	$\frac{1}{2}\frac{1}{2}$	1	0	$2\sqrt{5}/27$
$0s^4 0p^2$	$[f]_{cs}$	(λ_1, μ_1)	$S_1 I_1$	$[f_2]$	(λ_2, μ_2)	$S_2 I_2$	S	L	$\kappa_{orb} \times \kappa_{NN}^{0s^4 0p^2}$
4	$[2^2]$	(1,0)	1 0	[2]	(0,1)	0 0	1	0	$-1/6$
5	$[2, 1^2]$	(1,0)	1 1	$[1^2]$	(0,1)	1 1	1	0	$\sqrt{3}/18$
6	$[2, 1^2]$	(1,0)	1 1	$[1^2]$	(0,1)	1 1	1	0	$1/(9\sqrt{2})$

ii) $J = 1, I = 0$ case.
Now the coefficients are:

$$c = \{0.9997, -0.0016, 0.0111, -0.0178, -0.0096, -0.0019 \text{ for } \hbar\omega = 300\text{MeV}\}$$

and

$$A_{0s}[1, 0] = -0.167,\ A_{1s}[1, 0] = 0.0034 \tag{11}$$

Before concluding the section it is worth mentioning that components other than the two-nucleon ones can, of course, be examined as members of the cluster including colored ones. Thus in the case of the dominant configurations, i.e. $0s^6$ $J = 0$, $I = 1$ and $0s^6 J = 1, I = 0$ we obtain the expansion with the three-body coefficients found in Table 3.

The color singlet component in the case of the $J = 0, I = 1$ is about 13%, the color singlet nucleon-like part is even smaller, 11%. Again,[45] this can provide a strong enrichment of the high-momentum component for ^{6}He considered in $\alpha + q^6(I = 1, J = 0)$ cluster model.

Table 3. Three-body cfps, J=1, I=0 on the left and $J = 0, I = 1$ on the right. In each matrix the relevant quantum numbers $q^3[f_1](\lambda_1\mu_1)S_1\, I_1$, $\times$ $q^3[f_2](\lambda_2\mu_2)S_2\, I_2$ are understood in the first two columns.

$J = 1,\ I = 0$	cfp	$J = 0,\ I = 1$	cfp
$[21](11)\frac{1}{2}\frac{1}{2} \times [21](11)\frac{1}{2}\frac{1}{2}$	$-\frac{\sqrt{5}}{3\sqrt{2}}$	$[21](11)\frac{1}{2}\frac{1}{2} \times [21](11)\frac{1}{2}\frac{1}{2}$	$\frac{\sqrt{2}}{3}$
$[21](11)\frac{1}{2}\frac{1}{2} \times [21](11)\frac{3}{2}\frac{1}{2}$	$-\frac{\sqrt{5}}{3\sqrt{2}}$	$[21](11)\frac{1}{2}\frac{1}{2} \times [21](11)\frac{3}{2}\frac{1}{2}$	$\frac{1}{3}$
$[21](11)\frac{3}{2}\frac{1}{2} \times [21](11)\frac{1}{2}\frac{1}{2}$	$\frac{\sqrt{5}}{3\sqrt{2}}$	$[21](00)\frac{1}{2}\frac{1}{2} \times [21](00)\frac{1}{2}\frac{1}{2}$	$\frac{1}{3}$
$[21]](11)\frac{3}{2}\frac{1}{2} \times [21](11)\frac{3}{2}\frac{1}{2}$	$\frac{\sqrt{5}}{6}$	$[21](11)\frac{1}{2}\frac{1}{2} \times [111](11)\frac{1}{2}\frac{3}{2}$	$\frac{\sqrt{2}}{3}$
$[21](00)\frac{1}{2}\frac{1}{2} \times [21](00)\frac{1}{2}\frac{1}{2}$	$-\frac{1}{6}$	$[111](11)\frac{1}{2}\frac{3}{2} \times [21](11)\frac{1}{2}\frac{1}{2}$	$\frac{\sqrt{2}}{3}$
$[111](11)\frac{1}{2}\frac{3}{2} \times [1^3](11)\frac{1}{2}\frac{3}{2}$	$\frac{\sqrt{5}}{3}$	$[111](00)\frac{3}{2}\frac{3}{2} \times [1^3](00)\frac{3}{2}\frac{3}{2}$	$\frac{2}{3\sqrt{5}}$
$[111](00)\frac{3}{2}\frac{3}{2} \times [1^3](00)\frac{3}{2}\frac{3}{2}$	$-\frac{2}{3}$	$[111](11)\frac{1}{2}\frac{3}{2} \times [111](11)\frac{1}{2}\frac{3}{2}$	$\frac{1}{3\sqrt{5}}$

5. Mixing of the Cluster States with the Two-Particle Shell Model States

In evaluating mixing of the six-quark cluster states $\Phi(6q; J_c = 0, I_c = 1, (I_c)_3 = \pm 1) = |q^6\ 0, 1, \pm 1\rangle$ with the two-particle shell model states one would need the effective two nucleon interaction. For this purpose it is necessary that the cluster states, defined with the size parameter b_N, be expressed in terms of the shell model states usually expressed in terms of the harmonic oscillator parameter b_A, see,[38] section 8. For harmonic oscillator wave functions this is straightforward, bur such an expansion converges rather slowly and for $0s$ and $1s$ radial functions, which are of interest in this work, the number of radial quantum numbers $n_1\ n_2$ to a good approximation can be restricted to the 12 combinations n_1, n_2:

$$n_1 n_2 = \{(0,0), (0,1), (0,2), (1,1), (0,3), (1,2), (0,4), (1,3), (2,2), (0,5), (1,4), (2,3)\} \tag{12}$$

For light nuclei with $b_A/b_N \approx 3$ the radial overlaps in the above basis are of two types

$$C^{n_1,n_2}_{0s0s} = \{0.245178, 0.468387, 0.408421, 0.223701, 0.344056, 0.39012, .284612, 0.328642, 0.170088, 0.232808, 0.271861, 0.28656\} \tag{13}$$

$$C^{n_1,n_2}_{0s1s} = \{-0.234193, -0.127658, -0.0275665, -0.121938, 0.0473272, -0.026332, 0.097511, 0.045207, -0.022960, 0.01275, -0.074095, 0.039419\}.$$

So one needs 12 combinations for each combination of the two shell model J $= 0$ states. Such matrix elements are not available. This however is not necessary, if we are content to use one of the simpler interactions found in the literature.[46]

Then one can easily evaluate the matrix elements:

$$\begin{aligned}\mathrm{ME}_{(0s0s;n\ell),S,J,I} &= \langle 0s^2(b_N)L=0,S,JI|V(\mathbf{r})|(n\ell)^2L=0,S,J,I\rangle,\\ \mathrm{ME}_{(0s1s;n\ell),\ell,S,J,I} &= \langle 0s1s(b_N)L=0,S,JT|V(\mathbf{r})|(n\ell)^2L=0,S,J,I\rangle\end{aligned} \tag{14}$$

where in the ket the wave functions are defined for b_A and only the $L=0$ component of the two particles is relevant.

The simplest possibility to consider is a closed-shell nucleus with two nucleons outside the closed shell.

A = 6 nuclei, specifically ^{6}Li

We consider admixtures of the type $^6\mathrm{Li}{+}C_{1,0}\left|^4\mathrm{He}\times q^6\,1,0,0\right\rangle$. The mixing coefficient is given by

$$\begin{aligned}C_{1,0} &= \frac{1}{D}\langle ^4\mathrm{He};\times q^6\,1,0,0|V|^6\mathrm{Li}\;J=1,I=0\rangle,\\ D &= E(^6\mathrm{Li}-^4\mathrm{He})-(E_{6q}J=1,I=0-2E_N)\end{aligned} \tag{15}$$

Furthermore, for this relatively simple nucleus

$$\begin{aligned}C_{1,0} &= \frac{1}{D}\,\langle q^6\,1,0,0|V|0p^2,J=1,I=0\rangle\\ &= \frac{1}{D}\left(A_{0s}(1,0)\mathrm{ME}_{(0s0s;0p),1,1,0}+A_{1s}(1,0)\mathrm{ME}_{(0s1s;0p),1,1,0}\right)\end{aligned} \tag{16}$$

In shell model calculations one finds[47]

$$\psi(0p^2,10) = 0.7996\,|0p^2_{3/2}\,01\rangle - 0.258|0p^2_{1/2}\,01\rangle - 0.6004|0p_{3/2}0p_{1/2}\,01\rangle.$$

In the LS coupling scheme the component of interest is large with amplitude $0.93447\,|0p^2L=0,S=1\rangle$.

In a similar fashion we find the admixture of the type $^6\mathrm{He}{+}C_{0,1}\,|^4\mathrm{He}\times q^6 0,1,1\rangle$. Proceeding as above we find:

$$\begin{aligned}C_{0,1} &= \frac{1}{D}\langle q^6\,0,1,1|V|0p^2,J=0,I=1\rangle\\ &= \frac{1}{D}\left(A_{0s}(0,1)\mathrm{ME}_{(0s0s;0p),0,0,1}+A_{1s}(0,1)\mathrm{ME}_{(0s1s;0p),0,0,1}\right)\end{aligned} \tag{17}$$

In shell model calculations one finds

$$\psi(0p^2J=1,I=0) = 0.97841\,|0p^2_{3/2}\,01\rangle + 0.20670\,|0p^2_{1/2}\,01\rangle,$$

Similarly, the dominant LS component is large, $0.9182|0p^2L=0,S=0\rangle$. Both nuclei are therefore favorable subjects for the present study.

1s0d shell A = 18 The relevant wave functions for mass 18 are well studied, see *e.g.* Refs. 15, 29. The eigenvectors for the ^{18}O ground state in the Elliott SU(3) scheme are

$$0.81\,|(40)L=0\rangle - 0.46\,|(02)L=0\rangle + 0.36\,|(21)L=1\rangle$$

or 87% L=0. Further, most interactions and transitions favor proximity of the nucleons involved. The SU(3) representations ($\lambda 0$) (which are the leading representations for four or fewer particles) contain all the relative $0s(r_{12})$ components. Thus, just as the mass six nuclei should be favorable candidates to seek clusters, so are the mass 18 nuclei. Explicit calculations of the coupling of $6-q$ cluster to these nuclei are under way, but they will not be reported in this work.

The above calculations are straightforward. They can be extended to more complicated nuclear systems provided the wave functions are known. For the moment we provide some results for the p-shell nuclei for illustrative purposes. Twelve nuclear matrix elements are required in the coupling of the six-quarks to the above nuclei, They were obtained using the one-pion exchange potential given in Ref. 46. They are as follows:

$$\mathrm{ME}_{(0s0s;0p),0,0,1} = 0.296, \quad \mathrm{ME}_{(0s1s;0p),0,0,1} = 2.75,$$
$$\mathrm{ME}_{(0s0s;0p),1,1,0} = 0.301, \quad \mathrm{ME}_{(0s1s;0p),1,1,0} = 2.80,$$

The above mentioned fraction of the $L=0$ component in the nuclear wave function has been incorporated into the above ME. Thus the numerical vales of the expression in Eqs. 17 and 16 become:

$$\mathrm{Num(He)} = \left(A_{0s}(0,1)\mathrm{ME}_{(0s0s;0p),0,0,1} + A_{1s}(0,1)\mathrm{ME}_{(0s1s;0p),0,0,1}\right) = 0.0918$$
$$\mathrm{Num(Li)} = \left(A_{0s}(1,0)\mathrm{ME}_{(0s0s;0p),1,1,0} + A_{1s}(1,0)\mathrm{ME}_{(0s1s;0p),1,1,0}\right) = -0.0406$$

From nuclear mass tables we find that BE(^{4}He)=28.296, BE(^{6}Li) = 31.995 and our calculations indicate E(q^6(0,1)-2E_{3q}(1/2.1/2)=-7.95 MeV and E(q^6(1,0)-2 $E_{3q}(1/2,1/2)$=-4.71MeV. On the other hand β-decay experiments[48] indicate that the energy of ^{6}He is 5.5 MeV higher than that of ^{6}Li. Based on that information we find $D_{0,1} = 9.8$ MeV and $D(1,0) = 4.3$ MeV which results in $C_{0,1} = \frac{0.0918}{9.8} = 0.0101$ and $C_{1,0} = \frac{0.0406}{4.3} = -0.0088$. Thus the probability of finding the six-quark cluster is:

$$P(\mathrm{He}) = 1.02 \times 10^{-4}, \quad P(\mathrm{Li}) = 0.779 \times 10^{-4}. \tag{18}$$

Before embarking on more detailed calculations, however, ideally we would have a good experimental signature to enable us detect the presence of the six-quark clusters in nuclei. The most appropriate is the break up of the above nuclei into the closed-shell nucleus (^{4}He or ^{16}O) and a six-quark cluster to unravel the exotic structure given, *e.g,* in tables 3. For the moment we concentrate on double-beta and the possible effects of nuclear admixtures of six-quark clusters in nuclear transitions caused by very short range operators, not hindered by nuclear short-range correlations.

6. The Role of Six-Quark Admixtures in the the $0\nu\beta\beta$ Mediated by Heavy Majorana Neutrinos

We consider two states which can undergo neutrinoless double-beta decay:

$$(A, Z) \rightarrow (A, Z+2) + e^- + e^- \tag{19}$$

proceeding via the exchange of very Majorana neutrinos. This process is very important because, among other things, it will determine whether the massive neutrinos are of the Majorana type (the antineutrino coincides with the corresponding neutrino) or of the Dirac type. This process has extensively studied both theoretically and experimentally, see, e.g.,[49] for a review. The most interesting scenario is if $0\nu\beta\beta$-decay is mediated by a light Majorana neutrino. Other mechanisms exist whereby the process is mediated by mechanisms involving heavy particles and in particular very heavy Majorana neutrinos. In this case one must deal with a very short range operator and the relevant transition nuclear matrix element is greatly suppressed due the short range correlations. One, therefore, must go beyond the two-nucleon mode. One possibility is to consider the double-beta decay of pions in flight between two nucleons. This possibility has been considered[50] leading to an effective transition operator with a range of the pion mass. This way the short-range range correlations are under control leading to reasonable matrix elements. Another possibility is to assume that the nucleons are not point-like particles but have a size described by a form factor, *e.g.*, a dipole shape form factor.[49,51] This leads to an effective operator with a range of the proton mass and again the problem of short range correlations is circumvented.

The presence of six-quark clusters in nuclei opens a novel mechanism. The neutrinoless double beta decay can occur within the same hadron and the nuclear matrix element can be sizable even with δ function integration between quarks. In this case the nuclear transition takes the form:

$$|AJ_iT_z\rangle \rightarrow |AJ_f(T_z-2)\rangle \tag{20}$$

where we use here the nuclear convention of isospin $T_z|n\rangle = \frac{1}{2}|n\rangle$, $T_z|p\rangle = -\frac{1}{2}|p\rangle$. We do not exhibit the isospin, T, of the states, only their T_z.

We will also consider two additional nuclear states $(A-2)J_i'(T_z-1)$, $(A-2)J_f'(T_z-1)$ which via the six-quark cluster can couple to the A states. We begin with mixing to the initial state. *i.e.*

$$|i\rangle = |AJ_iT_z\rangle + C_{J_i,J_i'}|(A-2)J_i'(T_z-1)\times q^6 0,1,1\rangle.$$

The mixing coefficient is given by:

$$\begin{aligned} C_{J_i,J_i'} &= \frac{1}{D_i}\langle (A-2)J_i'(T_z-1)\times q^6 0,1,1|V|AJ_iT_z\rangle, \\ D_i &= E(AJ_iT_z) - E((A-2)J_i'(T_z-1)) - (E_B - 2E_N) \end{aligned} \tag{21}$$

In the energy denominator, in addition to the nuclear binding energies, the energy of the six-quark state relative to the two nucleons also appears.

Furthermore

$$\begin{aligned}&\langle (A-2)J_i'(T_z-1)\times q^6 0,1,1|V|AJ_iT_z\rangle\\&\quad=\sum_j \left(\langle (A-2)J_i'(T_z-1);j^2J=0|\}AJ_iT_z\rangle\right)\delta_{J,J'}\\&\qquad\langle q^6 0,1,1|V|j^2J=0\rangle\end{aligned}$$

where the coefficient $\langle (A-2)J_i'(T_z-1);j^2J=0|\}AJ_iT_z\rangle$ is the usual coefficient of fractional parentage (cfp) in symbolic form where $|j^2J=0\rangle$ is a short hand notation of the shell model states $|[(n\ell j)(n'\ell' j)]J=0\rangle$ $\langle\Phi(6q)J_c=0,I_c=1,(I_c)_3=1|V|j^2J=0\rangle$ is essentially in the $j-j$ scheme a short-hand notation of the matrix element given by Eq. (14).

$$\begin{aligned}C_{J_i}&=\frac{1}{D_i}\sum_j\langle (A-2)J_i(T_z-1);[(n\ell j)(n'\ell' j)]J=0|\}AJ_iT_z\rangle\\&\qquad\langle\Phi(6q)J_c=0,I_c=1,(I_c)_3=1|V|j^2J=0\rangle.\end{aligned}\tag{22}$$

Proceeding in an analogous fashion we can find the admixture to the final state:

$$|f\rangle=AJ_f(T_z-2)+C_{J_f,J_f'}(A-2)J_f'(T_z-1)\times q^6 0,1,-1\rangle.$$

The mixing coefficient is given by:

$$\begin{aligned}C_{J_f}&=\frac{1}{D_f}\sum_j\langle (A-2)J_i(T_z-1);[(n\ell j)(n'\ell' j)]J=0|\}(AJ_f(T_z-2))\rangle\\&\qquad\langle\Phi(6q)J_c=0,I_c=1,(I_c)_3=-1|V|j^2J=0\rangle\end{aligned}\tag{23}$$

with

$$\begin{aligned}D_i&=E(AJ_iT_z)-E((A-2)J_i(T_z-1))-(E_B-2E_N),\\D_f&=E(AJ_f(T_z-2))-E((A-2)J_f(T_z-1))-(E_B-2E_N)\end{aligned}\tag{24}$$

6.1. *Evaluation of the $0\nu\beta\beta$ Matrix Elements in the Six-Quark Cluster*

The $0\nu\beta\beta$ transition can now be computed once the nuclear wave functions are known. Here we will focus on the contribution coming from the six-quark contribution. The nuclear matrix element involved will be of the form:

$$|\mathrm{me}|^2=\left|\ \langle q^6(00)_c,01,1,-1|\Omega|q^6(00)_c,0,1,1\rangle\ \right|^2\tag{25}$$

where Ω is the transition operator at the quark level resulting from the diagram in Fig. 1. The amplitude $\mathcal{M}'$ associated with this diagram in momentum space is inversely proportional to the neutrino mass m_N. Transforming this into coordinate

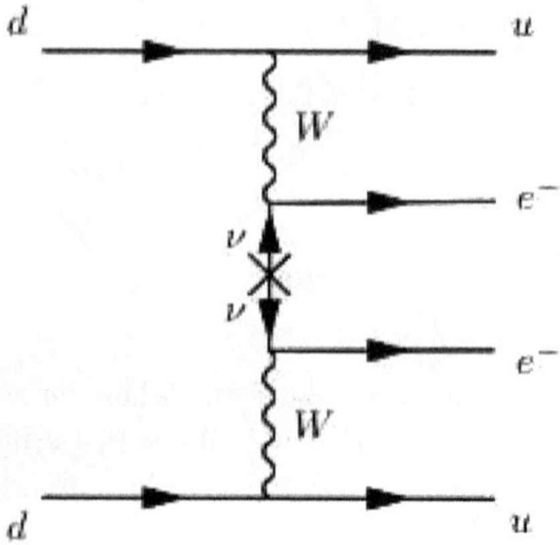

Fig. 1. The elementary diagram at the quark level leading to neutrinoless double-beta decay in the case of heavy Majorana neutrino mechanism. Two down quarks with momenta q_1 and q_2 are transform into two up quarks with momenta q_1' and q_2' with the simultaneous emission of two electrons with momenta p_1 and p_2.

space using the momentum conservation δ-function $\delta(\mathbf{q_1}+\mathbf{q_2}-\mathbf{q}_1'-\mathbf{q}_2'-\mathbf{p_1}-\mathbf{p_2})$ we find

$$\mathcal{M}=\frac{1}{m_N}\delta\left(\mathbf{r_1}-\mathbf{r_2}\right)e^{i(\mathbf{p_1}+\mathbf{p_2}).\mathbf{r_2}}=\frac{1}{b_N^3 m_N}\omega,\ \omega=b_N^3\delta\left(\mathbf{r_1}-\mathbf{r_2}\right)e^{i(\mathbf{p_1}+\mathbf{p_2}).\mathbf{r_2}} \tag{26}$$

where b_N is the size parameter of the six-quark cluster introduced to make ω dimensionless

Absorbing the factor $\frac{1}{b_N^3 m_N}$ into the lepton violating parameter and incorporating the standard the spin-isospin dependence of ω, see, *e.g.*, Ref. 51, in the long wavelength approximation we get:

$$\omega=\sum_{i\neq j}b_N^3\delta(\mathbf{r_i}-\mathbf{r_j})\tau_-(\mathbf{i})\tau_-(\mathbf{j})\left(\mathbf{1}-\boldsymbol{\sigma}_\mathbf{i}.\boldsymbol{\sigma}_\mathbf{j}\right) \tag{27}$$

(at the quark level $g_A=1$). The required matrix elements can be obtained in the context of the six-quark cluster.

Let us consider the leading $0s^6$ configuration. Using the techniques of Ref. 36 and employing the cfp's found in Table 13 of Ref. 38 we find

$$\langle q^6 1,1,-1|\Omega|q^6 1,1,1\rangle=\frac{1}{2}\times 6\times 5$$

$$\langle 0,0;1,1|1,1\rangle^2\left(\frac{1}{4}\mathrm{me}_0+\frac{1}{12}\mathrm{me}_1\right)+\langle 2,0;1,1|1,1\rangle^2\left(\frac{1}{10}\mathrm{me}_0+\frac{1}{15}\mathrm{me}_1\mathrm{me}_1)\right)$$

where the two-body matrix elements

$$\mathrm{me}_s=\langle 0s^2 I=1,I_3=-1,s|\omega|0s^2 I=1,I_3=1,s\rangle$$

and the two body spin is indicated by s.

We note that due to isospin conservation the two-body ME can only involve $I = 1$, *i.e.*, only the $[1^2]$ color-spin symmetry is relevant. The first factors outside the parenthesis are the relevant isospin Clebsch-Gordan coefficients taking the values of 1 and $\frac{1}{10}$. In the above expression

In the present case the two-body matrix elements are $\text{me}_0 = 4R_{0s}$ and $\text{me}_1 = 0$. Due to the presence of the δ function in the interaction the radial integral becomes

$$R_{0s} = \frac{1}{4\pi}\int_0^\infty 0s^4(x)x^2dx = \frac{1}{(2\pi)^{3/2}} \tag{28}$$

We thus obtain

$$\langle q^6 0, 1, -1|\omega|q^6 0, 1, 1\rangle = \frac{39}{10}\text{me}_0 = \frac{78}{5}\frac{1}{(2\pi)^{3/2}} \approx 0.99. \tag{29}$$

This value is actually quite large but it appears to be small compared to other values involving heavy neutrinos that appear in standard two nucleon calculations. The reason is that in the case of heavy neutrino the double-beta decay life time is expressed as follows:[51]

$$\left[T_{1/2}(0\nu)\right]^{-1} = \eta_N^2\,(\text{ME})^2_{\mathit{eff}}\,G_{0\nu}(E_0, Z), \quad \eta_N = \sum_j U_{ej}^2\frac{m_p}{m_{N_j}} \tag{30}$$

where η_N is the lepton violating parameter and $G_{0\nu}(E_0, Z)$ is the phase space function. The nuclear matrix element is usually normalized by the factor $f_{sc} = m_p/m_e$ with a compensating factor in the lepton violating parameter. The effective matrix element associated with Eq. (29) becomes

$$\langle q^6 0, 1, -1|\omega|q^6 0, 1, 1\rangle_{\mathit{eff}} \approx 1.82 \times 10^3 \tag{31}$$

This would be the case if the probability of finding the six-quark cluster in a nucleus were close to unity. Unfortunately, as we will see in the next section, the actual probability is very small.

6.2. *Evaluation of the Double-Beta Decay Matrix Element for A = 48*

We consider for simplicity the A=48 system within the $f_{7/2}$ shell assuming an initial state, $0f^8_{7/2}$ neutrons with $J_i = 0$. The allowed mass 46 $|J'_i, T'_i\rangle$ states are $J'_i = 0, 2, 4, 6$, $T'_i = 1$ with the cfp's in the spirit of Eq. (22) being $C_{J'} = \frac{\sqrt{2J'_i+1}}{2\sqrt{7}}$. Only the $J' = 0$ is relevant here with a cfp $= 1/(2\sqrt{7})$.

The final state is of very simple form, a product of $0f^6_{7/2}J'_f$ neutrons and $0f^2_{7/2}J'_f$ protons coupled to a total angular momentum $J_f = 0$. The corresponding amplitudes for $J'_f = 0, 2, 4, 6$ can be obtained by a simple shell model calculation, *e.g.* $0.845, -0.523, 0.109, -0.001$, respectively found in Ref. 52. The relevant amplitude is for $J'_f = 0$ or 0.845.

From nuclear mass tables we find the binding energies are for ^{48}Ca ^{46}Ca and ^{48}Ti 415.991, 398.769, *and* 420.191 MeV, respectively. Taking the quark energy difference the same as for ^{6}He, that is E(q^6(0,1)-2E_{3q}(1/2,1/2)=-7.95 MeV, the energy denominators become

$$D_i = -9.27\ \text{MeV}, \quad D_f = -16.71\ \text{MeV}. \tag{32}$$

Then the specialization of Eq. (14) in the case of ^{48}Ca yields:

$$\text{ME}_{(0s0s;0f),0,0,1} = -0.305, \text{ME}_{(0s1s;0f),0,0,1} = 0.537 \Rightarrow$$
$$\text{Num(Ca)} = \left(A_{0s}(0,1)\text{ME}_{(0s0s;0f),0,0,1} + A_{1s}(0,1)\text{ME}_{(0s1s;0f),0,0,1}\right) = -0.103$$

The fraction of $2/\sqrt{7}$ of the $L = 0$ component has been incorporated in the evaluation of the above ME. Thus

$$C_i = \frac{1}{2\sqrt{7}}\frac{-0.103}{D_i} = 0.00209, \quad C_f = \frac{-0.103}{D_f}0.845 = 00.00552\,. \tag{33}$$

Thus in the approximation of a purely $0f_{\frac{7}{2}}$ configuration, the probability of finding the $J = 0, I = 1$ six-quark cluster in the ^{48}Ca nucleus is extremely small

$$C_i^2 = 4.9 \times 10^{-6} \text{ for } ^{48}\text{Ca}, \quad C_f^2 = 2.7 \times 10^{-5} \text{ for } ^{48}\text{Ti}. \tag{34}$$

The resulting nuclear matrix element is small due to the above small probabilities. Indeed

$$ME_{0\nu\beta\beta} = C_i C_f 1.82 \times 10^3 = 0.020, \tag{35}$$

see Eq. (31). This is a tiny value compared with the standard phenomenological treatment:[51]

$$\text{ME}_{0\nu\beta\beta} = 72. \tag{36}$$

The A = 48 wave functions have large admixtures of $0p$ components which will increase the size of the result, but this requires more detailed calculations.

7. Concluding Remarks

In this paper we examined the structure of six-quark clusters in the context of the group $SU(6)_L \otimes SU(6)_{cs} \otimes SU(2)_I$ the first corresponding to orbital, the second to color-spin and the third to isospin symmetry. We considered a model involving harmonic oscillator single quark states within two $\hbar\omega$ with all color singlet states and with all allowed spin and isospin states, focusing on those of the two nucleons, *i.e.* color singlet states $J = 0, I = 1$ and $J = 1, I = 0$. The Hamiltonian matrix was diagonalized using the interaction derived from the one-gluon exchange potential. The wave functions were expanded into two q^3 clusters. To this end we computed all the one-particle, two-particle and three-particle cfp's for all the three symmetries involved.

The probability of two-nucleon components was about 1/9 and 1/36 for the $J = 0$, $I = 1$ and $J = 1$, $I = 0$, respectively. The remainder corresponds to other baryon-like states or colored three-quark components. Using only the two nucleon-like component, we obtained the coupling of the six-quarks to two-nucleon shell model states and thus the probability of finding the six-quark structure in a nucleus. This was straightforward for the nuclei that are considered as two particles outside a closed shell. For such simple cases this probability is about 0.01% in the case of ^{6}Li and ^{6}He, perhaps small for experimental searches. As the mass number increases the spin-orbit interaction becomes more important and this tends to destroy LS coupling. But isolated areas of the mass table can be found in which components like $(\ell j)^2, J = 0, 1$ can dominate the ground state nuclear wave functions and, in addition, they can couple strongly to the six-quark cluster. In general for more complicated nuclei the above mixing may suppressed due to the fact the $\langle A-2, (\ell j)^2 | \} A \rangle$ coefficients of fractional parentage may be small.

To test for a signature of the presence of six-quark cluster in nuclei we considered a process involving a very short-range transition operator, in particular the neutrinoless double-beta decay mediated by very heavy Majorana neutrinos. We found that indeed this proceeds even with a delta-function interaction between the quarks inside the six-quark bag, with a large nuclear matrix element. As a start we considered the simple double-beta decay ^{48}Ca to ^{48}Ti. The resulting neutrinoless double-beta decay nuclear matrix element was found, however, to be very small, mainly due to the smallness of the mixing of the six-quark cluster caused by size of the nuclear coefficients of fractional parentage involved.

Acknowledgments

This work was started during D.S's Alexander von Humboldt Fellow award at the Institute for Theoretical Physics, Goethe University, Frankfurt, hosted by Prof. Walter Greiner and Horst Stöcker.

References

1. D. Strottman, *Phys. Rev. D* **18**, 2716 (1978).
2. D. Strottman, *Phys. Rev. D* **20**, 748 (1979).
3. R. Jaffe, *Phys. Rev. Lett* **38**, 195 (1977).
4. R. Jaffe, *Phys. Rev.D* **15**, 281 (1977).
5. F. Wang, J. Ping, D. Qing, and T. Goldman, *Phys. Atom. Nucl.* **68**, 808 (2005). *Yad. Fiz.* 68:841-846:2005; arXiv:nucl-th/0406036.
6. R. Aaij *et al.*, *Phys. Rev. Lett.* **115**, 072001 (2015). arXiv:1507.03414.
7. R. Aaij *et al.*, *Phys. Rev. Lett.* **117**, 261801 (2016). The LHCb collaboration; arXiv:1608.06420 (astro-ph.CO).
8. R. Aaij *et al.*, *Phys. Rev. Lett.* **17**, 082003 (2016).
9. R. Ali, C. Hambrock, and M. J. Aslam, *Phys. Rev. Lett.* **104**, 162001 (2010).
10. R. Aaij *et al.*, *Phys. Rev. Lett.* **112**, 222002 (2014).
11. J. D. Vergados and C. Moustakidis, *Subatomic Physics (Nuclear and Particle Physics and Astrophyscs).* World Scientific (2021).

12. J. Vergados, *Nuc. Phys. B* **250**, 418 (1985).
13. E. P. Wigner, *Phys. Rev.* **51**, 106, 947 (1937).
14. E. P. Wigner, *Phys. Rev.* **56**, 519 (1939).
15. H. H. T. Inoue, T. Sebe, and A. Arima, *Nucl. Phys.* **59**, 1 (1964).
16. A. A. Y. Akiyama and T. Sebe, *Nucl. Phys.* **A138**, 273 (1969).
17. Y. Akiyama, *Nucl. Data* **A2**, 403 (1966).
18. D. Strottman and J. Vergados, *arXiv:* **1808.06042**, 1 (2018).
19. J. Richard, *arXiv:* **1205.43262**, 1 (2012).
20. H. A. Jahn and H. van Wieringen, *Proc. Roy. Soc. Ser. A* **209**, 502 (1951).
21. J. H. J.P. Elliott, and H. Jahn, *Phil. Trans. Roy. Soc.* **A249**, 241 (1953).
22. J. P. Elliott and B. F. Flowers, *Proc. Roy. Soc.* **A242**, 57 (1957).
23. J. Elliott, *Proc. Roy. Soc. Series A* **A245**, 128 (1958).
24. J. Elliott and M. Harvey, *Proc. Roy. Soc. Series* **A272**, 557 (1963).
25. M. Harvey, *Advances in Nuclear Physics* **1**, 67 (1968).
26. A. Arima and D. Strottman, *Nucl. Phys.* **A162**, 423 (1971).
27. D. Strottman and A. Arima, *Nucl. Phys.* **A162**, 605 (1971).
28. D. Strottman, *Phys. Letts.* **398**, 457 (1972).
29. D. Strottman, *Oxford Nuclear Physics Laboratory technical report.* (1973).
30. K. Hecht, *Nucl. Phys.* **62**, 1 (1965).
31. J. Vergados, *Nucl. Phys.* **111**, 681 (1968).
32. J. Draayer and Y. Akiyama, *J. Math. Phys.* **14**, 1904 (1973).
33. Y. Akiyama and J. Draayer, *Comput. Phys. Commun.* **5**, 405 (1973).
34. D. Millener, *J. Math. Phys.* **19**, 1513 (1978).
35. S. I. So and D. Strottman, *J. Math. Phys.* **20**, 153 (1976).
36. D. Strottman, *J. Math. Phys.* **20**, 1643 (1979).
37. H. Weyl, *The Theory of Groups and Quantum Mechanics* (1931).
38. J. Vergados and D. Strottman, Symmetries in subatomic six-quark systems, *arXiv* **nuc-th**, 1 (2022). arXiv:2202.04563 (nucl-th).
39. G. Racah, *Phys. Rev* **62**, 438 (1942).
40. A. de Shalit and I. Talmi, *Nuclear Shell Theory.* Academic Press (1962).
41. D. Strottman and J. Vergados, Quark cluster model for $\beta\beta$-decay, *arXiv* **nuc-th**, 1 (2018). arXiv:1808.06042 (nucl-th).
42. S. Ohta, M. Oka, A. A. Arima, and K. Yazaki, Shell model study of six quark system, *Phys. Lett B* **119**, 15 (1982).
43. T. Skyrme, *Proc. Roy. Soc.* **260**, 127 (1961).
44. K. Hecht, *Nucl. Phys.* **A170**, 34 (1971).
45. M. Kakenov, V. Kukulin, V. Pomerantsev, and O. Bayakhmetov, Properties of dibaryons in nuclear medium, *Eur. Phys. J. A* **56**, 266 (2020). arXiv:2010.10557 [nuc-th].
46. I. E. Lagaris and V. R. Pandaripande, *Nucl. Phys. A* **359**, 331 (1981).
47. J. D. Vergados, *Phys. Lett. B* **41**, 560 (1972).
48. A. Knecht *et al.*, Precision measurement of the ^{6}He half-life and the weak axial current in nuclei, *Phys. Rev. Lett.* **100**, 122502 (2012).
49. J. D. Vergados, H. Ejiri, and F. Simkovic, *Rep. Prog. Phys.* **75**, 106301. (2012).
50. J. D. Vergados, Pion double charge exchange contribution to neutrinoless double β-decay, *Phys. Rev. D* **25**, 914 (1982).
51. J. D. Vergados, A limit on neutrino mass from double β decay, *Phys. Lett. B* **109**, 96 (1982).
52. J. D. Vegados, *Nuc. Phys. B* **218**, 109 (1983).

https://doi.org/10.1142/9789819813353_0020

Back and Forth with Akito Arima

Larry Zamick and Castaly Fan
Rutgers, the State University of New Jersey, Piscataway, NJ 08854

In 1967 Akito Arima spent a year as a visiting professor in the physics department of Rutgers University. In this work we pay tribute to him by discussing topics that we worked on that were directly influenced buy his works or were closely related to his interests. These include nuclear Symmetries, magnetic and other moments, analytic expressions in the single j shell model and schematic interactions.

1. Introduction

Akito Arima was a visiting professor at Rutgers in 1967. He was not the only Japanese visitor. Also Shiro Yoshida and a bit later a very bright student of Arima-Koichi Yazaki. Yoshida stayed on as a professor for about 5 years and I wrote the following article with him. "Electromagnetic Moments and Transitions Annual Review of Nuclear Science — Vol 22". I developed lifelong friendships with all these people. I was invite to give one of the banquet speeches at a 1972 conference in Japan honoring Akito. I said diplomatically: "Akito Arima is the most respected nuclear theorist in the world and Shiro Yoshida is the most respected nuclear theorist in Japan".

Akito's work was not only a big influence on the topics that I chose but was also of importance to the experimental group at Rutgers — Noemie Koller, Gerfried Kumbartzki, and from Bonn, Karl Heinz Speidel. This group measured magnetic moments of excited states of even-even nuclei and the work of Arima and Hori, well known at that time, provided a solid theoretical background.

When I was postdoc at Princeton Gerry Brown suggested to me and to a then student Harry Mavromatis that we also work on magnetic moments. The Arima-Horie theory was basically first order perturbation theory so we were to do second order. This is important for a closed major shell plus (or minus) one nucleon because for such cases first order vanishes. Ichimura and Yazaki also had worked on this. By the way soon after Gerry Brown left Princeton for Stony Brook an the took Akito with him. I recall at conferences Akito would stand up to ask a question: He started with Akito Arima, Stony Brook and Tokyo and then the fearsome question.

Besides magnetic moments I will here discuss topics which paralleled the interest of Arima — at least I hope they did. These include Nuclear Symmetries, quadrupole moments, single j shell properties and schematic interactions. In the same time period — after 2000, Akito Arima and Yu-Min Zhao wrote many papers on the single j shell (e.g.number of states of identical particles), and I had written a few myself. I recall sending emails to Akito about this but got no replies. I related this to Igal Talmi who laughed and said "Akito does not answer anyone's emails." At the same Akito and Yu-Min were very generous in mentioning my works on these topics and I reciprocated in turn. I hope the next few sections will help to covey how important Arima's presence both at Rutgers and on the world scene helped to enlivened my life in physics.

2. Symmetries

Akito's Arima contributions to the subject of symmetries in nuclei is overwhelming. Here we show some work on this topic which we hope would have met with his approval.

2.1. *Isospin*

Note in Table 1 that the excitation energies of the even J states in ^{42}Ca, ^{42}Sc and ^{42}Ti are early the same. This is evidence of the charge independence of the nuclear force. The fact that odd J states appear only in ^{42}Sc shows the Pauli

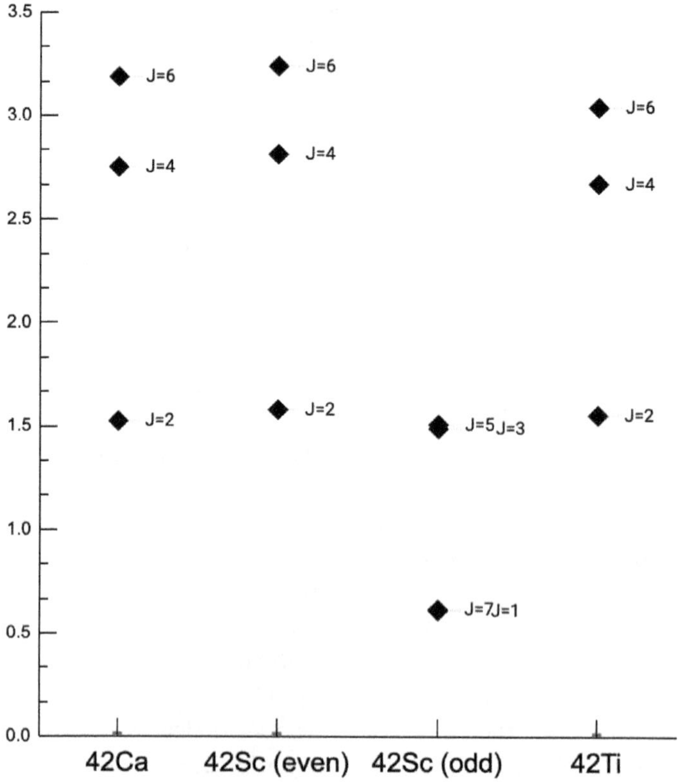

Fig. 1. Energy levels of ^{42}Ca, ^{42}Sc and ^{42}Ti shown in order to display the near charge independence of the nuclear force.

Table 1. The spectra of ^{42}Ca, ^{42}Sc, and ^{42}Ti.

	^{42}Ca	^{42}Sc	^{42}Ti
$J = 0$	0.0000	0.0000	0.0000
$J = 1$	—	0.6110	—
$J = 2$	1.5247	1.5803	1.5546
$J = 3$	—	1.4904	—
$J = 4$	2.7524	2.8153	2.6746
$J = 5$	—	1.5100	—
$J = 6$	3.1893	3.2420	3.0430
$J = 7$	—	0.6163	—

principle in action. In ^{42}Ca and ^{42}Ti we have 2 identical nucleons so we can only have antisymmetric states. This tells us that in the j^2 configuration states with even J are antisymmetric. In ^{42}Sc we do not have identical nucleons so we can have symmetric states. These are the odd J states. And the there is the multiplicity rule. Even J states occur 3 times so $(2T+1) = 3$ and so $T = 1$. The odd J states occur only once so $(2T+1) = 1$ and so $T = 0$.

2.2. *Signature*

In the $f_{7/2}$ paper of McCullen *et al.*[1,2] we also discuss briefly signature selection rules. For say ^{48}Ti we have a system of 2 protons and 2 neutron holes in the $f_{7/2}$ shell. We find that the wave functions are either even or odd under the interchange of protons and neutron holes. This is different from isospin in that a state of even signature and one of odd signature can have the same isospin.

It has been shown that this signature property leads to several selection rules. For example, for the electric quadrupole operator the B(E2) between states of opposite signature is proportional to (ep + en),and between states of the same signature to (ep - en). The quadrupole moment of the 2^+ state is proportional to (ep-en). It turns out that the $2{+}i$ state of Ti in the single j shell calculation has odd signature, but the 2+2 state has even signature. Hence B(E2) from the $J = 0$ ground state to 2(1) goes as $(\text{ep} + \text{en})^2$ whilst to 2(2) as $(\text{ep} - \text{en})^2$.

Consider next the double Gamow-Teller operator $(\sigma t_-)_i(\sigma t_-)_j$ connecting ^{48}Ca to ^{48}Ti. Zamick and Moya de Guerra[3] showed that in this single j shell model the transition to the 2(1) state (negative signature) vanishes because of the signature selection rule. On the other hand a transition to the 2(2) (positive signature) state is allowed. This simple model shows that there might be surprises when calculating double beta decay transitions.

2.3. *Seniority*

Some of the well-known statements and theorems concerning states of good seniority are

1. The seniority is roughly the number of identical particles not coupled to zero. Hence, for a single nucleon the seniority v is equal to 1. For two nucleons in a

$J = 0$ state we have $v = 0$, but for $J = 2, 4, 6$, etc., $v = 2$. For three nucleons there is one state with seniority $v = 1$, which must have $J = j$; all other states have seniority $v = 3$.

2. The number of seniority-violating interactions is $[(2j - 3)/6]$, where the square brackets mean the largest integer contained therein. For $j = 7/2$ there are no seniority-violating interactions, while for $j = 9/2$ there is one.
3. With seniority-conserving interactions, the spectra of states of the same seniority is independent of the particle number.
4. At midshell we cannot have any mixing of states with seniorities v and $v + 2$; one can mix v and $v + 4$ states.

III. SPECIAL BEHAVIORS FOR $I = 4^+$ AND 6^+ STATES OF THE $g^4_{9/2}$ CONFIGURATION

For a system of four identical nucleons in the $g_{9/2}$ shell, the possible seniorities are $v = 0$, 2, and 4, with $v = 0$ occurring only for a state of total angular momentum $I = 0$. There is also a $v = 4$ state with $I = 0$.

For $I = 4$ and 6, we can have three states, one with seniority $v = 2$ and two with seniority $v = 4$. For the two $v = 4$ states we have at hand, we can construct different sets of $v = 4$ states by taking linear combinations of the original ones. If the original ones are $(4)_1$ and $(4)_2$, we can form

$$(4)_A = a(4)_1 + b(4)_2,$$
$$(4)_B = -b(4)_1 + a(4)_2, \quad (2)$$

with $a^2 + b^2 = 1$. The set $(4)_A$, $(4)_B$ is as valid as the original set.

However, we here note that if we perform a matrix diagonalization with any two-body interaction—seniority conserving or not—one state emerges which does not depend on what the interaction is. The other two states are, in general, mixtures of $v = 2$ and $v = 4$ which do depend on the interaction. The values of the coefficients of fractional parentage (cfp's) of this unique state of seniority 4 are shown in Table I. The states with $J_0 \neq 4.5$ all have seniority $v = 3$. Note that in this special $v = 4$ state there is no admixture of states with $J_0 = j = 9/2$, be they $v = 1$ or $v = 3$. Again, no matter what

Fig. 2. We quote from the paper.[4]

TABLE I. A unique $J = 4$, $v = 4$ cfp for $j = 9/2$.

J_0	$(j^3 J_0 j \| \} j^4 I = 4, v = 4)$
1.5	0.1222
2.5	0.0548
3.5	0.6170
4.5 ($v = 1$)	0.0000
4.5 ($v = 3$)	0.0000
5.5	−0.4043
6.5	−0.6148
7.5	−0.1597
8.5	0.1853

two-body interaction is used, this $I = 4$ state remains a unique state.

Amusingly, this state *does not appear* in the compilation of seniority-classified cfp's of Bayman and Lande [20] or de Shalit and Talmi [5]. We should emphasize that, although different, the Bayman–Lande cfp's are perfectly correct (as are the ones of de Shalit and Talmi, whose cfp's are also different from those of Bayman and Lande [20]). But then, why do they not obtain the unique state that we have shown above? Bayman and Lande use group theoretical techniques to obtain the cfp's, diagonalizing the following Casimir operator for $Sp(2j + 1)$:

$$G(Sp_{2j+1}) = \frac{1}{2j+1} \sum_{\text{odd } k=1}^{2j} (-1)^k (2k+1)^{3/2} [U^k U^k]^0_0, \quad (3)$$

where $U^k_q \equiv \sum_{i=1}^{N} U^k_q(i)$ and U is the Racah unit tensor operator

$$\langle \Psi^{j'}_{m'} | U^k_q | \Psi^j_m \rangle = \delta_{jj'} (kjqm | j'm'). \quad (4)$$

The two seniority $v = 4$ states are degenerate with such an interaction and, since there is no seniority mixing, we can have arbitrary linear combinations of the 4^+ states. Only by using an interaction which removes the degeneracy and violates seniority, do we learn about the special state in Table I.

Fig. 2. (*Continued*)

3. Nuclear Moments

3.1. *Magnetic moments*

With n nucleons of one kind there are simple formulas for nuclear moments in a single j shell. For example all g factors should be the same. From this it follows that for states of the same J the magnetic moments should be the same. The magnetic moment of a free neutron (in units of nuclear magnetons) is $\mu_n = -1.913$ and that of a free proton is $\mu_p = +2.793$. In a single j shell of neutrons with $j = l + 1/2$ the magnetic moments are predicted to be the same as those of a free neutron-namely -1.913 ; for protons it is $(2.793 + l)(\mu_n)$. Here L is the orbital angular momentum.

The single particle magnetic moments, commonly called the Schmidt moments are given here:

1. for an odd proton:
   - $\mu = j - 1/2 + \mu_p$ for $j = l + 1/2$
   - $\mu = j/(j+1)[j + 3/2 - \mu_p]$ for $j = l - 1/2$
2. for an odd neutron:
   - $\mu = \mu_n$ for $j = l + 1/2$
   - $\mu = -j/(j+1)\mu_n$ for $j = l - 1/2$.

We can discuss these in a more physical manner. The magnetic moment of a free neutron (in units of nuclear magnetons) is $\mu_n = -1.913$ and that of a free proton is $\mu_p = +2.793$. In a single j shell of n neutrons with $j = l + 1/2$ the magnetic moments are predicted to be the same as those of a free neutron-namely -1.913; for n protons it is $(2.793 + l)$. Here l is the orbital angular momentum. For a $j = l - 1/2$ neutron we have a quantum effect so that the magnetic moment is only minus that of a free neutron in the large j limit. In general it is $-j/(j+1)$ that of a free neutron.

In the sixties Arima was already famous for the Arima-Horie theory for quenching magnetic moments which is basically first order perturbation theory.[5,6] I am showing a figure that I like (Fig. 3) because it has both theorists and experimentalists at Rutgers and Bonn testing out the Arima-Horie theory of quenching.

3.1.1. *Magnetic moments of isotones and isotopes*

In the single shell model, all factors (magnetic moment/angular momentum) are the same. Many people are under the impression that in Arima-Horie that is also true of the quenched g factors. But that is not the case. Rather they are predicted to lie on a straight line with a negative slope. Experimental confirmation of this is shown beautifully in the figure. The slope for states of even nuclei is different than for the ground states of odd nuclei. This is at it should be. One might think one could also apply this to the Calcium isotopes but there intrude states come in to spoil the picture, especially of the g factors of the 2^+ states.[7,8] For example, for the

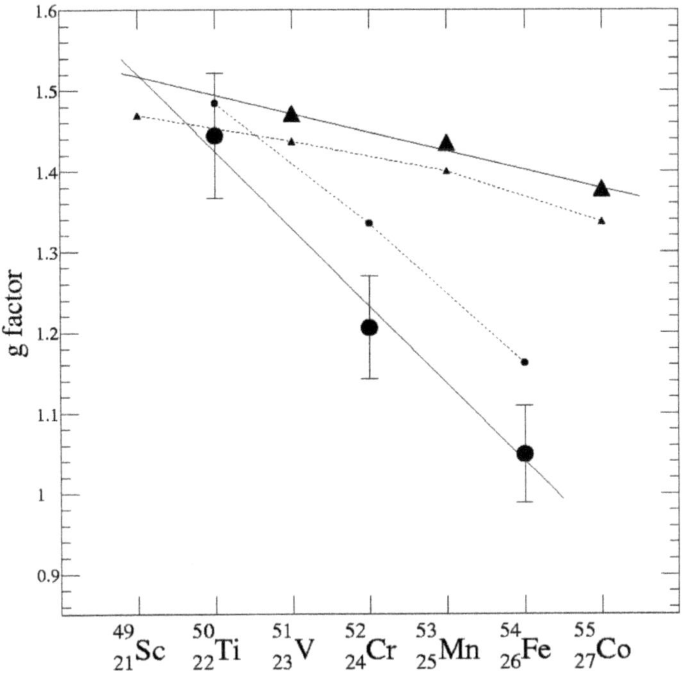

Fig. 3. Triangles: factors of ground states of odd isotopes. Circles: g factors of 2^+ states of even isotopes. Light lines: theory.

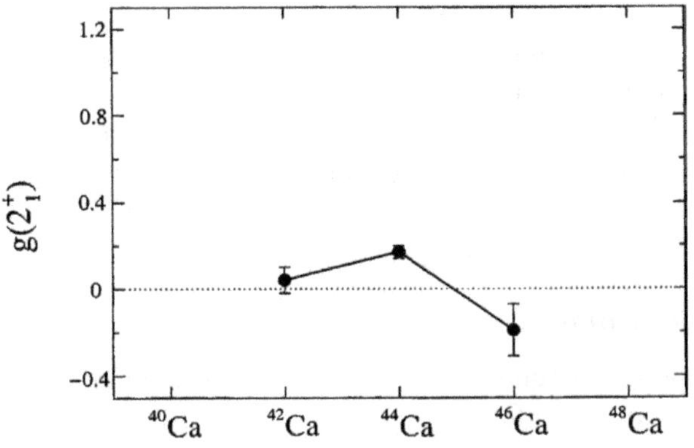

Fig. 4. Source: Phys. Rev C 68, 061302(R) (2003).

state of ^{44}Ca the g factor in the $f_{7/2}$ model is about -0.5, whilst that of a highly deformed intruder state is about $+0.5$. We explain the measured result of close to zero for this state by assuming a 50% admixture of the shell model and intruder state (see Fig. 4).[9]

3.1.2. *Second order perturbation theory*

My first foray into this subject of magnetic moments was actually not on the work above — i.e. first order perturbation theory. Rather with Gerry Brown and a Princeton student Harry Mavromatis we dealt with cases where first order perturbation theory was zero and we had to go to second order — much more complicated.[10–12] Work on second order was also done in Japan by Ichimura and Yazaki.[13] We deal with a closed major shell plus a nucleon e.g. ^{17}O, ^{17}F, ^{41}Ca, ^{41}Sc. These calculations involved a lot of complicated Feynman diagrams. The results were in the right direction to remove the discrepancy from theory and experiment.

The calculations were done first with the Kallio-Koltveit(KK) interaction[14] which does not contain a central interaction and then with the more realistic Hamada Johnson (HJ) interaction.[15] Note that with KK the corrections for mirror pairs are equal and opposite i.e. there is no isoscalar correction. However with HJ, which contains a tensor interaction we do get an isoscalar correction in second order perturbation theory. This result was proved by the authors. The above interactions were soon superseded by the Kuo-Brown matrix elements.[16]

3.1.3. *Isoscalar magnetic moments*

Isoscalar magnetic moments have been extensively discussed by S.S. Yeager, L. Zamick, Y.Y. Sharon and S.J.Q. Robinson[18] Isoscalar magnetic moments are much closer to the Schmidt values than the isovector ones. Nevertheless, there are small but systematic deviations. It was noted by Talmi[19] that "The experimental values of $\langle S\rangle$ seems to follow a simple rule. They are always smaller in absolute value than the values calculated in jj coupling." Arima, however, noted[20] that the smallness of the isoscalar deviation is due to the small isoscalar spin coupling (0.44) relative to that of the isovector coupling (2.353). If one divides the deviation by the lowest order result one can get a rather large ratio even in the isoscalar case, even up to 50%.

3.2. *Quadrupole Moments*

For quadrupole moments there is also an n dependent simple formula for ground states of odd nuclei in a single j shell

$$Q(n) = [(2j + 1 - 2n)/(2j - 1)]Q(sp)$$

Note that for a single hole $n = 2j$. The formula becomes $Q = -Q(sp)$. I.e. the quadruple moment of a hole is minus that of a particle. We can understand this another way. A nuclear moment is the expectation value of a moment operator in a state with $M = J$,

$$Q^2 = \langle\Psi_J^J|Q_0^2|\Psi_J^J\rangle.$$

Table 2. The calculated sums for the two mirror pairs and the results for the second order corrections to the magnetic moments of the eight nuclei.[17]

	1	2	3	4	5	6	7	8	9
	Nucleus	Ground state	Spin	Experimental moment in nucleon magnetons	Schmidt moment in nuclear magnetons	Exp. -Sch. (column 4-column 5)	2nd order correction calculated with H.-J.	2nd order correction calculated with K.-K.	Calculated (with H.J.) moments (columns 5,7)
1	${}^{8}_{7}N^{15}$	${}^{1}p^{-1}_{\frac{1}{2}}$ proton	$\frac{1}{2}$	−0.28	−0.26	−0.02	−0.10	−0.10	−0.36
2	${}^{7}_{8}O^{15}$	${}^{1}p^{-1}_{\frac{1}{2}}$ neutron	$\frac{1}{2}$	0.72	0.64	+0.08	+0.14	+0.10	+0.78
	Total	rows 1 and 2		0.44	0.38	+0.06	+0.04	0.00	+0.42
3	${}^{8}_{9}F^{17}$	${}^{1}d_{\frac{5}{2}}$ proton	$\frac{5}{2}$	4.72	4.79	−0.07	−0.30	−0.16	4.49
4	${}^{9}_{8}O^{17}$	${}^{1}d_{\frac{5}{2}}$ neutron	$\frac{5}{2}$	−1.89	−1.91	+0.02	+0.28	+0.16	−1.63
	Total	rows 3 and 4		2.83	2.88	−0.05	−0.02	0.00	2.86
5	${}^{20}_{19}K^{39}$	${}^{1}d^{-1}_{\frac{3}{2}}$ proton	$\frac{3}{2}$	0.39	0.12	+0.27	—	−0.26	—
6	${}^{19}_{20}Ca^{39}$	${}^{1}d^{-1}_{\frac{3}{2}}$ neutron	$\frac{3}{2}$	—	1.15	—	—	+0.26	—
7	${}^{20}_{21}Sc^{41}$	$\mathbf{I}f_{\frac{7}{2}}$ proton	$\frac{7}{2}$	—	5.79	—	—	−0.28	—
8	${}^{21}_{20}Ca^{41}$	$\mathbf{1}f_{\frac{1}{2}}$ neutron	$\frac{2}{2}$	−1.59	−1.91	0.32	—	0.28	—

Table 3. In Table 2 of the work of Yeager *et al.*[18] we show a table of empirical magnetic moments as well as shell model calculations and Schmidt estimates. The close agreement of the latter with experiment should be noted.

Nuclei	J	Measured g-factors	Large-scale shell model	Single-j model
^{2}H	1^+	0.857438228(9)	0.88[a]	0.88 ($s_{1/2}$)
^{6}Li	1^+	0.8220473(6)	0.87[b]	0.63 ($p_{3/2}$)
^{10}B	3^+	0.60021493(2)	0.61[b]	0.63 ($p_{3/2}$)
	1^+	0.63(12)	0.77[b]	0.63 ($p_{3/2}$)
^{14}N	1^+	0.40376100(6)	0.32[b]	0.37 ($p_{1/2}$)
^{18}F	3^+	0.59(4)	0.62[c]	0.58 ($d_{5/2}$)
	5^+	0.572(6)	0.58[c]	0.58 ($d_{5/2}$)
^{22}Na	3^+	0.582(1)	0.59[c]	0.58 ($d_{5/2}$)
	1^+	0.523(11)	0.52[c]	0.58 ($d_{5/2}$)
^{26}Al	5^+	0.561(8)	0.57[c]	0.58 ($d_{5/2}$)
^{38}K	3^+	0.457(2)	0.41[c]	0.42 ($d_{3/2}$)
^{46}V	3^+	0.55(1)	0.58[d]	0.55 ($f_{7/2}$)
^{58}Cu	1^+	0.52(8)	0.63[e]	0.63 ($p_{3/2}$)

[a]See table 1.
[b]With PJI interaction;[16] full p shell.
[c]With USDA interaction;[17] full sd shell.
[d]With GXPF1 interaction;[18] full fp shell.
[e]With GXPF1 interaction, up to 4 particles excited from $f_{7/2}$ orbit.

To create a hole nucleus in a state with $M = J$ we have to remove a nucleon from a closed shell with $M = -J$. The value of Q for a closed shell is zero-this is the the sum of Q for the hole nucleus and the nucleon removed. The value of Q^2 in a state with $M = J$ is the same as it is for $M = -J$ — namely $Q(sp)$. So we have $Q(\text{hole}) + Q(sp) = 0$ or $Q(\text{hole}) = -Q(sp)$. For magnets moments we have the opposite the value for $-J$ is minus that for $+J$. Thus we have 2 minus signs and $\mu(\text{hole}) = \mu(sp)$.

As a first example of the sturdiness of the shell model we look at the work of Ruiz *et al.*[21] on measurements and theoretical analysis quadrupole moments of odd A nuclei in the "f-p" region. They measured the quadruple moments of the $J = 7/2^-$ ground states of Calcium isotopes with $A = 43, 45, 47$ which have ground state spins $J = 7/2^-$. They did not do $A = 41$ but this case could be obtained from another source. They also obtained results for $A = 49, 51$ with $J = 3/2^-$ spins.

A starting point for $A = 41$ to 47 would be the $f_{7/2}$ shell while for $A = 49, 51$ it would be the $p_{3/2}$ shell. The theoretical calculations were performs with many interactions and different model spaces. The latter include complete pf space, (pf + $g_{9/2}$) and breaking the ^{40}Ca core by allowing 2p-2h admixtures. They use effective charges of 1.5 for the protons and 0.5 for the neutrons. In general the calculations are in excellent agreement with the measurements. We will not go into further details about the calculations except to say that they involve an enormous number of shell model configurations.

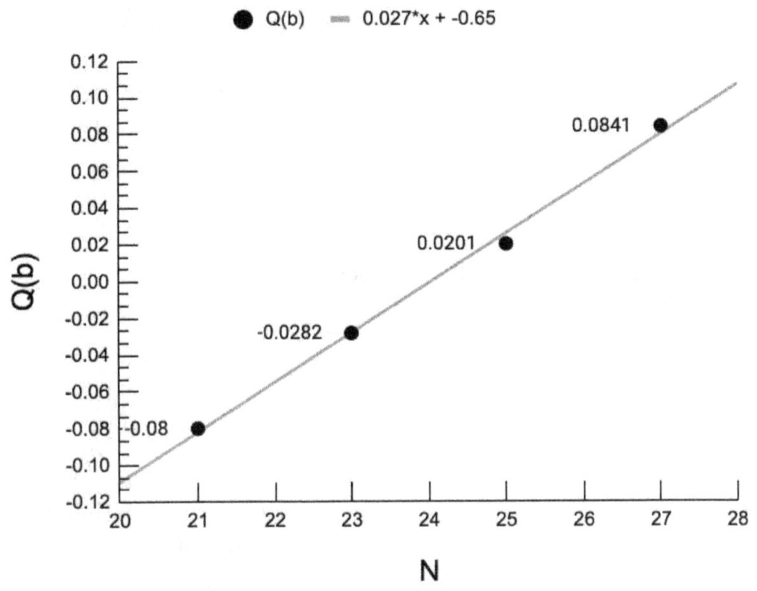

Fig. 5. The quadrupole moments vs. A from $A = 41$ to 47.

Rather in Fig. 5 we show the quadrupole moments vs. A and show the the remarkable result that the measured moments from $A = 41$ to $A = 47$ lie, to an excellent approximation on a straight line. As noted in the introduction this is exactly what a single j calculation predicts. To repeat $Q = (2j - 1 - 2n)/(2j - 1) * Q(s.p.)$ This simple result seems to survive the large shell attack. For $A = 51$ the measured quadrupole moment $Q = +0.04$ b. It is nearly equal and opposite of that for $A = 49$ $Q = -0.04$ b. This is the prediction of the simplest shell model in which $A = 49$ consist of a single $p_{3/2}$ neuron and $A = 51$ of a $p_{3/2}$ hole.

Before leaving this section we should mention that a purist might say that the real prediction of single j is that all the charge quadrupole moments are zero because the neutrons have no charge. We have to assign an effective charge to the neutrons, popular choice being $e_{\text{eff}} = 0.5$. But note that even the large space calculations including those of[21] require effective charges in order to get agreement with experiment. In first order perturbation theory the effective charge comes from $\Delta N = 2$ excitations. For example for ^{41}Ca excitations from 0p to 1p; from 0d to 0g, 1d, and 2s. As large as model spaces are in[21] and in nearly all other calculations these configurations are not present and one needs to insert effective charges.

We next consider an "empirical" limit for expectation of the isoscalar spin operator $\langle \sigma \rangle$. In the single particle model (i.e. Schmidt) the value for $j = L - 1/2$ is $-j/(j+1)$. While the value for$j = L + 1/2$ is one. For the most part the measured values lie between these 2 limits and this has been called an empirical rule. Occasionally some one comes up with an exception. In the work of Kramer *et al.*,[22] the magnetic moment of ^{21}Mg is measured, which when combined with the moment of ^{21}F yields an isoscalar magnetic moment and an expectation value of the spin operator.

They find a value of $\langle\sigma\rangle = 1.15(2)$. They call this an anomalous result. We pointed out however that the "empirical rule" is not a theoretical rule. In LS coupling the value of the spin operator is

$$\langle\sigma\rangle = [S(S+1) + J(J+1) - L(L+1)]/(J+1). \tag{1}$$

The smallest value is $-2SJ/(J+1)$. And the largest value is 2S. So we can get in principle get values of $\langle\sigma\rangle$ that are greater than one. The empirical rule is not a theoretical rule.[32]

3.2.1. *Empirical rule*

As shown in Fig. 6 from the work of P.W. Zhao *et al.*[23] we have the strange case where without pairing we get a complicated behavior of Q vs N but when pairing is included one gets a linear behavior as in the single j shell. However the linear curve has more entries than are present for an $h_{11/2}$ shell.

3.3. *Isotope Shifts*

In Fig. 7 we show measured values of isotope shifts in the Argon Isotopes by Blau *et al.*[24] (open circles). Also shown are spherical Hartree-Fock calculations in closed triangles, as well as a formula by Zamick[25] and by Talmi[26,27] which will soon be discussed.

Note that the data shows a lot of even-odd staggering but the HF calculations do not. The Zamick-Talmi calculations yield excellent fits to the data and have the even-odd scattering features well under control. In order to get the even-odd

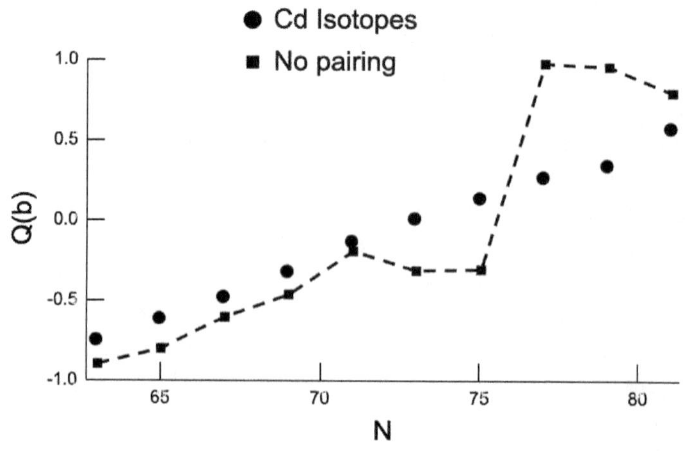

Fig. 6. Q vs. N.

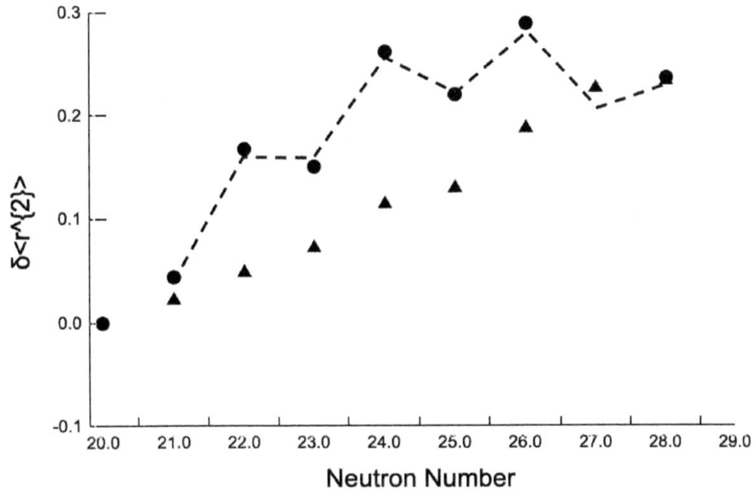

Fig. 7. Isotope shifts in the odd N Ar Isotopes. The filled circles correspond to the experiment of Blaum *et al.*[24] and the dashed line the Zamick formula.[25] The triangles correspond to spherical Hartee-Fock calculation[26,27]

staggering Zamick[25] introduced a 2 body effective radius operator in addition to the one body term.

We simply make the assumption that the effective charge radius operator has a two-body part as well as one body part

$$\delta r^2_{\text{eff}} = \sum_i O(i) + \sum_{i<j} V(i,j) \tag{2}$$

where the symbol V for the two-body part has been written to suggest the similarity with the two-body potential, since both are scalars.

The problem of evaluating this operator for n particles in the $j = f_{7/2}$ shell is exactly the same problem as calculating the binding energies of nuclei whose configuration consists of several nucleons in a single j shell. This problem has been solved and used with great success by the "Israeli group" including de-Shalit, Racah, Talmi, Thieberger, and Unna.[28] In analogy with their binding energy formula we get for the change in charge radius

$$\delta r^2(40+n) = nC + \frac{n(n-1)}{2}\alpha + \left[\frac{n}{2}\right]\beta, \tag{3}$$

where

$$\left[\frac{n}{2}\right] = \begin{cases} \dfrac{n}{2} & \text{for even } n \\[2ex] \dfrac{n-1}{2} & \text{for odd } n. \end{cases} \tag{4}$$

The parameter C comes from the one-body part and is equal to $\delta r^2(41)$, the difference in charge radius of ^{41}Ca and ^{40}Ca. The quantities α and β come from the two-body part

$$\alpha = -\frac{2(j+1)\bar{E}_2 - E_0}{2j+1}, \qquad \beta = \frac{2(j+1)(\bar{E}_2 - E_0)}{2j+1}, \tag{5}$$

where

$$E_0 = \langle j^2 J = 0 | V | j^2 J = 0 \rangle, \qquad \bar{E}_2 = \frac{\sum_{J \neq 0} (2J+1) \langle j^2 J | V | j^2 J \rangle}{\sum_{J \neq 0} (2J+1)}. \tag{6}$$

4. Redmond Modifications and Counting Pairs

Akito Arima and Yu-Min Zhao have written many papers concerning relations of states in the single j shell. One example is "Number of States for Nucleons in a

A recursion formula for cfp's due to Redmond [1] is presented in the books of de Shalit and Talmi [2] on p. 528, and Talmi [3] on p. 274. It can be written as follows

$$\begin{aligned}(n+1)&[j^n(\alpha_0 J_0)jJ|\}j^{n+1}[\alpha_0 J_0]J][j^n(\alpha_1 J_1)jJ|\}j^{n+1}[\alpha_0 J_0]J] \\ &= \delta_{\alpha_1\alpha_0}\delta_{J_1 J_0} + n(-1)^{J_0+J_1}\sqrt{(2J_0+1)(2J_1+1)}\sum_{\alpha_2 J_2}\begin{Bmatrix} J_2 & j & J_1 \\ J & j & J_0 \end{Bmatrix} \\ &\quad \times [j^{n-1}(\alpha_2 J_2)jJ_0|\}j^n\alpha_0 J_0][j^{n-1}(\alpha_2 J_2)jJ_1|\}j^n\alpha_1 J_1].\end{aligned} \tag{7}$$

In the above, square bracket designates the principal parent used to calculate the cfp. Actually, the principal parent sometimes looses its significance because in some cases more than one principal parent can yield the same cfp. In tables of cfp's, the principal parent is usually not listed. The quantities in parentheses $(\alpha_0 J_0)$ are listed. The cfp with $(\alpha_0 J_0)$ is the probability amplitude that a system of $(n+1)$ identical particles can be separated into a system of n particles with quantum numbers $(\alpha_0 J_0)$ and a single nucleon.

We here present the equivalent of the Redmond recursion relation, but for cfp's classified by the seniority quantum number v and for which there are no redundancies. Here is our formula

$$\begin{aligned}(n+1)&\sum_{v_s}[j^n(v_0 J_0)jI_s|\}j^{n+1}v_s I_s][j^n(v_1 J_1)jI_s|\}j^{n+1}v_s I_s] \\ &= \delta_{J_0 J_1}\delta_{v_0 v_1} + n(-1)^{J_0+J_1}\sqrt{(2J_0+1)(2J_1+1)}\sum_{v_2 J_2}\begin{Bmatrix} J_2 & j & J_1 \\ I_s & j & J_0 \end{Bmatrix} \\ &\quad \times [j^{n-1}(v_2 J_2)jJ_0|\}j^n v_0 J_0][j^{n-1}(v_2 J_2)jJ_1|\}j^n v_1 J_1].\end{aligned} \tag{19}$$

This differs from the Redmond formula inasmuch as there is now a sum on the left-hand side of the equation over v_s. Note that I_s is fixed. Basically, then, the sum is over all states that are present which have angular momentum I_s for the $(n+1)$-particle system.

Fig. 8. We quote from the paper.[30]

single j shell".[29] I was involved in this kind of business as well and was delighted by the generous references to my works by Arima and Zhao.

Let me give one example "New Relations for coefficients of fractional parentage", where we simplify a recursion relation due to Redmond.[30]

5. Closing Remarks

Already in 1967, when Akito Arima was at Rutgers he was a big name on the world scene.Indeed when Gerry Brown went from Princeton to Stony Brook he took Arima with him as well as Tom Kuo. But as they say, the best was yet to come. I am sure the summary of all his accomplishments will appear somewhere in this compendium so I won't mention them. Well maybe a couple — pseudo spin and the interacting boson model. And in service-president of the University of Tokyo and head of Ricken. Rather I would like to dwell on the fun time it was to be a nuclear physicist in New Jersey around 1967. Princeton and Rutgers had a joint seminar called Nuclear News run by Rubby Sherr. Amongst the faculty, post-docs, senior grad students and visitors at that time were the following:

- Princeton: Gerry Brown, Tom Kuo, Tony Green, Chun WA Wong, George Bertsch, Felix Wong, Alex Lande Yitzhak Sharon, Julian Noble, Harry Mavromatis;
- Rutgers: Joe Ginocchio, Aldo Covello, Giovanni Sartoris, George Ripka and oh yes me.

And for icing on the cake. Shiro Yoshida, Akito Arima and Koichi Yazaki. Arima's discussions were appreciated not only by the theorists buy also the experimentalists — Noemie Koller, George Temmer, Rubby Sherr and others. Rutgers had a major program of measuring magnetic moments of excited states. Those were great times and Akito was a major contributor to the fun we all had.

References

1. B. F. Bayman, J. D. McCullen, and Larry Zamick, *Phys. Rev. Lett.* **11**, 215 (1963) — Published 1 September 1963.
2. J. D. McCullen, B. F. Bayman, and Larry Zamick, *Phys. Rev.* **134**, B515 (1964) — Published 11 May 1964.
3. L. Zamick and E. Moya de Guerra, *Phys. Rev. C* **34**, 290 (1986) — Published 1 July 1986.
4. A. Escuderos and L. Zamick, Seniority conservation and seniority violation in the g9/2 shell, *Phys Rev C* **73**, 044302 (2006).
5. A. Arima and H. Horie, Configuration mixing and magnetic moments of nuclei, *Progress of Theoretical Physics*, 11,509 (1954); A. Arima and H. Horie, Configuration mixing and magnetic moments of odd nuclei, *Progress of Theoretical Physics*, 12,623 (1954).
6. H. Noya, A. Arima and H. Horie, Nuclear moments and configuration mixing, *Progress of Theoretical Physics Supplement* **8**, 33 (1958).

7. Evidence for 40 Ca core excitations from g factor and B(E2) measurements on the 2+ states of 42,44 Ca, S. Schielke, D. Hohn, K. H. Speidel, O. Kenn, J. Leske, N. Gemein, M. Offer, J. Gerber, P. Maier-Komor, O. Zell, F. Nowacki, Y. Y. Sharon, and L. Zamick, *Phys. Lett. B* **571**, Oct. 2003, 29–35.
8. Competing Core and Single Particle excitations in the 2+ State in 44Ca, M. J. Taylor, N. Benczer-Koller, G. Kumbartzki, T. J. Mertzimekis, S. J. Q. Robinson, Y. Y. Sharon, L. Zamick, A. E. Stuchbery, C. Hutter, C. W. Beausang, J. J. Ressler, and M. A. Caprio, *Phys. Lett.* **B559**, (2003) 187.
9. Core polarization in the light of new experimental g factors of fp shell, N=28, isotones, K. H. Speidel, R. Ernst, O. Kenn, J. Gerber, P. Maier-Komor, N. Benczer-Koller, G. Kumbartzki, L. Zamick, M. S. Fayache, and Y. Y. Sharon, *Physical Review C* **62**, September 2000, 031301.
10. First and Second Order Corrections to the Magnetic Moments of Nuclei Using Realistic Interactions, L. Zamick, H. A. Mavromatis, and G. E. Brown, *Nuclear Phys.* **80**, 545 (1966).
11. Magnetic Moments of Nuclei with Closed J-J Shells Plus One or Minus One Nucleon, L. Zamick and H. A. Mavromatis, *Nucl. Phys.* **A104**, 17 (1967).
12. H. A. Mavromatis and Larry Zamick, *Physics. Letters* **20**(2), 171 (1966).
13. M. Ichumira, K. Yazaki, *Nuclear Phys.*, **63** (1965), p. 401.
14. A. Kallio and K. Kolltveit, *Nuclear Phys.* **53** (1964) 87.
15. B. T. Hamada and I. D. Johnston, *Nuclear Phys.* **34** (1962) 382.
16. T. T. S. Kuo and G. E. Brown, *Nucl. Phys.* **85**, 40 (1966).
17. H. A. Mavromatis, L. Zamick. Magnetic moment corrections to second order in perturbation theory, *Phys. Rev.* **20**(2) (1966).
18. S. Yeager, L. Zamick, Y. Y. Sharon, S. J. Q. Robinson, *Europhys. Lett.* **88**, 52001, 2009.
19. I. Talmi, Hyperfine Excited States of Nuclei, *Gordon and Breach*, New York, 1971.
20. A. Arima, Hyperfine Interactions, volume 4, 1978.
21. R. F. Garcia Ruiz, *et al.*, *Phys. Rev. C* **91**, 041304(R) (2015).
22. J. Kramer *et al.*, *Phys. Lett. B* **678**, 465 (2009).
23. P. W. Zhao, S. Q. Zhang and J. Meng, *Physical Review* **C89**, 01130 (R) 2014.
24. K. Blaum, *et al.* Nuclear moments and charge radii of argon isotopes between the neutron-shell closures N = 20 and N = 28, *Nuclear Physics A* **799**, 30–45 (2008).
25. L. Zamick, *Annals of Physics*, **66**(2), 784–789 (1971).
26. I. Talmi, *Nucl. Phys.* **A423** (1984) 189.
27. A. Klein, B. A. Brown, U. Georg, M. Keim, P. Lievens, R. Neugart, M. Neuroth, R. E. Silverans, L. Vermeeren, ISOLDE Collaboration, *Nucl. Phys.* A 607 (1996) 1.
28. I. Talmi and R. Thieberger, *Phys. Rev.* **103** (1956), 718; S. Goldstein and I. Talmi, *Phys. Rev.* **105** (1957), 995; R. Thieberger and A. de-Shalit, *Phys. Rev.* **108** (1957), 378; I. Talmi and I. Unna, *Ann. Rev. Nucl. Sci.* **10** (1960), 353; I. Talmi, *Rev. Mod. Phys.* **34** (1962), 104.
29. Y.M. Zhao and A. Arima, *Phys. Rev C* **72**, 064333 (2005).
30. The Redmond Recursion Relation with Seniority. L. Zamick and A. Escuderos, *Annals of Physics* **321** (2006) 987–998.
31. P. J. Redmond, *Proc. R. Soc. London* **A222** (1954) 84.
32. L. Zamick, *Phys. Rev. C* **84**, 017302 (2011).

https://doi.org/10.1142/9789819813353_0021

Simple Relations of Atomic Masses

Y. M. Zhao

School of Physics and Astronomy, Shanghai Jiao Tong University, Shanghai 200240, China

This paper is a mini-review of the "local" mass formulas of neighboring nuclei performed in last decade. The results related to the Garvey-Kelson mass relations and their generalizations, the mass relations related to the proton-neutron interactions and Coulomb energies, as well as the parametrizations of symmetry energy and Wigner energy terms, are discussed in this context.

1. Introduction

It is a duty and meanwhile an honor for me to contribute to the Volume in memory of Arima Akito Sensei, with whom I have been collaborating for more than two decades. I am very grateful to Arima Sensei for his constant instructions, collaborations and encouragements. He was a charming, warm-hearted, ingenious, brave, innovative and far-sighted man, gifted in multi-dimensions.

Arima Sensei and I collaborated on four topics, including nuclear structure in the presence of random interactions,[1] nucleon-pair approximation of the nuclear shell model,[2] decomposition of the shell model configuration space,[3] and atomic mass formulas. We co-authored more than one hundred papers in peer-reviewed journals. This paper is a mini review on the fourth topic, local and simple relations of atomic masses.

Atomic mass is a sum of nuclear mass plus the masses of all electrons [a free electron mass equals 0.51099895000(15) MeV/c^2, where c is the speed of light in vacuum] in the atom, subtracted by the binding energy of electrons. As the atomic structure is well studied, there have been a number of approaches to evaluate the binding energy of electrons rather accurately. Thus the problem of atomic mass is reduced to the evaluation of nuclear mass. The mass of given nucleus at its ground state with Z protons and N neutrons, denoted by $M(Z,N)$, equals $(Zm_\mathrm{p}+Nm_\mathrm{n})c^2$[$m_\mathrm{p}$ and m_n is the mass of a free proton and a free neutron, equals 938.2720813(58) and 939.5654133(58) MeV, respectively] subtracted by the total binding energy [denoted by $B(Z,N)$] of all nucleons. Thus the nuclear mass involve of complicated interactions among all nucleons in the nucleus.

In the last three or four decades, the nucleosynthesis problem in astrophysics is in need of nuclear mass database (in particular some nuclear masses far from the

β-stability line) as one of key inputs, and furthermore, it has been also realized that nuclear mass database provides us with a convenient and very rich information of nuclear structure. Therefore to perform experimental measurements and to construct theories of nuclear masses have attracted more and more attention.

There are model-dependently 7000–9000 bound systems of nucleons, however, only about 2000 nuclei have experimental values of masses so far; different versions of the atomic mass evaluations have been compiled every four or five years, and the latest versions are the AME 2012,[4] AME2016,[5] and AME2020[6] databases. Clearly, both evaluation and theoretical predictions of nuclear masses are necessary. One of comments of this brief review is that, understanding the nuclear masses is as difficult as understanding other quantities of atomic nuclei, or any other complicated many-body systems. There are various contributions from different origins, each of which is challenging to pin down. For nuclei not very far from the known borders, extrapolations based on local mass relations have been found to be quite reliable, and for those very far from the known borders, in particular for those in the neutron-rich side, one has to resort to calculated results, although those predicted results [e.g., Refs. 7–9] are largely model dependent. Besides local mass relations and theoretical models, there are simple and practical formulas of nucleon-separation energy, which work either for large regions[10–13] or in given shells,[14–16] on which the readers should go to literature.

This paper is organized as follows. In Sec. 2 we discuss mass relations and related systematics related to proton-neutron interactions; in Sec. 3 we discuss the Coulomb energy and its relation with the nuclear mass; in Sec. 4, we comment on the coefficients of the symmetry energy and Wigner energy; in Sec. 5 say a few words in memory of Arima Sensei; and finally in Sec. 6 we we summarize this paper.

2. Local Mass Relations and Proton-Neutron Interactions

For a long time since 1960's, the best known "local" mass relations of atomic nuclei are two Garvey-Kelson relations,[17] one is called longitudinal and the other called transversal, abbreviated as the GKL and the GKT, given as follows, respectively.

$$\begin{aligned} D_{\rm L}(N,Z) &= B(N+1,Z+1) + B(N-1,Z) + B(N,Z-1) \\ &\quad - B(N+1,Z) - B(N,Z+1) - B(N-1,Z-1) \approx 0, \end{aligned} \tag{1}$$

$$\begin{aligned} D_{\rm T}(N,Z) &= B(N+1,Z-1) + B(N,Z+1) + B(N-1,Z) \\ &\quad - B(N+1,Z) - B(N,Z-1) - B(N-1,Z+1) \approx 0, \end{aligned} \tag{2}$$

A simple explanation of the above relations is as follows. Assuming that the interactions between nucleons are of one-body and two-body, and the interactions among nucleons of given nucleus are essentially constants in comparison with its neighboring nuclei. With these assumptions, all interactions are canceled out in Eqs. (1)–(2). The GKL and GKT have been found to be very accurate for realistic nuclei, with the RMSD values around 170 keV for mass number $A \geq 100$.

There have been many discussions and generalizations of these formulas.[18–23] These formulas can be written in a unified form, called $S^{i,j}_{N',Z'}(N,Z)=0$, where

$$\begin{aligned} S^{i,j}_{N',Z'}(N,Z) &= S_{N',Z'}(N,Z)+S_{N',Z'}(N-i,Z-j) \\ &\quad - S_{N',Z'}(N-i,Z)-S_{N',Z'}(N,Z-j) \\ &= \delta V^{i,j}_{0,0}-\delta V^{(i,j)}_{-N',-Z'}. \end{aligned}$$

Here $S_{N',Z'}(N,Z)$ is N'-neutron-Z'-proton separation energy, i.e., $S_{N',Z'}(N,Z)=B(N,Z)-B(N-N',Z-Z')$. It is a simple generalization of the usual one or two nucleon separation energies, for example, the two-neutron separation energy $S_{2,0}(N,Z)=S_{2n}(N,Z)$. According to numerical experiments in Ref. 23, the RMSD value of the relation

$$\begin{aligned} S^{(1,1)}_{2,0}(N,Z) &= M(N-3,Z-1)-M(N-2,Z-1) \\ &\quad - M(N-1,Z-1)+M(N,Z-1)-M(N-3,Z) \\ &\quad + M(N-2,Z)+M(N-1,Z)-M(N,Z)\cong 0 \end{aligned} \tag{3}$$

is the smallest.

The two GK relations, GKL and GKT, can be written in terms of proton-neutron interactions. The empirical proton-neutron interaction between the last i protons and j neutrons is defined by[24, 25]

$$\begin{aligned} \delta V_{i\mathrm{p}-j\mathrm{n}}(N,Z) &= B(N,Z)+B(N-j,Z-i) \\ &\quad - B(N-j,Z)-B(N,Z-i). \end{aligned} \tag{4}$$

We note that $\delta V_{2\mathrm{p}-1\mathrm{n}}$ and $\delta V_{1\mathrm{p}-2\mathrm{n}}$ can be decomposed in terms of $\delta V_{1\mathrm{p}-1\mathrm{n}}$, the residual interaction between the last proton and the last neutron. By using $\delta V_{1\mathrm{p}-1\mathrm{n}}$, the GKL and GKT is rewritten as below.

$$\begin{aligned} D_{\mathrm{L}}(N,Z) &= \delta V_{1\mathrm{p}-1\mathrm{n}}(N,Z)-\delta V_{1\mathrm{p}-1\mathrm{n}}(N+1,Z+1), \\ D_{\mathrm{T}}(N,Z) &= \delta V_{1\mathrm{p}-1\mathrm{n}}(N,Z+1)-\delta V_{1\mathrm{p}-1\mathrm{n}}(N+1,Z). \end{aligned}$$

Figure 1 plots the results with $(i,j)=(1,1),(1,2),(2,1)$. For nuclei with the same mass number A, we calculate the average value of $\delta V_{i\mathrm{p}-j\mathrm{n}}(A)$, denoted by $\overline{\delta V_{i\mathrm{p}-j\mathrm{n}}}(A)$. Here we did not include the results in which nuclei with $Z=N$ are involved, because $\delta V_{1\mathrm{p}-1\mathrm{n}}$ for nuclei with $Z=N$ is exceptionally large due to the Wigner energy, which we shall come to later. In Fig. 1(a), the results are classified into two cases, even number of A and odd number of A. Based on a special form of pairing interaction, it was shown in Ref. 26 that

$$\overline{\delta V_{1\mathrm{p}-1\mathrm{n}}}(\text{even } A)-\overline{\delta V_{1\mathrm{p}-1\mathrm{n}}}(\text{odd } A)\simeq a_{\text{pairing}} I A^{-1/3},$$

where $I=|N-Z|/A$. This leads to larger values of $-\delta V_{1\mathrm{p}-1\mathrm{n}}$ for nuclei with even A (open circles in black) than those with odd A (open circles in blue). According

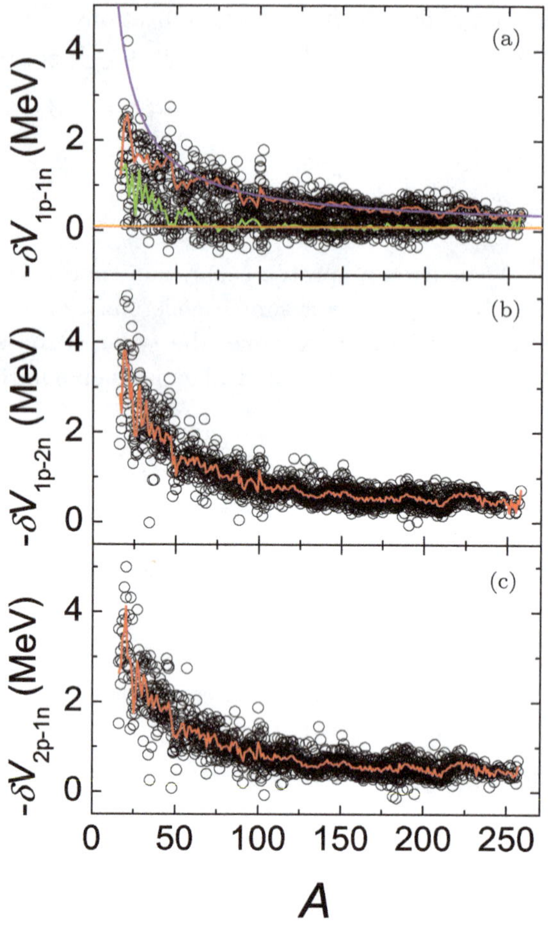

Fig. 1. (Color online) Proton-neutron interactions for nuclei with mass number $A \geqslant 16$. A minus sign is added to δV_{ip-jn} in this figure so that most results are positive. In panel (a) $-\delta V_{1p-1n}$ are discriminated nuclei with even A (black circles) from those with odd-A (blue circles), and the curve in red (green) is plotted by $\overline{\delta V_{ip-jn}}(A)$ for nuclei with A even (odd). The straight line in orange is plotted with $-\delta V_{1p-1n}(N, Z) = 74\,\text{keV}$, and the curve in violet-blue is plotted with $-\delta V_{1p-1n}(N, Z) = 74 + 69861/A\,\text{keV}$. In panels (b) and (c), circles in black correspond to $-\delta V_{1p-2n}$ and $-\delta V_{2p-1n}$, respectively; and curves in red are plotted by using $\overline{\delta V_{ip-jn}}(A)$. The results with $Z = N$, $-\delta V_{1p-2n}$ for nuclei with $Z = N$ or $Z = N - 1$, or $-\delta V_{2p-1n}$ for nuclei nuclei with $Z = N$ or $Z = N + 1$ are excluded.

to Fig. 1(a), the $\overline{\delta V_{1p-1n}}(A)$ (see the solid curve in green) for odd-A nuclei with $A \geq 100$ is almost a constant (close to zero, see the solid line in orange), i.e.,

$$B(N,Z) + B(N-1,Z-1) - B(N-1,Z) - B(N,Z-1) \simeq \overline{\delta V_{1p-1n}}(A) \simeq -74\,\text{keV}, \tag{5}$$

where $N + Z = A$ is odd and $A \geq 100$. Numerical experiments show that the RMSD of the above relation is only 132 keV for $A \geq 100$; this relation is more precise than

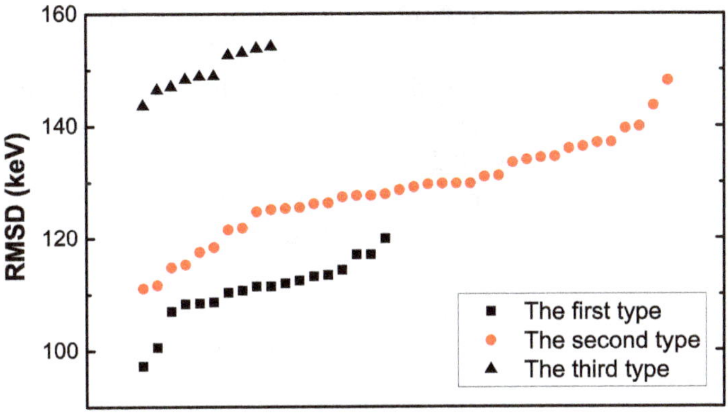

Fig. 2. (Color online) The RMSD of the GKs for $n = 2$. There are totally sixty-six combinations, among which there are eighteen cases belong to the first case, thirty-six belong to the second case, and twelve belong to the third case. One sees that the first case gives the smallest RMSD, and the third gives the largest RMSD, as expected. The RMSD in the second case is intermediate between the first and the third cases. See the text for details.

the Garvey-Kelson mass relations, the RMSD of which is 170 keV for the same set of the AME database. We empirically have

$$\overline{\delta V_{1p-1n}}(\text{odd}-A) - \overline{\delta V_{1p-1n}}(\text{even}-A) \simeq \frac{69861}{A}\text{keV},$$

as shown by using the solid curve in violet-blue in Fig. 1(a). From this relation, one obtains

$$\begin{aligned} &B(N,Z) + B(N-1,Z-1) - B(N-1,Z) - B(N,Z-1) \\ &\simeq \overline{\delta V_{1p-1n}}(A) \simeq -74 - \frac{69861}{A}\text{keV}, \end{aligned} \tag{6}$$

where $N + Z = A$ is even and $A \geq 100$. The RMSD of this relation is 168 keV, very close to that of the Garvey-Kelson mass relations (170 keV). As these simple relations are more accurate than the famous Garvey-Kelson mass relations, they were applied to predict unknown nuclear masses by extrapolations.[27]

Scrutinizing more carefully, one sees that the values of $\delta V_{1p-1n}(N,Z)$ for odd-even and even-odd (odd-A) nuclei, or for even-even and odd-odd (even-A) nuclei deviate from zero,[28] although the differences between them are very small on average. This suggests that the deviations D of the GK relations from the experimental database are not statistically random with respect to zero; if the values of D were zero on average and were randomly distributed, one would expect that an average of predictions based on the multiple applications of the GK relations yield more accurate predictions of nuclear masses, and numerical experiments seem to confirm this picture.[20] Yet, the fact that statistical but systematic deviations from zero for the GKL and GKT actually leads to correlations of various Garvey-Kelson relations for given nucleus. For example, there are in total twelve different Garvey-Kelson relations for given nucleus, if all neighboring nuclear masses are available; thus there are totally sixty-six combinations if one predict the mass by two of them. We classify these sixty-six approaches into three cases: in the first, one of the relations yield a deviation $D = \delta V(\text{ee}) - \delta V(\text{oo})$, and the other leads to $D = \delta V(\text{oo}) - \delta V(\text{ee})$, or one yields to $\delta V(\text{oe}) - \delta V(\text{eo})$, and the other yields to $D = \delta V(\text{eo}) - \delta V(\text{oe})$; in the third case, both GK relations give $D = \delta V(\text{ee}) - \delta V(\text{oo})$ or $D = \delta V(\text{eo}) - \delta V(\text{oe})$. The second case is simply between the first and third cases, namely, one of the two R_i corresponds to a deviation between δV of an even-even nucleus and that of an odd-odd nucleus, while the other corresponds to a deviation between δV of an even-odd nucleus and that of an odd-even nucleus. Apparently, the cancelation of δV values is the best done in the first case, thus one expects the smallest RMSD. In Ref. 29 such correlations were discerned, with the recommendation of the optimal combinations.

Clearly, the proton-neutron interaction is the strongest when valence protons and neutrons are in the same orbit, and is small if valence protons and valence neutrons are in very different orbits. Therefore one expect small values of $\delta V_{1\text{p}-1\text{n}}(Z, N)$ in the latter case, and the best example corresponds to the major shells $82 < Z, 104 < N < 126$. Indeed, one finds that the $\delta V_{1\text{p}-1\text{n}}(Z, N)$ is relatively small and compact. Minor modifications on the $\delta V_{1\text{p}-_{1\text{n}}}(Z, N)$ yields nice mass relations. With statistical correlations into consideration, one obtain the RMSD values below 40 keV for mass relations [involved of $\delta V_{1\text{p}-1\text{n}}(Z, N)$] of this region.[30]

For heavy nuclei, α decay energy [denoted by Q_α] is an important source to extract the binding energy of nucleus. The relation between Q_α and nuclear binding energies is as below.

$$\begin{aligned} Q(N, Z) &= -B(N, Z) + B(N-2, Z-2) + B({}^4_2\text{He}) \\ &= M(N, Z) - M(N-2, Z-2) - M({}^4_2\text{He}), \end{aligned}$$

where M is the atomic mass. From Eq. (1), one sees that the GKL relation is the difference of $\delta V_{1\text{n}-1\text{p}}(N, Z)$ and $\delta V_{1\text{n}-1\text{p}}(N+1, Z+1)$, of two neighboring nuclei with fully different parity of N and Z. Suppose that both N and Z are even. $D_{\text{L}}(N, Z)$ equals $\delta V_{1\text{n}-1\text{p}}$ of an even-even nucleus subtracted by that of its odd-odd neighbor with $N+1$ neutrons and $Z+1$ protons, and $D_{\text{L}}(N-1, Z-1)$ equals $\delta V_{1\text{n}-_{1\text{p}}}$ of an odd-odd nucleus subtracted by that of its even-even neighbor with N neutrons and Z protons. Therefore, one arrives at statistical correlation

$$\overline{D}_{\text{L}}(\text{ee}) \approx -\overline{D}_{\text{L}}(\text{oo}), \quad \overline{D}_{\text{L}}(\text{eo}) \approx -\overline{D}_{\text{L}}(\text{oe}).$$

Here we use abbreviations "ee" for even-even, "oo" for odd-odd, "eo" for even-odd, and "oe" for odd-even.

We rewrite the GKL by using the definition of Q value defined in Eq. (7) for $D_{\text{L}}(N, Z)$ and $D_{\text{L}}(N-1, Z-1)$ as follows.

$$\begin{aligned} D_{\text{L}}(N, Z) = B(N-1, Z) &+ B(N, Z-1) - B(N+1, Z) \\ &- B(N, Z+1) - Q(N+1, Z+1) + B({}^4_2\text{He}), \end{aligned} \tag{7}$$

$$D_{\rm L}(N-1,Z-1) = B(N-2,Z-1) + B(N-1,Z-2) - B(N,Z-1) - B(N-1,Z) - Q(N,Z) + B({}^4_2{\rm He}), \quad (8)$$

$D_{\rm L}(N,Z) + D_{\rm L}(N-1,Z-1) \simeq 0$ immediately leads to the following relation

$$Q(N+1,Z) + Q(N,Z+1) - Q(N,Z) - Q(N+1,Z+1) \approx 0. \quad (9)$$

This relation was found to be accurate in predicting the Q values of α decays[31] of heavy nuclei.

Usually, the integrated proton-neutron interactions V_{NP} between valence nucleons are rather large (e.g., a few dozens of MeV), while the RMS-D values of extrapolated masses are well below 1 MeV, it is reasonable to apply such predicted masses as surrogates of experimental data in predicting the V_{NP}. It was shown that such extracted V_{NP} are in linear of N_pN_n, but the ratio exhibits particle-hole discrimination which is a reflection of differences of average neutron-proton interaction strength for different regions in one major shell, as discussed in reference.[32] These differences are consistent with asymmetries of $E_{2_1^+}$ results versus proton number or neutron number, for the particle-particle, particle-hole, and hole-hole regions. The collective observables of even-even nuclei such as $E+$ evolve individually in the $N_{\rm p}$ $N_{\rm n}$ scheme for particle-particle, particle-hole, and hole-hole regions, were found to follow *one unified and compact* trajectory[33] in terms of the values of $V_{\rm NP}$. We note that this interesting correlation is robust of theoretical mass models in the market.

3. Parametrizations of Symmetry Energy and Wigner Energy

The systematics of $\delta V_{i{\rm p}-j{\rm n}}$ is not only useful in constructing the mass relations, but also helps in refining the important physical parameters. For instance, we come to the symmetry energy coefficient in the Bethe-Weizsäcker mass formula,

$$B(N,Z) = a_vA + a_sA^{2/3} + a_{\rm Coul}\frac{Z^2}{A^{1/3}}[1 - Z^{-2/3}] + a_{\rm sym}I^2A + a_{\rm pair}A^{-1/3}\delta_{np} + E_W. \quad (10)$$

Here the coefficients a_v, a_s, $a_{\rm Coul}$ and $a_{\rm pair}$ correspond to the volume, surface, Coulomb, and the pairing energy terms, respectively. $a_{\rm sym}$ corresponds to the symmetry energy term. $E_{\rm W}$ is the Wigner term, assuming here that $E_{\rm W} = C_0I^2A(\frac{2-|I|}{2+|I|A})$.

We now define our "experimental" symmetry energies $e(Z,A) = B(N,Z) - [a_vA + a_sA^{2/3} + a_{\rm Coul}\frac{Z^2}{A^{1/3}}[1 - Z^{-2/3}] + a_{\rm pair}A^{-1/3}\delta_{np} + E_{\rm W}]$, where the parameters taken from Ref. 34 Similar to the definition of $\delta V_{i{\rm p}-j{\rm n}}$, we define

$$R_{i{\rm p}-j{\rm n}}(Z,A) = e(Z,A) + e(Z-i,A-i-j) - e(Z,A-j) - e(Z-i,A-i).$$

The resultant values of $R_{i{\rm p}-j{\rm n}}(Z,A)/(ij)$ are plotted in Fig. 3, for $(i,j) = (1,2)$, (2,1), (2,2), (2,3) and (3,2). One sees that the correlation between $R_{i{\rm p}-j{\rm n}}(Z,A)$ and A in Fig. 3 is very compact. Very similar to the case of $\delta V_{i{\rm p}-j{\rm n}}(Z,A)$ with even

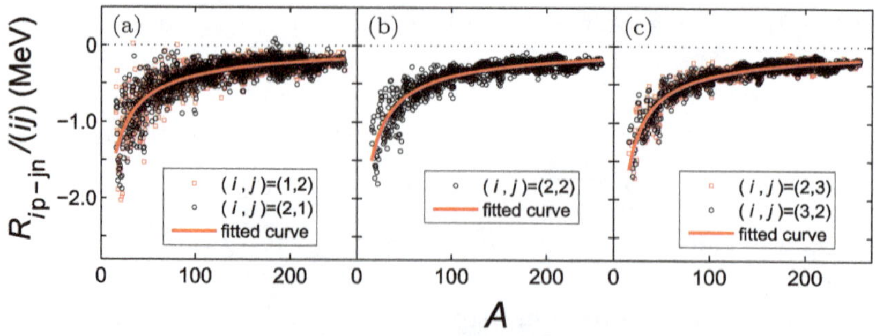

Fig. 3. (Color online) $R_{ip-jn}/(ij)$ (MeV) versus mass number A. R_{ip-jn} is given by double difference of "experimental" symmetry energies of four neighboring nuclei. Refer to the text for details. Panel (a): $(i,j) = (1,2)$ and (2,1); (b): $(i,j) = (2,2)$; and (c): $(i,j) = (2,3)$ and (3,2). The solid lines in red color are used to guide eyes. Equation (11) is used to describe $R_{ip-jn}/(ij)$.

values of ij, no noticeable difference is seen between even-odd and odd-even nuclei, or between even-even and odd-odd nuclei, for $R_{ip-jn}(Z,A)$ with even values of ij.

Assuming that $e(Z,A) = a_{\rm sym}I^2A = (c_{\rm sym}^{(V)} - c_{\rm sym}^{(S)}A^{-1/3})I^2A$, one obtains

$$\begin{aligned}
R_{ip-jn}(Z,A) = c_{\rm sym}^{(V)} \Bigg\{ & -\frac{j^2}{A-j} + \frac{(i-j)^2}{A-i-j} - \frac{i^2}{A-i} \\
& + \left[\frac{-1}{A-j} + \frac{1}{A-i-j} + \frac{1}{A} - \frac{1}{A-i}\right] I^2A^2 \\
& + \left[\frac{2i-2j}{A-i-j} - \frac{2i}{A-i} + \frac{2j}{A-j}\right] IA \Bigg\} \\
& + c_{\rm sym}^{(S)} \Bigg\{ \left[\frac{1}{(A-j)^{4/3}} - \frac{1}{(A-i-j)^{4/3}} - \frac{1}{A^{4/3}} + \frac{1}{(A-i)^{4/3}}\right] I^2A^2 \\
& + \left[-\frac{2i-2j}{(A-i-j)^{4/3}} + \frac{2i}{(A-i)^{4/3}} - \frac{2j}{(A-j)^{4/3}}\right] IA \\
& + \frac{j^2}{(A-j)^{4/3}} - \frac{(i-j)^2}{(A-i-j)^{4/3}} + \frac{i^2}{(A-i)^{4/3}} \Bigg\}. \qquad (11)
\end{aligned}$$

An optimization of the above R_{ip-jn} yields[35]

$$C_{\rm sym}^{(V)} = 31.26 \pm 0.32\,\text{MeV}, \qquad (12)$$

$$C_{\rm sym}^{(S)} = 52.15 \pm 1.12\,\text{MeV} (\kappa_{S/V} = 1.67 \pm 0.05). \qquad (13)$$

The uncertainties of such optimal values are sizably smaller than previous parametrizations in the market. Therefore one might be curious why this simple practice yields higher accuracy of $c_{\rm sym}^{(V)}$ and $c_{\rm sym}^{(S)}$. The main advantage of this procedure is that other contributions than the symmetry energy in the binding energy can be well canceled out in double differences for neighboring nuclei; For example, the consideration of the shell correction [which is difficult to evaluate precisely, meanwhile, is an important term in the binding energy] on $R_{ip-jn}(Z,A)$ leads to very minor modifications (less than 2%) on the parameters of $c_{\rm sym}^{(V)}$ and $c_{\rm sym}^{(S)}$.

The same idea was adopted to study the possible I^4 dependence of the symmetry energy based on a few popular mass models. There are a number of studies along this line: in Ref. 36 it was suggested based on the AME2012 database that the magnitude of I^4-volume coefficient is sizable, while the I^4-surface term seems unfavorable in binding energy. In Refs. 37, 38 it was shown that the coefficients of the I^4 term in finite nuclei extracted from different mass models correlate strongly with the corresponding RMSD from the experimental binding energies and also with the number of bound nuclei in the nuclear landscape predicted by theoretical models, and the resultant I^4-volume coefficient is in good agreement with those of Ref. 36, i.e., sizable magnitude of I^4-volume coefficient (~8–10 MeV). Clearly, both the possible existence of the I^4 term and a more accurate parametrization of I^2 term in symmetry energy are very interesting.

An unexpected result in this study is as follows. The R_{ip-jn} values for odd values of i and j shows an odd-even feature, as shown in Fig. 4. In Ref. 26, the odd-even difference of δV_{1p-1n}, δV_{1p-3n} and δV_{3p-1n} was explained to be dominantly originated from the pairing interaction. In $R_{ip-jn}(Z,A)$ the contribution from pairing interaction is subtracted with the updated "best" parameters. Yet, one sees in Fig. 4 for $(i,j) = (1,1)$ that the odd-even difference of $R_{ip-jn}(Z,A)$ is striking for $A \leq 150$. Namely, the magnitude of $R_{ip-jn}(Z,A)$ with even A are systematically larger than

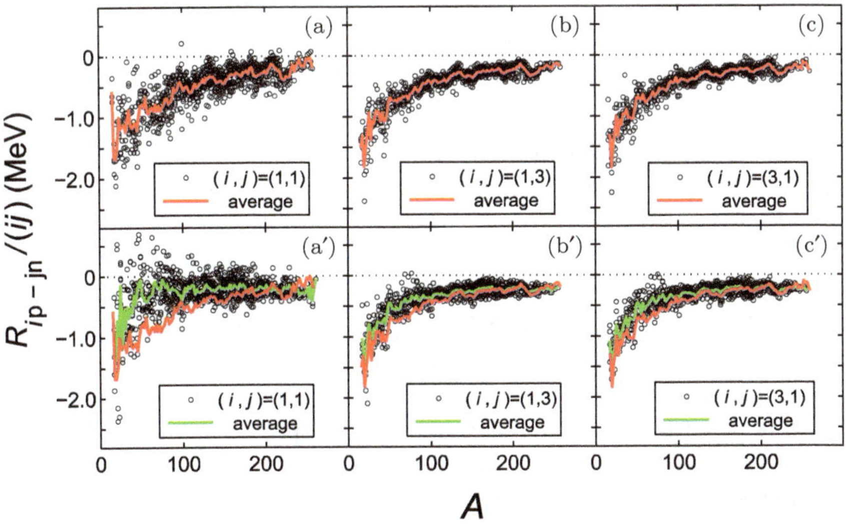

Fig. 4. (Color online) $R_{ip-jn}/(ij)$ (MeV) versus mass number A for $(i,j) = (1,1)$, (1,3) and (3,1). The upper three panels correspond to nuclei with even A and the lower with odd A. The lines in red color correspond to the average values of $R_{ip-jn}/(ij)$ (MeV) with even A(they are also plotted in the lower panels for comparison), and those in green color correspond to those with odd A. The feature are independent of the form of the Wigner energy adopted.

that of $R_{ip-jn}(Z, A)$ with odd A, when $A \leq 150$. Because there is no odd-even effect in the symmetry energy, this odd-even difference must be originated from improper approximation made in extracting the "experimental" symmetry energies. In other words, this observation demonstrates the disadvantage of these mass models for the pairing correlation, and should be considered in further refinement of the mass formulas. Indeed, in Refs. 39, 40 the odd-even staggering of deviations for theoretical δV_{1n-1p} values in the Duflo-Zuker model and the improved Weizsäcker-Skyrme model with respect to those extracted from experimental data was found to be well represented by an isospin-dependent term. The consideration of this odd-even staggering improves our description of binding energies and one-neutron separation energies in both the Duflo-Zuker model and the improved Weizsäcker-Skyrme model.

The parametrization of the Wigner energy is another interesting application of local mass relations. Due to the Wigner effect, these GK relations are restricted to cases in which all nuclei involved in the mass relations have $N \neq Z$ and locate on the same side with respect to the $N = Z$ line. Assuming that deviations of the local mass relations from experimental data along the N = Z line are originated solely from the Wigner energy, one may extract the Wigner energy from these anomalous deviations.

The Wigner energy is usually rewritten as follows.

$$B_{\rm w}(N, Z) = -W(A)|N - Z| - d(A)\delta_{N,Z}\pi_{np}. \tag{14}$$

Here π_{np} equals to 1 for odd-odd nuclei and vanishes for other (N, Z) parities. For short, the first term of Eq. (14) is called "W term", and the second term "d term". We note that the binding-energy difference between the lowest $T = 0$ and $T = 1$ states in odd-odd $N = Z$ nuclei was given by the competition between the symmetry energy and the isovector neutron-proton pairing energy.

The strategy was applied in Ref. 41 to extract the values of $B_{\rm w}(N, Z)$, and $d(A)$ of odd-odd $N = Z$ nuclei. In Fig. 5 we summarize the values of $W(A)$ with even A and $d(A)$ of odd-odd $N = Z$ nuclei. In Ref. 42 it was expected from the SU(4) symmetry that the values of W and d are positive and in similar magnitudes. However, here the values of $d(A)$ deviate from $W(A)$ for $A > 30$ and becomes negative for $A > 40$. This "anomaly" is accompanied by the evolution of ground state isospin from $T = 0$ to $T = 1$ for odd-odd $N = Z$nuclei, as pointed out in Ref. 43 by Satula and collaborators. The ground states of the odd-odd self-conjugate nuclei with $A > 40$ (also the^{34}Cl nucleus) have isospin $T = 1$ rather than $T = 0$ (except for ^{58}Cu). We therefore extract the value of d in these anomalous cases [denoted now by using $d_{T=0}(A)$] by the lowest $T = 0$ state energies compiled in Ref. 44, instead of the ground state energy with $T = 1$ in the odd-odd $N = Z$ nuclei. One sees the values of $d_{T=0}(A)$ are all positive. As mass number A increases, W and $d_{T=0}$ decrease with fluctuations. Assuming that

$$W(A) = a_W/A, \quad d_{T=0}(A) = a_d/A.$$

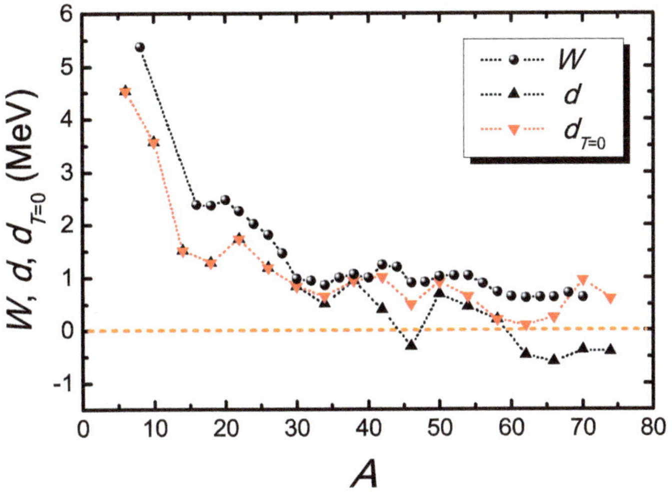

Fig. 5. (Color online) W (even-mass), d and $d_{T=0}$ (odd-odd $N = Z$) versus mass number A. The values of d extracted by using binding energies of the ground states deviate substantially from W for $A > 30$ and becomes negative for $A > 40$, while $d_{T=0}$ extracted by using binding energies of the lowest $T = 0$ states (not the ground states for ^{34}Cl and nuclei with $A > 40$ except for ^{58}Cu) are in similar magnitudes with W (with fluctuations).

A χ^2 fitting procedure with the AME2013 database yielded that

$$a_{\rm W} = 42.7 \pm 1.2\,\text{MeV}, \tag{15}$$

$$a_{\rm d} = 28.7 \pm 1.5\,\text{MeV}. \tag{16}$$

These values are close to those given in Ref. 45 ($a_{\rm W} = 43 \pm 5$ MeV and $d(A)/W(A) = 0.56 \pm 0.27$, and here $a_{\rm W} \simeq 42.7$ MeV and $d(A)/W(A) = 0.67$). In addition, $a_{\rm W}$ here is also very close to that given in Ref. 46 ($a_{\rm W} = 42.5 \pm 1.2$ MeV), which was extracted by using the Q values of double-β decay through the nuclear chart.

Adopting parameters for Eq. (11) and parameters of the Wigner energy coefficients of Eqs.(15)–(16), a χ^2 fitting for $R_{\rm 2p-2n}$, $R_{\rm 2p-3n}$ and $R_{\rm 3p-2n}$ for $10 \le A \le 80$ yields

$$C_{\rm sym}^{(V)} = 32.367 \pm 0.317\,\text{MeV}, \quad \kappa_{S/V} = 1.6743 \pm 0.0303. \tag{17}$$

They are well consistent with those obtained in Eqs. (12)–(13).

4. Mass Relations of Mirror Nuclei

It has been long known that the isospin is a good quantum number for nuclear interaction. This means that the wave functions of ground states and excited states for two mirror nuclei would be the same if the Coulomb interaction did not exist; or in other words, the mass difference of ground states for two mirror nuclei is given by the Coulomb energy [denoted by E_c] and the neutron-proton mass difference [denoted by $\Delta E_{\rm np}^{(0)}$].

Coulomb energy term is usually assumed to take the value of a uniformly charged sphere with Z protons, i.e.,

$$E_{\rm c} = \frac{3}{5}\cdot\frac{1}{4\pi\epsilon_0}\cdot\frac{Z^2e^2}{A^{\frac{1}{3}}r_0} = a_{\rm c}^{(0)}\frac{Z^2e^2}{A^{\frac{1}{3}}}, \tag{18}$$

where $r_0 = 1.2$ fm is adopted. The value of $a_{\rm c}^{(0)}$

$$a_{\rm c}^{(0)} = \frac{3}{5}\cdot\frac{e^2}{4\pi\epsilon_0 r_0} \simeq 0.720\,\text{MeV}.$$

Instead of N and Z, in this section we use $M(K-k,K)$ and $M(K,K-k)$ to represent the masses of mirror nuclei, with $M(K-k,K)$ the proton-rich and $M(K,K-k)$ the neutron-rich, respectively, for the sake of convenience. From Eq. (18), one has

$$\begin{aligned}\Delta_{\rm m}(K-k,K) &\equiv M(K-k,K) - M(K,K-k).\\ &= a_{\rm c}^{(0)}kA^{2/3} - k\Delta E_{\rm np}^{(0)},\end{aligned} \tag{19}$$

where $\Delta E_{\rm np}^{(0)}$ represents the mass difference of a proton and a neutron for atomic masses,

$$\Delta E_{\rm np}^{(0)} = M_n - M_p - m_e = 0.782\,\text{MeV}.$$

In practice one might use Eq. (19) and treat $a_{\rm c}^{(0)}$ and $\Delta E_{\rm np}^{(0)}$ as free parameters, denoted by $a_{\rm c}$ and $\Delta E_{\rm np}$, respectively, namely,

$$\Delta_{\rm m}(K-k,K) = a_{\rm c}kA^{2/3} - k\Delta E_{\rm np}, \tag{20}$$

One similarly obtains

$$\begin{aligned}&\Delta_{\rm n}(K-k,K)\\ &\quad\equiv M(K-1-k,K) - M(K-k,K) - M(K,K-k-1) + M(K,K-k)\\ &\quad= a_{\rm c}\delta_{\rm c}^{\rm n} + (M_{\rm p} - M_{\rm n}) = a_{\rm c}\delta_{\rm c}^{\rm n} - \Delta E_{\rm np}, \end{aligned}\tag{21}$$

$$\begin{aligned}&\Delta_{\rm p}(K-k,K)\\ &\quad\equiv M(K-k,K-1) - M(K-k,K) - M(K-1,K-k) + M(K,K-k)\\ &\quad= a_{\rm c}\delta_{\rm c}^{\rm p} + (M_{\rm n} - M_{\rm p}) = a_{\rm c}\delta_{\rm c}^{\rm p} + \Delta E_{\rm np}, \end{aligned}\tag{22}$$

where $\Delta_{\rm n}$ is the difference of one-neutron separation energy between the nucleus with $(N,Z) = (K-k,K)$ and one-proton separation energy of the nucleus with $(N,Z) = (K,K-k)$, and $\Delta_{\rm p}$ is the difference of one-proton separation energy of the nucleus with $(N,Z) = (K-k,K)$ and one-neutron separation energy of the nucleus with $(N,Z) = (K,K-k)$. According to Eqs. (17)–(18) of Ref. 47,

$$\delta_{\rm c}^{\rm n} = (k+1)(A-1)^{2/3} - kA^{2/3}, \tag{23}$$

$$\delta_{\rm c}^{\rm p} = (k-1)(A-1)^{2/3} - kA^{2/3}, \tag{24}$$

with $A = 2K - k = N + Z$. The Δ_n-δ_c^n formula of Eq. (21) and the Δ_p-δ_c^p relation of Eq. (22) have been exemplified to the AME2016 database, with their RMSD equal to 113 keV and 132 keV, respectively. Clearly, both Δ_n and Δ_p are double mass differences for two successive mirror nuclei. If the isospin symmetry is conserved, then the mass difference of mirror nuclei should be interpreted in terms of Coulomb energy and neutron-proton mass difference. By these unified notations of Δ for Δ_p and Δ_n, and δ_c for δ_c^n and δ_c^p, one has

$$\Delta = a_c \delta_c \mp \Delta E_{np}, \tag{25}$$

where we take $-$ sign for Δ_n and $+$ sign for Δ_p. The Δ-δ_c plot is presented in Fig. 6, where the results for Δ_p with odd proton numbers and for Δ_n with odd neutron numbers are denoted by solid circles in red, and by solid squares in black otherwise. To be compact and convenient, the plot for Δ_p is presented by using $|\Delta_p| = -\Delta_p$ and $|\delta_c^p| = -\delta_c^p$ (both Δ_p and δ_c^p are negative). One sees a clear odd-even feature of Δ_n and Δ_p for small $|\delta_c^n|$ and $|\delta_c^p|$, as shown in Fig. 6, which is a reflection of the short-range attractive pairing correlation between protons.[48–50] Taking this factor into account, the resultant RMSD value of Δ-δ_c relation of Eq. (25) was found to be 93 keV in ref.[51] for 111 pairs of mirror nuclei. The advantage of this odd-even improvement is that the difference between Coulomb energies of two successive nuclei is small, and thus the effect of ambiguity in theoretically evaluated Coulomb energy becomes much smaller — as demonstrated in numerical experiments[52] on

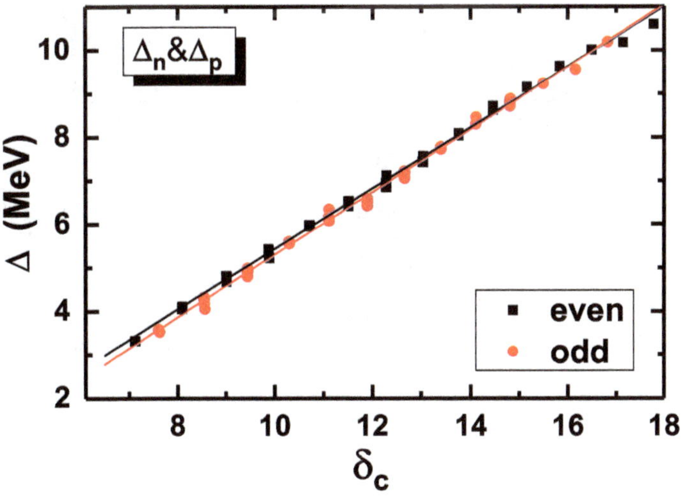

Fig. 6. (color online) Δ_n-δ_c^n and Δ_p-δ_c^p correlations for nuclei with $N, Z \geq 10$. Both Δ_p and δ_c^p are multiplied by -1 (see the text for details). Solid squares in black corresponds to $\Delta_n(\Delta_p)$ with even neutron (proton) numbers, and circles in red corresponds to those with odd neutron (proton) numbers. The results are extracted based on the AME2016, except that the mass excess of the ^{4}V nucleus is replaced by that measured recently.[?] The straight lines are plotted by using optimized a_c and ΔE_{np}.

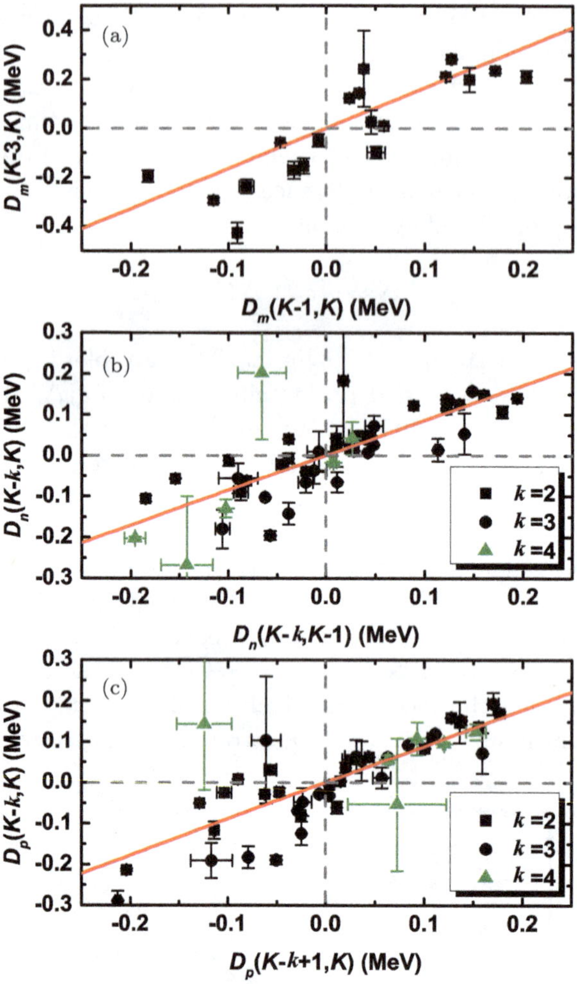

Fig. 7. (color online) Correlations between deviations between theoretical results and experimental data. The deviations of Eqs. (20), (21) and (22) are denoted by D_m, D_n and D_p respectively. Panel (a): Correlation between $D_m(K-3,K)$ and $D_m(K-1,K)$, (b) Correlation between $D_n(K-k,K)$ and $D(K-k,K-1)$, and (c) Correlation between $D_p(K-k,K)$ and $D_p(K-k+1,K)$. The red lines are plotted by a slope $y=\lambda x$ and used to guide eyes. We put the error bars for cases which involves of large experimental uncertainties.

various calculations of Coulomb energy, there is considerably large uncertainties in theoretical Coulomb energy.

There exists strong statistical correlations between the deviations of theoretically predicted results and experimental data, denoted by D_m, D_p and D_n. Figure 7 shows the correlations of $D_m(K-k,K)$ versus $D_m(K-k+2,K)$, $D_n(K-k,K)$ versus $D_n(K-k,K-1)$, $D_p(K-k,K)$ versus $D_p(K-k+1,K)$, respectively, where linear correlation is clearly seen. Taking such linear correlation into account, the RMSD values are reduced to 51 keV.[53]

Yet, the above treatment of the Coulomb force has one advantage, one has to resort the value of $\Delta E_{\rm np} \simeq 1.5$ MeV, which is much larger than $\Delta E_{\rm np}^{(0)} = 0.782\,{\rm MeV}$. This deviation of the parameter is remedied by using a more sophisticated Coulomb energy of atomic nuclei, as follows.

$$\begin{aligned} E_{\rm c} &= E_{\rm c}^{\rm d} + E_{\rm c}^{\rm e} + E_{\rm c}^{\rm s} \\ &= a_{\rm c}\frac{Z^2}{A^{1/3}} - \frac{5}{4}\left(\frac{3}{2\pi}\right)^{2/3} a_{\rm c}\frac{Z^{4/3}}{A^{1/3}} - a_{\rm c}\frac{Z}{A^{1/3}}, \end{aligned} \tag{26}$$

where $E_{\rm c}^{\rm d}, E_{\rm c}^{\rm e}, E_{\rm c}^{\rm s}$ are called the direct, the exchange and the self-energy terms, respectively. It has also been well known that $E_{\rm c}$ exhibits an odd-even staggering. With such considerations, Eq. (19) is improved, and take the following form

$$\Delta_{\rm m}(K-k, K) = a_{\rm c} k \delta_{\rm c}^{\rm m} - k\Delta E_{\rm np} + \frac{1-(-)^k}{2}(-)^K \frac{\beta}{A}, \tag{27}$$

where the last term represents an empirical odd-even staggering in Coulomb energy,[54] and

$$\delta_{\rm c}^{\rm m} = A^{2/3} - 0.808 - A^{-1/3}.$$

In Eq. (27) both $a_{\rm c}$ and $\Delta E_{\rm np}$ are adjustable parameters,

$$a_{\rm c} = a_{\rm c}^{(0)} + a' = 0.720 + a'({\rm MeV}), \tag{28}$$

$$\Delta E_{\rm np} = \Delta E_{\rm np}^{(0)} + \alpha = 0.782 + \alpha({\rm MeV}), \tag{29}$$

where a' and α are expected to be small values (below 100 keV). In the first row (labeled by $\Delta_{\rm m}$) of Table 1, we present the optimized values of a', α and β based on the AME2020 database. One sees that the values of $a_{\rm c}$ and $\Delta E_{\rm np}$ are close to their "conventional" values, 0.72 MeV and 0.782 MeV. The RMSD of Δ involving of only three parameters, $a_{\rm c}$ and $\Delta E_{\rm np}$ and β, is below 100 keV (96 keV). Namely, in this improved version of mass relations of mirror nuclei, there are only three parameters, which are very close to their generally adopted values and describe very satisfactorily experimental data. A simple generalization of this model to the unbound nuclear systems yields a RMSD of 151 keV for the 32 unbound systems beyond the proton drip line.[54]

Table 1. The RMSD (in keV) of Eqs. (27), (21)–(22), and the parameters a', α and β (in unit of keV) optimized based on AME2020 database. It is noted that the resultant RMSD and the values of these three parameters are very robust, namely, one obtains almost the same values if one adopts the AME2012 or AME2016 database.

	RMSD	a'	α	β
$\Delta_{\rm m}$	147	-20 ± 2	67 ± 24	1276 ± 329
$\Delta_{\rm n}, \Delta_{\rm p}$	96	-16 ± 2	105 ± 24	1473 ± 150

5. Arima Sensei as a Great Man

I would like to take this opportunity to say a few words about Arima Sensei. For a few times I heard a comment that Arima Sensei is the right person, or perhaps the only right person, whose contribution could represent Asians in the international nuclear physics society in the last forty or fifty years, and I agreed to this comment.

Arima Sensei made great achievements not only in science, but also in education, poetry of haiku, and administrations. Many of his former students have become big professors in Japan and overseas. He was one of the most famous haiku poets for many years in Japan. His vision is not confined inside Japan. He knew both the eastern and western cultures, and pushed forward many international exchanges between Japan and other countries, in particular, China. He visited China more than 150 times, and hosted numerous Chinese guests from different fields.

Very very luckily in my life, I met Arima Sensei and had collaborated with him for more than twenty years. One of his post-docs in late 1990's, Jie Meng (now a distinguished professor at Peking University), said to me very loudly after he was drunk in a party in RIKEN, perhaps in June of 1998 or 1999, "Yu-Min, it is a great great privilege for us to be taken care of by Arima Sensei." Indeed, Jie was right, I was too much lucky to have Arima Sensei in my life. Through hundreds of discussions, Arima Sensei gradually trained me from a weed into a small tree. Apparently and definitely, to say "thank you" before Arima Sensei is too shallow. For a long time we treated each other like kin. About ten days before he passed away suddenly, he told me in a phone call that he would visit me right after the COVID epidemic finishes. Therefore the sudden news of his death threw me into a deep and long sorrow. At the Winter solstice ceremony in late December of 2020, my wife and I went to a local temple, put Arima Sensei's name together with my father's name and names of my parents-in-law who passed away in recent years, among the temple's ceremonial tablets, burned joss papers one by one, and prayed together with other Buddhists for souls of our respected parentages in the Heaven.

On April 8th, 2021, Rick Casten, Da Hsuan Feng, and myself, organized a memorial zoom session in memory of Arima Sensei, and it was very successful. There were about 400 participants worldwide who joined this session, with twenty one presentations. We compiled a vedio of this session and distributed it over the international nuclear physics society and Japanese haiku poetry society. In December of 2021, again the three of us (Rick, Da Hsuan and I), edited a book in English, *Akito Arima: Scientist, Educator and Poet*, in Shanghai Jiao Tong University Press. We hope that more and more people, generation after generation, will have a wide vision like Arima Sensei, to make efforts for and to contribute to, the happiness and peace of the whole humanity.

6. Summary

To summarize, in this paper we discussed a few empirical approaches to evaluate and to predict nuclear masses, including the Garvey-Kelson mass relations, the formulas involved of the proton-neutron interactions, the parametrizations of symmetry energy and the Wigner energy, and mass relations of mirror nuclei. Numerical experiments showed that these relations are not only simple but also accurate in extrapolations for the unknown data close to the known border.

As shown in Ref. 52, the Coulomb energy, which is considered to be the simplest part in nuclear binding energy, is vulnerable to rather large uncertainties in numerical calculations. Indeed, none of the phenomenological contributions, the volume, surface, Coulomb terms, the Wigner energy, or the symmetry-energy term, has been parameterized accurately. This is easy to understand; for instance, one cannot perform direct measurements of total Coulomb energy of atomic nuclei, thus it is impossible to decouple the Coulomb energy from the binding energy. Because of such difficulties, cancellation of complicated phenomenological terms (such as the shell effect or Coulomb energy) in mass relations is very useful in refining the parametrizations of mass formulas, as discussed in Sec. 3 of this review. Local and similar relations of other quantities are beyond the scope of this review. Here we would mention a few results of nuclear charge radii[55–59] for references.

Clearly, theoretical and experimental studies of atomic mass will remain to be challenging and important in nuclear physics in the long future.

Acknowledgement

This work is financially supported by the National Natural Science Foundation of China (Grants No. 11975151 and No. 11961141003), and MOE Key Lab for Particle Physics, Astrophysics and Cosmology.

References

1. For a review, see Y. M. Zhao, A. Arima, and N. Yoshinaga, *Phys. Rep.* **400**, 1 (2004).
2. For a review, see Y. M. Zhao and A. Arima, *Phys. Rep.* **545**, 1 (2014).
3. E.g., Y. M. Zhao and A. Arima, *Phys. Rev. C* **68**, 044310 (2003); *Phys. Rev. C* **71**, 047304 (2005); *Phys. Rev. C* **72**, 064333 (2005).
4. M. Wang, G. Audi, A. H. Wapstra, and F. G. Kondev *et al.*, *Chin. Phys. C* **36**, 1603 (2012).
5. W. J. Huang, G. Audi, and M. Wang *et al.*, *Chin. Phys. C* **41**, 030002 (2017); M. Wang, G. Audi, and F. G. Kondev *et al.*, *Chin. Phys. C* **41**, 030003 (2017).
6. F. G. Kondev, M. Wang, W. J. Huang, S. Naimi, and G. Audi, *Chin. Phys. C* **45**, 030001 (2021); W. J. Huang, M. Wang, F. G. Kondev, G. Audi, and S. Naimi, *Chin. Phys. C* **45**, 030002 (2021).
7. P. Möller, W. D. Myers, H. Sagawa, and S. Yoshida, *Phys. Rev. Lett.* **108**, 052501.
8. N. Wang and M. Liu, *Phys. Rev. C* **84**, 051303(R) (2011).
9. J. Duflo and A. P. Zuker, *Phys. Rev. C* **52**, 23(R) (1995).
10. K. Vogt, T. Hartmann, and A. Zilges, *Phys. Lett. B* **517**, 255 (2001).
11. M. Bao, Z. He, Y. M. Zhao, and A. Arima, *Phys. Rev. C* **87**, 044313 (2013).
12. Z. He, M. Bao, Y. M. Zhao, and A. Arima, *Phys. Rev. C* **90**, 054320 (2014).
13. H. C. Yu, M. Q. Lin, M. Bao, Y. M. Zhao, and A. Arima, *Phys. Rev. C* **100**, 014314 (2019).

14. G. Streletz, A. Zilges, N. V. Zamfir, R. F. Casten, D. S. Brenner, and Benyuan Liu, *Phys. Rev. C* **54**, 2815(R) (1996).
15. H. Jiang, G. J. Fu, M. Bao, Z. He, Y. M. Zhao, and A. Arima, *Phys. Rev. C* **86**, 014327 (2012).
16. G. J. Fu, M. Bao, Z. He, H. Jiang, Y. M. Zhao, and A. Arima, *Phys. Rev. C* **86**, 054303 (2012).
17. G. T. Garvey and I. Kelson, *Phys. Rev. Lett.* **16**, 197 (1966).
18. G. T. Garvey, W. J. Gerace, R. L. Jaffe, I. Talmi, and I. Kelson, *Rev. Mod. Phys.* **41**, S1 (1969).
19. J. Barea, A. Frank, J. G. Hirsch, and P. Van Isacker, *Phys. Rev. Lett.* **94**, 102501 (2005).
20. J. Barea, A. Frank, J. G. Hirsch, P. Van Isacker, S. Pittel, and V. Velzquez, *Phys. Rev. C* **77**, 041304(R) (2008).
21. M. Bao, Z. He, Y. Lu, Y. M. Zhao, and A. Arima, *Phys. Rev. C* **88**, 064325 (2013).
22. J. L. Tian, N.Wang, C. Li, and J. J. Li, *Phys. Rev. C* **87**, 014313 (2013).
23. Y. Y. Cheng, Y. M. Zhao, and A. Arima, *Phys. Rev. C* **90**, 064304 (2014).
24. J. Y. Zhang, R. F. Casten, and D. S. Brenner, *Phys. Lett. B* **227**, 1 (1989); D. S. Brenner, C. Wesselborg, R. F. Casten, D. D. Warner, and J. Y. Zhang, *Phys. Lett. B* **243**, 1 (1990).
25. G. J. Fu, H. Jiang, Y. M. Zhao, and A. Arima, *Phys. Rev. C* **82**, 014307 (2010).
26. G. J. Fu, Y. Lei, H. Jiang, Y. M. Zhao, B. Sun, and A. Arima, *Phys. Rev. C* **84**, 034311 (2011).
27. H. Jiang, G. J. Fu, B. Sun, M. Liu, N. Wang, M. Wang, Y. G. Ma, C. J. Lin, Y. M. Zhao, Y. H. Zhang, Zhongzhou Ren, and A. Arima, *Phys. Rev. C* **85**, 054303 (2012).
28. Z. He, M. Bao, Y. M. Zhao, and A. Arima, *Phys. Rev. C* **87**, 057304 (2013).
29. Y. Y. Cheng, Y. M. Zhao, and A. Arima, *Phys. Rev. C* **89**, 061304(R) (2014).
30. X. Yin, R. Shou, and Y. M. Zhao, to be published.
31. M. Bao, Z. He, Y. M. Zhao, and A. Arima, *Phys. Rev. C* **90**, 024314 (2014).
32. R. B. Cakirli, D. S. Brenner, R. F. Casten, and E. A. Millman, *Phys. Rev. Lett.* **94**, 092501 (2005).
33. M. Q. Lin, C. Ma, Y. M. Zhao, to be published.
34. N. Wang, Z. Y. Liang, M. Liu, and X. Z. Wu, *Phys. Rev. C* **82**, 044304 (2010).
35. H. Jiang, G. J. Fu, Y. M. Zhao, and A. Arima, *Phys. Rev. C* **85**, 024301 (2012).
36. H. Jiang, M. Bao, Lie-Wen Chen, Y. M. Zhao, and A. Arima, *Phys. Rev. C* **90**, 064303 (2014).
37. H. Jiang, N. Wang, L. W. Chen, Y. M. Zhao, and A. Arima, *Phys. Rev. C* **91**, 054302 (2015).
38. H. Jiang, Y. Y. Cheng, N. Wang, L. W. Chen, Y. M. Zhao, and A. Arima, *Phys. Rev. C* **94**, 064301 (2016).
39. Y. Lu, Y. M. Zhao, and A. Arima, *Phys. Rev. C* **89**, 017301 (2014).
40. Y. Y. Cheng, Y. M. Zhao, and A. Arima, *Phys. Rev. C* **91**, 024314 (2015).
41. Y. Y. Cheng, M. Bao, Y. M. Zhao, and A. Arima, *Phys. Rev. C* **91**, 024313 (2015).
42. D. D. Warner, M. A. Bentley, and P. Van Isacker, *Nature Physics* **2**, 311 (2006).
43. W. Satula, D. J. Dean, J. Gary, S. Mizutori, and W. Nazarewicz, *Phys. Lett. B* **407**, 103 (1997).
44. http://www.nndc.bnl.gov/ensdf/.
45. A. S. Jensen, P. G. Hansen, and B. Jonson, *Nucl. Phys. A* **431**, 393 (1984).
46. M. W. Kirson, *Phys. Lett. B* **661**, 246 (2008).
47. Y. Y. Zong, M. Q. Lin, M. Bao, Y. M. Zhao, and A. Arima, *Phys. Rev. C* **100**, 054315 (2019).

48. B. C. Carlson and I. Talmi, *Phys. Rev.* **96**, 436 (1954).
49. S. Sengupta, *Nucl. Phys.* **21**, 542 (1960).
50. J. A. Nolen and J. P. Schiffer, *Annu. Rev. Nucl. Sci.* **19**, 471 (1969).
51. Y. Y. Zong, C. Ma, Y. M. Zhao, and A. Arima, *Phys. Rev. C* **102**, 024302 (2020).
52. C. Ma, Y. Y. Zong, S. Q. Zhang, J. Li, K. Wang, Y. M. Zhao, and A. Arima, *Phys. Rev. C* **103**, 054326 (2021).
53. C. Ma, Y. Y. Zong, Y. M. Zhao, and A. Arima, *Phys. Rev. C* **102**, 024330 (2020).
54. Y. Y. Zong, C. Ma, M. Q. Lin, and Y. M. Zhao, to be published.
55. J. Piekarewicz, M. Centelles, X. Roca-Maza, and X. Vinas, *Eur. Phys. J. A* **46**, 379 (2010).
56. B. H. Sun, Y. Lu, J. P. Peng, C. Y. Liu, and Y. M. Zhao, *Phys. Rev. C* **90**, 054318 (2014).
57. M. Bao, Y. Lu, Y. M. Zhao, and A. Arima, *Phys. Rev. C* **94**, 064315 (2016).
58. M. Bao, Y. Y. Zong, Y. M. Zhao, and A. Arima, *Phys. Rev. C* **102**, 014306 (2020).
59. C. Ma, Y. Y. Zong, Y. M. Zhao, and A. Arima, *Phys. Rev. C* **104**, 014303 (2021).

www.ingramcontent.com/pod-product-compliance
Lightning Source LLC
Chambersburg PA
CBHW071801130925
32297CB00004B/19

* 9 7 8 9 8 1 9 8 1 3 3 4 6 *